Guidelines for
Preventing Human Error
in Process Safety

Publications Available from the

CENTER FOR CHEMICAL PROCESS SAFETY

of the

AMERICAN INSTITUTE OF CHEMICAL ENGINEERS

Guidelines for Preventing Human Error in Process Safety
Guidelines for Evaluating the Characteristics of Vapor Cloud Explosions, Flash Fires, and BLEVEs
Guidelines for Implementing Process Safety Management Systems
Guidelines for Safe Automation of Chemical Processes
Guidelines for Engineering Design for Process Safety
Guidelines for Auditing Process Safety Management Systems
Guidelines for Investigating Chemical Process Incidents
Guidelines for Hazard Evaluation Procedures, Second Edition with Worked Examples
Plant Guidelines for Technical Management of Chemical Process Safety
Guidelines for Technical Management of Chemical Process Safety
Guidelines for Chemical Process Quantitative Risk Analysis
Guidelines for Process Equipment Reliability Data, with Data Tables
Guidelines for Vapor Release Mitigation
Guidelines for Safe Storage and Handling of High Toxic Hazard Materials
Guidelines for Use of Vapor Cloud Dispersion Models
Safety, Health, and Loss Prevention in Chemical Processes: Problems for Undergraduate Engineering Curricula
Safety, Health, and Loss Prevention in Chemical Processes: Problems for Undergraduate Engineering Curricula—Instructor's Guide
Workbook of Test Cases for Vapor Cloud Source Dispersion Models
Proceedings of the International Process Safety Management Conference and Workshop, 1993
Proceedings of the International Conference on Hazard Identification and Risk Analysis, Human Factors, and Human Reliability in Process Safety, 1992
Proceedings of the International Conference/Workshop on Modeling and Mitigating the Consequences of Accidental Releases of Hazardous Materials, 1991.
Proceedings of the International Symposium on Runaway Reactions, 1989
CCPS/AIChE Directory of Chemical Process Safety Services

Guidelines for Preventing Human Error in Process Safety

Center for Chemical Process Safety

INTER-SCIENCE

To the Memory of John Embrey, 1937–1993

Copyright © 1994
American Institute of Chemical Engineers
345 East 47th Street
New York, New York 10017

Library of Congress Cataloging-in Publication Data
Guidelines for preventing human error in process safety.
    p.   cm.
Includes bibliographical references and index.
ISBN 13: 978-0-8169-0461-7
    1. Chemical processes—Safety measures. 2. Human engineering.
I. American Institute of   Chemical Engineers. Center for Chemical
Process Safety.
TP155.5.G778   1994
660' .2804—dc20   94–2481
                                       CIP

Contents

List of Figures and Tables   ix
Preface   xii
Glossary and Acronyms   xvii

1. Introduction: The Role of Human Error in Chemical Process Safety   1

1.1. Introduction   1
1.2. The Role of Human Error in System Accidents   4
1.3. Why Is Human Error Neglected in the CPI?   10
1.4. Benefits of Improved Human Performance   10
1.5. The Traditional and System-induced Error Approach   12
1.6. A Demand–Resource Mismatch View of Error   15
1.7. A Case Study Illustrating the System-Induced Error Approach   17
1.8. From Theory to Practice   19
1.9. Appendix: Case Studies   22

2. Understanding Human Performance and Error   39

2.1. Purpose of the Chapter   39
2.2. Concepts of Human Error   39
2.3. An Overview of the Four Perspectives on Human Error   44
2.4. The Traditional Safety Engineering Approach to Accidents
      and Human Error   47
2.5. The Human Factors Engineering and Ergonomics Approach (HF/E)   55
2.6. The Cognitive Engineering Perspective   67
2.7. The Sociotechnical Perspective   85
2.8. Summary   93
2.9. Appendix 2A: Process Plant Example of the Stepladder Model   94
2.10. Appendix 2B: Flowcharts for Using the Rasmussen Sequential Model
      for Incident Analysis (Petersen, 1985)   96
2.11. Appendix 2C: Case Study Illustrating the Use of the Sequential Model   100

3. Factors Affecting Human Performance in the Chemical Industry   **103**

3.1. Introduction   103
3.2. Applications of Performance-Influencing Factors   105
3.3. A Classification Structure for Performance-Influencing Factors   107
3.4. Operating Environment   109
3.5. Task Characteristics   120
3.6. Operator Characteristics   133
3.7. Organizational and Social Factors   142
3.8. Interaction of Performance-Influencing Factors   148
3.9. Variability of Human Performance during Normal
      and Emergency Situations   149
3.10. Summary   151

4. Analytical Methods for Predicting and Reducing Human Error   **153**

4.1. Introduction   153
4.2. Data Acquisition Techniques   154
4.3. Task Analysis   161
4.4. Human Error Analysis Techniques   189
4.5. Ergonomics Checklists   196
4.6. Summary   199

5. Qualitative and Quantitative Prediction of Human Error in Risk Assessment   **201**

5.1. Introduction   201
5.2. The Role of Human Reliability in Risk Assessment   202
5.3. System for Predictive Error Analysis and Reduction (SPEAR)   207
5.4. Critical Task Identification and Screening Analysis   209
5.5. Qualitative Human Error Analysis   211
5.6. Representation   219
5.7. Quantification   222
5.8. Summary   240
5.9. Appendix 5A: Influence Diagram Calculations   241

6. Data Collection and Incident Analysis Methods   **247**

6.1. Introduction   247
6.2. An Overview of Data Collection Systems   249
6.3. Types of Data Collection Systems   251

6.4.   Organizational and Cultural Aspects of Data Collection   255
6.5.   Types of Data Collected   260
6.6.   Methods of Data Collection, Storage, and Retrieval   266
6.7.   Data Interpretation   268
6.8.   Root Cause Analysis Techniques   270
6.9.   Implementing and Monitoring the Effectiveness
       of Error Reduction Measures   287
6.10. Setting Up a Data Collection System in a Chemical Plant   288
6.11. Summary   290

7. Case Studies   291

7.1.   Introduction   291
7.2.   Case Study 1: Incident Analysis of Hydrocarbon Leak from Pipe 292
7.3.   Case Study 2: Incident Investigation: Mischarging of Solvent
       in a Batch Plant   307
7.4.   Case Study 3: Design of Standard Operating Procedures
       for Task in Case Study 2   314
7.5.   Case Study 4: Design of Visual Display Units
       for Computer-Controlled Plant   325
7.6.   Case Study 5: Audit of Offshore Emergency Blowdown Operations 334

8. Implementing an Integrated Error and Process Safety Management System at the Plant   345

8.1.   Introduction   345
8.2.   Managing Human Error By Design   346
8.3.   Setting Up an Error Management System in an Existing Plant   357
8.4.   Summary   363

References   365

Bibliography   376

Index   379

List of Figures and Tables

FIGURES

FIGURE 1.1.   Production System Structure.   7
FIGURE 1.2.   Conditions Conducive to Accidents .   9
FIGURE 1.3.   The Dynamics of Incident Causation.   11
FIGURE 1.4.   Accident Causation Sequence.   14
FIGURE 1.5.   System-Induced Error Approach.   15
FIGURE 1.6.   A Demand–Resource View of Human Error.   16
FIGURE 1.7.   Overview of the Systems Approach.   20
FIGURE 1.8.   The Light Shows That the Solenoid Is Deenergized, Not
              That the Oxygen Flow Has Stopped.   29
FIGURE 1.9.   Valve B was Operated by Different Workers.   34
FIGURE 2.1.   Arrangement of Bypass Pipe at Flixborough.   43
FIGURE 2.2.   The Human–Machine Interface.   58
FIGURE 2.3.   Modes of Interacting with the World.   71
FIGURE 2.4.   The Continuum between Conscious and Automatic
              Behavior.   71
FIGURE 2.5.   Dynamics of Generic Error Modeling System (GEMS).   72
FIGURE 2.6.   Classification of Human Errors.   75
FIGURE 2.7.   Decision-Making Model including Feedback.   77
FIGURE 2.8.   Flow Chart for Classifying Skill-, Rule-, and
              Knowledge-Based Processing.   80
FIGURE 2.9.   Sequential Model of Error Causation Chain.   82
FIGURE 2.10. TRIPOD Failure-State Profiles of Two Production
              Platform.   87
FIGURE 2.11. Factors in Human Factors Assessment Methodology.   88
FIGURE 2.12. Example of use of HFAM Tool for Evaluation.   89
FIGURE 2.13. Classification of Causal Factors..   91
FIGURE 2.14. Sociotechnical Model Underlying Audit Tool..   92

ix

FIGURE 2.15. Flowchart for Determining Initiating Events.   97

FIGURE 2.16. Flowchart for Determining Internal Error Mechanisms.   98

FIGURE 2.17. Flowchart for Determining Internal Error Modes.   99

FIGURE 3.1.   Circadian Variations in Oral Temperatures and Alertness
              for Six Process Workers.   117

FIGURE 3.2.   Circadian Variations in Performance on High- and Low-
              Memory Load Tasks.   118

FIGURE 3.3.   Circadian Variations in Errors Made by Process Workers
              Compared with Body Temperature Changes.   118

FIGURE 3.4.   Risk Homeostasis Model.   139

FIGURE 3.5   Individual and Cognitive Phenomena under Stress.   151

FIGURE 4.1.   Activity Analysis for the Control of "Substance" in Paper
              Making.   159

FIGURE 4.2.   HTA Diagram of Isolating a Level Transmitter for
              Maintenance   164

FIGURE 4.3.   Tabular HTA Showing How to Optimize a High Pressure
              in a Distillation Column   166

FIGURE 4.4.   Event Tree for a Gas Leak from a Furnace.   168

FIGURE 4.5.   Decision/Action Flow Diagram of a Furnace Start-Up
              Operation.   171

FIGURE 4.6.   Decision/Action Diagram for Fault Diagnosis in a Crude
              Distillation Plant.   173

FIGURE 4.7.   Temporal Operational Sequence Diagram in a Slitting
              Coated Fabric Operation.   174

FIGURE 4.8.   Partitioned Operational Sequence Diagram for Loading a
              Recipe to a Computer Controlled Reactor.   175

FIGURE 4.9.   Block Diagram and Signal-Flow Graph for "Substance"
              Control System in Paper Making.   178

FIGURE 4.10. Decision/Action Elements of the Rasmussen Model.   181

FIGURE 4.11. CADET Analysis of a Fault-Diagnostic Task in an Oil
              Refinery.   182

FIGURE 4.12. Murphy Diagram for "Planning" Element of Rasmussen
              Model.   184

FIGURE 4.13. Example of a Mental Model Elicited by IMAS.   185

FIGURE 4.14. Criteria for Evaluating the Suitability of Various
              TA Methods   188

FIGURE 4.15. How to Use Various TA Methods in Human Factors
              Application   189

FIGURE 4.16. Error Classification used in Predictive Error Analysis   192

FIGURE 4.17. Documentation of the Results of Human Error Analysis   193

FIGURE 5.1.   Flammable Liquid Storage Tank P&ID.   203

FIGURE 5.2.   Fault Tree Analysis of Flammable Liquid Storage Tank.   204

FIGURE 5.3.   Sample of HAZOP Worksheet (CCPS, 1985).   206

FIGURE 5.4.   System for Predictive Error Analysis and Reduction.   208

FIGURE 5.5.   Relationship of SPEAR to Human Reliability Assessment Methodology   210

FIGURE 5.6.   Chlorine Tanker Task Analysis.   213

FIGURE 5.7   Error Classification.   215

FIGURE 5.8   Results of Predictive Human Error Analysis.   219

FIGURE 5.9.   Error Reduction Recommendations Based on PHEA   220

FIGURE 5.10. Offshore Drilling Blowout Fault Tree Subtree, "Fail to use shear rams to prevent blowout."   221

FIGURE 5.11. Operator Action Tree for ESD Failure Scenario.   223

FIGURE 5.12. THERP Event Tree.   227

FIGURE 5.13. Propane Condenser Schematic.   229

FIGURE 5.14. HRA Event Tree for Improper Condenser Isolation.   231

FIGURE 5.15. Influence Diagram.   239

FIGURE 6.1.   Overall Structure of Data Collection System   250

FIGURE 6.2.   Accident Causation Model.   258

FIGURE 6.3.   Onion Model of Accident Causation.   263

FIGURE 6.4.   Data Interpretation, Remedial Strategy Generation, and Implementation.   269

FIGURE 6.5.   The Spanish Campsite Disaster Described Using the Tree of Causes Diagram.   272

FIGURE 6.6.   Management Oversight and Risk Tree.   274

FIGURE 6.7.   STEP Diagram for the Spanish Campsite Disaster.   276

FIGURE 6.8.   Root Cause Tree.   279

FIGURE 6.9.   HPIP Flowchart.   283

FIGURE 6.10. The Six Steps of Change Analysis.   284

FIGURE 7.1.   Simplified Process Diagram: Hydrocarbon Leak from Pipe.   293

FIGURE 7.2.   STEP Diagram of Hydrocarbon Leak from Pipe   300

FIGURE 7.3.   Statements of Witnesses   304

FIGURE 7.4.   Investigating Engineer's Report   306

FIGURE 7.5.   Data for Process Data Recording System.   306

FIGURE 7.6.   Simplified Schematic Plant Diagram.   308

FIGURE 7.7.   Charging Manifold.   309

FIGURE 7.8.   Variation Tree for Mischarging of Solvent.   310

FIGURE 7.9.   Events and Causal Factors Chart.   313

FIGURE 7.10. Diagram Showing the Flow of Solvents from the Storage Tanks to the Blenders and Reactors.   315

FIGURE 7.11. HTA of Pumping Solvent to Blender.   316

FIGURE 7.12. Extract of PHEA for the "pumping solvent" Task.   318

FIGURE 7.13. Example of Step-by-Step Procedure for Pumping
Solvents.   322

FIGURE 7.14. Example of Checklist for Pumping Solvents.   327

FIGURE 7.15. Original Graphic Display for Furnace A.   329

FIGURE 7.16. Hierarchical Task Analysis of the Task of Increasing
Furnace Load.   331

FIGURE 7.17 Recommended Graphic Display for Furnace A.   332

FIGURE 7.18. Overview Display of the Four Furnaces of
the Distillation Unit.   333

FIGURE 7.19. Decision Flow Chart for Manual Blowdown in Gas
Release Scenario.   339

FIGURE 7.20. Task Analysis of Operator Response to a Significant
Unignited Gas Leak in MSM.   341

FIGURE 8.1. The Phases of a Capital Project.   349

FIGURE 8.2. General Error Management Structure.   357

FIGURE 8.3. Stages in Setting Up an Error Management Program.   358

TABLES

TABLE 1.1   Studies of Human Error in the CPI: Magnitude of the
Human Error Problem   6

TABLE 2.1   Comparisons between Various Perspectives on Human
Error   45

TABLE 2.2   Effect of Different Motivational Schemes on Use of PPE
(adapted from Pirani and Reynolds, 1976)   51

TABLE 2.3   Example Error Reduction Recommendations Arising
from the SRK Model   83

TABLE 3.1   Examples of PIF Scales   106

TABLE 3.2   A Classification Structure of Performance Influencing
Factors   108

TABLE 4.1   A Checklist on Procedures Extracted from the "Short
Guide to Reducing Human Error"   198

TABLE 5.1   Events Included in the HRA Event Tree   232

TABLE 5.2   Human Reliability Analysis Results   232

TABLE 5.3   PIF Ratings   235

TABLE 5.4   Rescaled Ratings and SLIs   236

TABLE 5.5   Effects of Improvements in Procedures on Error
Probabilities Calculated Using SLIM   238

TABLE 7.1   Data Collection Techniques in the Human Factors Audit   336

TABLE 7.2   Major Human Errors Affecting Time to Blowdown   342

Preface

The Center for Chemical Process Safety (CPS) was established in 1985 by the American Institute of Chemical Engineers (AIChE) for the express purpose of assisting the Chemical and Hydrocarbon Process Industries in avoiding or mitigating catastrophic chemical accidents. To achieve this goal, CCPS has focused its work on four areas:

- establishing and publishing the latest scientific and engineering guidelines (not standards) for prevention and mitigation of incidents involving toxic and/or reactive materials;
- encouraging the use of such information by dissemination through publications, seminars, symposia and continuing education programs for engineers;
- advancing the state-of-the-art in engineering practices and technical management through research in prevention and mitigation of catastrophic events; and
- developing and encouraging the use of undergraduate education curricula that will improve the safety knowledge and awareness of engineers.

It is readily acknowledged that human errors at the operational level are a primary contributor to the failure of systems. It is often not recognized, however, that these errors frequently arise from failures at the management, design, or technical expert levels of the company. This book aims to show how error at all of these levels can be minimized by the systematic application of tools, techniques and principles from the disciplines of human factors, ergonomics, and cognitive psychology. The book is the result of a project in which a group of volunteer professionals from CCPS sponsor companies prepared a project proposal and then worked with the successful contractor, Dr. David Embrey of Human Reliability Associates, to produce this book. The ensuing dialogue has resulted in a book that not only provides the underlying principles and theories of the science of human factors, but also goes on to show their application to process safety problems and to the CCPS technical management of process safety system.

ACKNOWLEDGMENTS

The American Institute of Chemical Engineers (AIChE) wishes to thank the Center for Chemical Process Safety (CCPS) and those involved in its operation, including its many sponsors, whose funding made this project possible; the members of its Technical Steering Committee who conceived of and supported this *Guidelines* project and the members of its Human Reliability Subcommittee for their dedicated efforts, technical contributions, and enthusiasm.

This book was written by Dr. David Embrey of Human Reliability Associates, with the assistance of the CCPS Human Reliability Subcommittee. Section 8.2, Managing Human Error by Design, which deals with the application of human factors principles in the process safety management system, was written by the Human Reliability Subcommittee.

- *The main authors of the text of the book were the following staff members of Human Reliability Associates:*
  Dr. David Embrey
  Dr. Tom Kontogiannis
  Mark Green
- *Other contributions from the following individuals are gratefully acknowledged:*
  Dr. Trevor Kletz
  Dr. Deborah Lucas
  Barry Kirwan
  Andrew Livingston
- *The members of the Human Reliability Subcommittee were:*
  Gary A. Page, American Cyanamid Co., (Chairman)
  Joseph Balkey, Union Carbide Corp.
  S. Barry Gibson, DuPont
  Mark D. Johnson, Eastman Kodak Co.
  Joseph B. Mettalia, Jr., CCPS Staff Consultant
  Gary Van Sciver, Rohm and Haas Co.
  Joseph C. Sweeney, ARCO Chemical Co.
- *Reviewers were:*
  Daniel A. Crowl, Mich. Tech. University
  Randolph A. Freeman, Monsanto Co.
  Thomas O. Gibson, The Dow Chemical Co.
  William N. Helmer, Hoechst Celanese Corp.
  Michele M. Houser, Martin Marietta Energy Systems
  Trevor A. Kletz, Process Safety Consultant
  Donald K. Lorenzo, Process Safety Institute
  Denise B. McCafferty, DNV Technica, Inc.
  Michael T. McHale, Air Products and Chemicals, Inc.
  David Meister, Consultant

Robert W. Ormsby, Air Products and Chemicals, Inc.
Wayne A. Pennycook, Exxon
John D. Snell, OxyChem
Marvin F. Specht, Hercules Incorporated
Donald Turner, CH2M Hill
Lester H. Wittenberg, CCPS

The Human Reliability Subcommittee wishes to express its appreciation to Lester Wittenberg, Thomas Carmody, and Bob G. Perry of CCPS for their enthusiastic support.

Glossary and Acronyms

GLOSSARY

Active Errors An active human error is an intended or unintended action that has an immediate negative consequence for the system.

Cognitive "tunnel vision" A characteristic of human performance under stress. Information is sought that confirms the initial hypothesis about the state of the process while disregarding information that contradicts the hypothesis.

Encystment A characteristic of human performance under stress. Encystment occurs when minor problems and details are focused on to excess while more important issues are ignored.

External Error Mode The observable form of an error, for example, an action omitted, as distinct from the underlying process

Externals Psychological classification of individuals who assume (when under stress), that the problem is out of their immediate control and therefore seek assistance.

Human Error Probability The probability that an error will occur during the performance of a particular job or task within a defined time period. *Alternative definition:* The probability that the human operator will fail to provide the required system function within the required time.

Human Information-Processing A view of the human operator as an information-processing system. Information-processing models are conventionally expressed in terms of diagrams which indicate the flow of information through stages such as perception, decision-making, and action.

Human Reliability The probability that a job will be successfully completed within a required minimum time.

Human–Machine Interface The boundary across which information is transmitted between the process and the worker, for example, analog displays, VDUs.

Internal Error Mechanism The psychological process (e.g., strong stereotype takeover) that underlies an external error mode.

Internal Error Mode The stage in the sequence of events preceding an external error mode at which the failure occurred (e.g., failed to detect the initial signal).

Internals Individuals who, when under stress, are likely to seek information about a problem and attempt to control it themselves.

Knowledge-Based Level of Control Information processing carried out consciously as in a unique situation or by an unskilled or occasional user

Latent error An erroneous action or decision for which the consequences only become apparent after a period of time when other conditions or events combine with the original error to produce a negative consequence for the system.

Locus of Control The tendency of individuals to ascribe events to external or internal causes, which affects the degree of control that they perceive they have over these events. (See also *Externals* and *Internals*.)

Manual Variability An error mechanism in which an action is not performed with the required degree of precision (e.g., time, spatial accuracy, force).

Mindset Syndrome A stress-related phenomenon in which information that does not support a person's understanding of a situation is ignored. (See also *Cognitive tunnel vision*.)

Mistakes Errors arising from a correct intentions that lead to incorrect action sequences. Such errors may arise, for example, from lack of knowledge or inappropriate diagnosis.

Performance-Influencing Factors Factors that influence the effectiveness of human performance and hence the likelihood of errors.

Population Stereotype Expectations held by a particular population with regard to the expected movement of a control or instrument indicator and the results or implications of this movement

Reactance Occurs when a competent worker attempts to prove that his or her way of doing things is superior in response to being reassigned to a subordinate position.

Recovery Error Failure to correct a human error before its consequences occur.

Risk Assessment A methodology for identifying the sources of risk in a system and for making predictions of the likelihood of systems failures.

Risk Homeostasis The theory that an operator will attempt to maintain a stable perception of risk following the implementation of new technology that increases the safety of a human–machine system. The theory predicts that operators will take greater risks where more safety devices are incorporated into the system.

Role Ambiguity Exists when an individual has inadequate information about his or her roles or duties.

Role Conflict Exists when there is the simultaneous occurrence of two or more sets of responsibilities or roles such that compliance with one is not compatible with compliance with the other(s).

Root Causes The combinations of conditions or factors that underlie accidents or incidents.

Rule-Based Level of Control In the context of chemical industry tasks, the type of human information processing in which diagnoses are made and actions are formulated on the basis of rules (e.g., "if the symptoms are X then the problem is Y").

Rule Book Culture An organization in which management or workers believe that all safety problems can be resolved by rigid adherence to a defined set of rules.

Skill-Based Level of Control A mode of information processing characterized by the smooth execution of highly practiced, largely physical actions requiring little conscious monitoring.

Slips Errors in which the intention is correct but failure occurs when carrying out the activity required. Slips occur at the skill-based level of information processing.

Stereotype Fixation Occurs when an individual misapplies rules or procedures that are usually successful.

Stereotype Takeover Occurs when an incorrect but highly practiced action is substituted for a correct but less frequently occurring action in a similar task. Also called a *strong habit intrusion*.

Traditional Safety Engineering A safety management policy that emphasizes individual responsibility for system safety and the control of error by the use of motivational campaigns and punishment.

Vagabonding Stress-related phenomenon in which a person's thoughts move rapidly and uncontrollably among issues, treating each superficially.

Verbal Protocol Analysis Technique in which the person is asked to give a "self-commentary" as he or she undertakes a task.

Violation An error that occurs when an action is taken that contravenes known operational rules, restrictions, and/or procedures. The definition of violations excludes actions taken to intentionally harm the system (i.e., sabotage).

ACRONYMS

AT	Area Technician
CADET	Critical Action and Decision Evaluation Technique

CADs	Critical Actions or Decisions
CCPS	Center for Chemical Process Safety
CCR	Central Control Room
CCTV	Closed-Circuit Television
CHAP	Critical Human Action Profile
CPI	Chemical Process Industry
CPQRA	Chemical Process Quantitative Risk Assessment
CR	Control Room
CRT	Cathode Ray Tube
CSE	Cognitive Systems Engineering
CT	Critical Tasks
CTI	Critical Task Identification
CV	Current Values
DA chart	Decision Action Chart
ECFC	Events and Causal Factors Charting
ERS	Error Reduction Strategies
FMECA	Failure Modes and Effects of Criticality Analysis
GEMS	Generic Error Modeling System
HAZOP	Hazard and Operability Study
HEA	Human Error Analysis
HEP	Human Error Probability
HFAM	Human Factors Assessment Methodology
HFE/E	Human Factors Engineering and Ergonomics Approach
HMI	Human–Machine Interface
HPES	Human Performance Evaluation System
HPIP	Human Performance Investigation Process
HRA	Human Reliability Analysis
HRAM	Human Reliability Assessment Method
HRP	Hazard Release Potential
HSP	Hazard Severity Potential
HTA	Hierarchical Task Analysis
IDA	Influence Diagram Approach
IMAS	Influence Modeling and Assessment System
IRS	Incident Reporting Systems
ISRS	International Safety Rating ⏐Systems
LTA	Less Than Adequate
MAST	Memory and Search Test

MORT	Management Oversight and Risk Tree
MSM	Molecular Sieve Module
NIOSH	National Institute of Occupational Safety and Health
NMRS	Near Miss Reporting System
NRC	US Nuclear Regulatory Commission
OAET	Operator Action Event Tree
OSD	Operational Sequence Diagram
P&ID	Piping and Instrumentation Diagram
PA	Public Address
PCS	Process Control System
PDCC	Program Development and Coordination Committee
PHEA	Predictive Human Error Analysis
PIF	Performance Influencing Factors
PORV	Pilot-Operated Relief Valve
PPE	Personal Protective Equipment
PRV	Pressure Relief Valve
PSA	Probabilistic Safety Analysis
PSF	Performance Shaping Factors
QRA	Quantitative Risk Assessment
RCAS	Root Cause Analysis System
RHT	Risk Homeostasis Theory
SFG	Signal Flow Graphs
SLI	Success Likelihood Index
SLIM	Success Likelihood Index Method
SM	Separator Module
SOP	Standard Operating Procedure
SORTM	Stimulus Operation Response Team Performance
SP	Set Points
SPEAR	System for Predictive Error Analysis and Reduction
SRK	Skill–Rule–Knowledge-Based Model
STAHR	Sociotechnical Approach to Human Reliability
STEP	Sequentially Timed Events Plotting Procedure
TA	Task Analysis
THERP	Technique for Human Error Rate Prediction
TQM	Total Quality Management
TSE	Traditional Safety Engineering
VDU	Visual Display Unit

Guidelines for Preventing Human Error in Process Safety

1

Introduction: The Role of Human Error in Chemical Process Safety

1.1. INTRODUCTION

1.1.1. Objective

This book has been written to show how the science of human factors can be applied at the plant level to significantly improve human performance and reduce human error, thus improving process safety.

1.1.2. Scope and Organization

The application of the science of human factors to eliminating error in all aspects of process design, management, operation, and maintenance is the focus of this work. Human error has been a major cause of almost all of the catastrophic accidents that have occurred in the chemical process industries (CPI). If one adopts the broad view of human error as being the result of a mismatch between human capabilities and process demands, then clearly management's role is critical in the following areas:

- Defining the process
- Providing the resources to manage, operate, and maintain the process
- Setting up the feedback systems to monitor the processes which are critical to ensuring safe operation

The book begins with a discussion of the theories of error causation and then goes on to describe the various ways in which data can be collected, analyzed, and used to reduce the potential for error. Case studies are used to teach the methodology of error reduction in specific industry operations. Finally, the book concludes with a plan for a plant error reduction program and a discussion of how human factors principles impact on the process safety management system.

1

The book is organized as follows:

Chapter 1, *The Role of Human Error in Chemical Process Safety*, discusses the importance of reducing human error to an effective process safety effort at the plant. The engineers, managers, and process plant personnel in the CPI need to replace a perspective that has a blame and punishment view of error with a systems viewpoint that sees error as a mismatch between human capabilities and demands.

Chapter 2, *Understanding Human Performance and Error*, provides a comprehensive overview of the main approaches that have been applied to analyze, predict, and reduce human error. This chapter provides the reader with the underlying theories of human error that are needed to understand and apply a systems approach to its reduction.

Chapter 3, *Factors Affecting Human Performance in the Chemical Industry*, describes how a knowledge of "performance-influencing factors" (PIFs), can be used to identify and then eliminate error-causing conditions at the plant.

Chapter 4, *Analytical Methods for Predicting and Reducing Human Error*, contains a discussion and critique of the various methods that are available for analyzing a process for its potential for human error.

Chapter 5, *Quantitative and Qualitative Prediction of Human Error in Safety Assessments*, describes a systematic process for identifying and assessing the risks from human error, together with techniques for quantifying human error probabilities.

Chapter 6, *Data Collection and Incident Analysis Methods*, examines the pitfalls involved in collecting data on human error and suggests possible approaches to improving the quality of the data.

Chapter 7, *Case Studies*, uses examples that illustrate the application of the various error analysis and reduction techniques to real world process industry cases.

Chapter 8, *A Systematic Approach to the Management of Human Error*, explains how the manager and safety professional can use human factors principles in the management of process safety. This chapter also provides a practical plan for a plant human error reduction program that will improve productivity and quality as well.

1.1.3. Purpose of This Book

The objectives of this book are ambitious. It is intended to provide a comprehensive source of knowledge and practical advice that can be used to substantially reduce human error in the CPI. The following sections describe how this is achieved.

1.1.3.1. Consciousness Raising

A major objective is to provide engineers, managers, and process plant personnel in the CPI with an entirely new perspective on human error. In particular, the intention is to change the attitudes of the industry such that human error is removed from the emotional domain of blame and punishment. Instead, a systems perspective is taken, which views error as a natural consequence of a mismatch between human capabilities and demands, and an inappropriate organizational culture. From this perspective, the factors that directly influence error are ultimately controllable by management. This book is intended to provide tools, techniques, and knowledge that can be applied at all levels of the organization, to optimize human performance and minimize error. One of the major messages of this book, with regard to implementing the ideas that it contains, is that methods and techniques will only be effective in the long term if they are supported by the active participation of the entire workforce. To this extent, the consciousness raising process has to be supported by training. The primary focus for raising the awareness of approaches to human error and its control is in Chapters 2 and 7.

1.1.3.2 Provision of Tools and Techniques

This book brings together a wide range of tools and techniques used by human factors and human reliability specialists, which have proved to be useful in the context of human performance problems in the CPI. Although many human factors practitioners will be familiar with these methods, this book is intended to provide ready access to both simple and advanced techniques in a single source. Where possible, uses of the techniques in a CPI context are illustrated by means of case studies.

Chapter 4 focuses on techniques which are applied to a new or existing system to optimize human performance or qualitatively predict errors. Chapter 5 shows how these techniques are applied to risk assessment, and also describes other techniques for the quantification of human error probabilities. Chapters 6 and 7 provide an overview of techniques for analyzing the underlying causes of incidents and accidents that have already occurred.

1.1.3.3 Provision of Solutions to Specific Problems

In addition to raising consciousness and acquainting the reader with a selection of tools for error reduction, this book is also intended to provide assistance in solving specific human error problems that the reader may be experiencing at the plant level. It should be emphasized that no textbook can substitute for appropriate training in human factors techniques or for the advice of human factors specialists. Readers requiring advice should contact professional bodies such as the Human Factors and Ergonomics Society (USA) or the Ergonomics Society (England) who have lists of qualified consultants.

However, given appropriate training, it is quite feasible for personnel such as engineers and process workers to apply techniques such as task analysis (Chapter 4) and audit methods (Chapter 3) to reducing error potential in the workplace.

1.1.3.4. Provision of a Database of Case Studies

The book provides a comprehensive set of examples and case studies that cover a wide variety of process plant situations. Some of these are intended to illustrate the range of situations where human error has occurred in the CPI (see Appendix 1). Other examples illustrate specific techniques (for example, Chapter 4 and Chapter 5). Chapter 7 contains a number of extended case studies intended to illustrate techniques in detail and to show how a range of different techniques may be brought to bear on a specific problem.

1.1.3.5 Cross-Disciplinary Studies

Although this book is primarily written for chemical process industry readers, it also provides a sufficiently wide coverage of methods, case studies and theory to be of interest to behavioral scientists wishing to specialize in process industry applications. Similarly, it is hoped that the a comprehensive description of current theory and practice in this area will stimulate interest in the engineering community and encourage engineers to gain a more in-depth knowledge of the topic. Overall, the intention is to promote the cross-disciplinary perspective that is necessary for effective problem solving in the real world environment.

1.1.3.6. A Complement to Other CCPS Publications

A final objective of this book is to complement other books in this series such as *Guidelines for Chemical Process Quantitative Risk Assessment* (CCPS, 1989b), *Guidelines for Investigating Chemical Process Incidents* (CCPS, 1992d), and *Plant Guidelines for the Technical Management of Chemical Process Safety* (CCPS, 1992a). In the latter volume, human factors was identified as one of twelve essential elements of process safety management. The application to this area of the concepts described in this book is addressed in Chapter 8.

1.2. THE ROLE OF HUMAN ERROR IN SYSTEM ACCIDENTS

After many years of improvements in technical safety methods and process design, many organizations have found that accident rates, process plant losses and profitability have reached a plateau beyond which further improvements seem impossible to achieve. Another finding is that even in organizations with good general safety records, occasional large scale disasters occur which shake public confidence in the chemical process industry. The common factor in both of these areas is the problem of human error. The purpose of this book is to provide a coherent strategy, together with appropriate knowledge and tools, to maximize human performance and minimize human error.

Human rror is probably the major contributor to loss of life, injury to personnel and property damage in the CPI. Human error also has a significant impact on quality, production, and ultimately, profitability. The publication: *One Hundred Large Losses: A Thirty Year Review of Property Damage Losses in the Hydrocarbon Chemical Industries* (Garrison, 1989), documents the contribution of operational errors to the largest financial losses experienced in the CPI up to 1984. This showed that human errors (defined as errors made on-site that have directly given rise to the losses) account for $563 million of these losses and as such are the second highest cause. If this analysis included off-site errors (e.g., Flixborough, due to an engineering error) human error would be the predominant contributor to these losses. A more recent analysis from the same source, Garrison (1989), indicates that in the period 1985–1990, human error was a significant factor in more than $2 billion of property damage in the CPI. These results are not confined to companies in the West. A study by Uehara and Hasegawa of fire accidents in the Japanese chemical industry between 1968 and 1980 indicated that of a total of 120 accidents, approximately 45% were attributed to human error. If the improper design and materials categories are also assumed to be due to human error, this figure rises to 58%. Little change was observed in this proportion over the twelve years examined. Further details of the study, together with others which indicate the central importance of human error in CPI safety, are given in Table 1.1.

In addition to these formal studies of human error in the CPI, almost all the major accident investigations in recent years, for example, Texas City, Piper Alpha, the Phillips 66 explosion, Feyzin, Mexico City, have shown that human error was a significant causal factor at the level of design, operations, maintenance or the management of the process.

One of the central principles presented in this book is the need to consider the organizational factors that create the preconditions for errors, as well as their immediate causes. Figure 1.1 (adapted from Reason, 1990) illustrates the structure of a general industrial production system. In the context of the CPI, this diagram can be interpreted as representing a typical plant. The plant and corporate management levels determine conditions at the operational level that either support effective performance or give rise to errors. Some of the factors that influence these conditions are given in Figure 1.1.The safety beliefs and priorities of the organization will influence the extent to which resources are made available for safety as opposed to production objectives. Attitudes towards blame will determine whether or not the organization develops a blame culture, which attributes error to causes such as lack of motivation or deliberate unsafe behavior. Factors such as the degree of participation that is encouraged in the organization, and the quality of the communication be-

TABLE 1.1
Studies of Human Error in the CPI: Magnitude of the Human Error Problem

STUDY	RESULTS
Garrison (1989)	Human error accounted for $563 million of major chemical accidents up to 1984
Joshchek (1981)	80–90% of all accidents in the CPI due to human error
Rasmussen (1989)	Study of 190 accidents in CPI facility: Top 4 causes: • insufficient knowledge 34% • design errors 32% • procedure errors 24% • personnel errors 16%
Butikofer (1986)	Accidents in petrochemical and refinery units • equipment and design failures 41% • personnel and maintenance failures 41% • inadequate procedures 11% • inadequate inspection 5% • other 2%
Uehara and Hoosegow (1986)	Human error accounted for 58% of the fire accidents in refineries • improper management 12% • improper design 12% • improper materials 10% • misoperation 11% • improper inspection 19% • improper repair 9% • other errors 27%
Oil Insurance Association Report on Boiler Safety (1971)	Human error accounted for 73% and 67% of total damage for boiler start-up and on-line explosions, respectively.

tween different levels of management and the workforce, will have a major impact on the safety culture. The existence of clear policies that will ensure good quality procedures and training will also impact strongly on error likelihood.

The next level represents the organizational and plant design policies, which will also be influenced by senior management. The plant and corporate management policies will be implemented by line management. This level of management has a major impact on the conditions that influence error. Even if appropriate policies are adopted by senior management, these policies may be ineffective if they do not gain the support of line management. Factors that

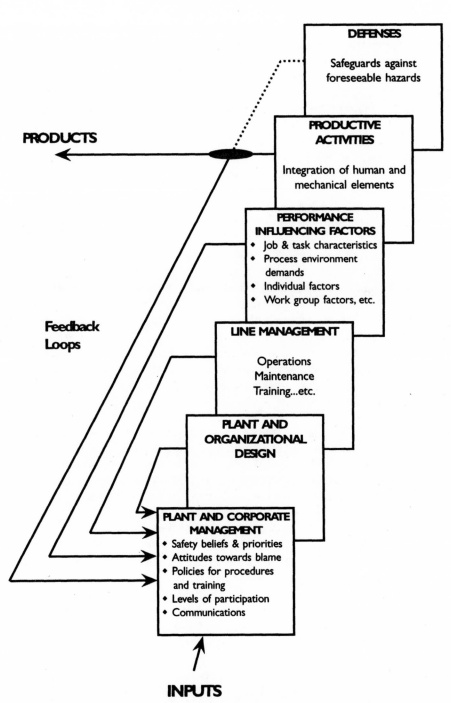

FIGURE 1.1   **Production System Structure (adapted from Reason 1990).**

7

directly affect error causation are located at the next level. These factors, which include the characteristics of the job performed by the worker (complexity, mental versus physical demands, etc.), and individual factors such as personality, and team performance factors, are called collectively performance-influencing factors, or PIFs. These factors are described in detail in Chapter 3.

The next layer in the production system structure represents the activities carried out at the plant level to make the product. These include a wide range of human interactions with the hardware. Physical operations such as opening and closing valves, charging reactors and carrying out repairs will be prominent in traditional, labor intensive, plants such as batch processing. In modern, highly automated plants, particularly those involving continuous production, there is likely to be a greater proportion of higher level "cognitive" skills involved such as problem solving, diagnosis, and decision making in areas such as process and production optimization. In all facilities, human involvement in areas such as maintenance and repairs is likely to be high.

The final elements of a production system represented in Figure 1.1 are the defenses against foreseeable hazards. These defenses exist in many forms. They may include engineered system features such as emergency shutdown systems, relief valves, bursting disks and valves or trips that operate on conditions such as high pressures or low flows. In addition to these hardware systems, the defenses also include human systems such as emergency response procedures, and administrative controls, such as work permits and training designed to give workers the capability to act as another line of defense against hazards.

The various feedback loops depicted in Figure 1.1 represent the information and feedback systems that should (but may not) exist to inform decision makers of the effectiveness of their policies. In Figure 1.2 the structure of Figure 1.1 is represented from the negative perspective of the conditions that can arise at various levels of the organization that will allow errors to occur with potentially catastrophic consequences. Inappropriate policies at the corporate level or inadequate implementation of correct policies by line management will create conditions at the operational level that will eventually result in errors. The term "latent failures" is used to denote states which do not in themselves cause immediate harm, but in combination with other conditions (e.g., local "triggers" such as plant disturbances) will give rise to active failures (e.g., "unsafe acts" such as incorrect valve operations or inadequate maintenance). If the system defenses (hardware or software) are also inadequate, then a negative or even catastrophic consequence may arise.

This model of accident causation is described further in Figure 1.3. This represents the defenses against accidents as a series of shutters (engineered safety systems, safety procedures, emergency training, etc.) When the gaps in these shutters come into coincidence then the results of earlier hardware or human failures will not be recovered and the consequences will occur. Inap-

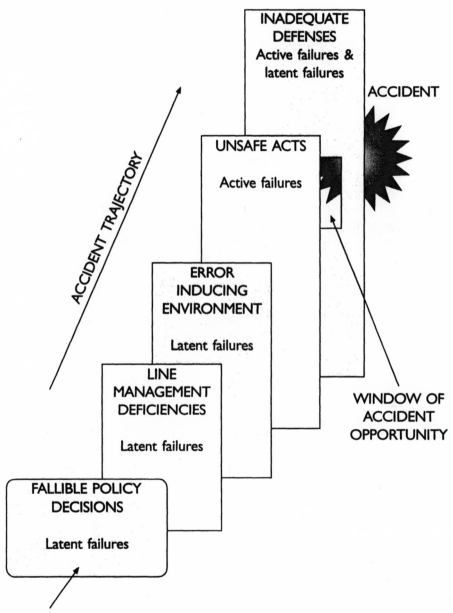

FIGURE 1.2 **Conditions Conducive to Accidents (adapted from Reason, 1990).**

9

propriate management policies create inadequate PIFs, which in turn give rise to a large number of opportunities for error, when initiated by local triggers or unusual conditions.

1.3. WHY IS HUMAN ERROR NEGLECTED IN THE CPI?

The evidence presented in the preceding section makes it clear that human performance problems constitute a significant threat to CPI safety. Despite this evidence, the study of human error has, in the past, been a much neglected area in the industry. There are several reasons for this neglect. Part of the problem is due to a belief among engineers and managers that human error is both inevitable and unpredictable. In subsequent chapters this assumption will be challenged by showing that human error is only inevitable if people are placed in situations that emphasize human weaknesses and do not support human strengths.

Another barrier to a systematic consideration of human error is the belief that increasing computerization and automation of process plants will make the human unnecessary. The fallacy of this belief can be shown from the numerous accidents that have arisen in computer controlled plants. In addition, considerable human involvement will continue to be necessary in the critical areas of maintenance and plant modification, even in the most automated process (see Chapter 2 for a further discussion of this issue).

Human error has often been used as an excuse for deficiencies in the overall management of a plant. It may be convenient for an organization to attribute the blame for a major disaster to a single error made by a fallible process worker. As will be discussed in subsequent sections of this book, the individual who makes the final error leading to an accident may simply be the final straw that breaks a system already made vulnerable by poor management.

A major reason for the neglect of human error in the CPI is simply a lack of knowledge of its significance for safety, reliability, and quality. It is also not generally appreciated that methodologies are available for addressing error in a systematic, scientific manner. This book is aimed at rectifying this lack of awareness.

1.4. BENEFITS OF IMPROVED HUMAN PERFORMANCE

The major benefits that arise from the application of human factors principles to process operations are improved safety and reduced down time. In addition, the elimination of error has substantial potential benefits for both quality and productivity. There is now a considerable interest in applying quality management approaches in the CPI. Many of the major quality experts em-

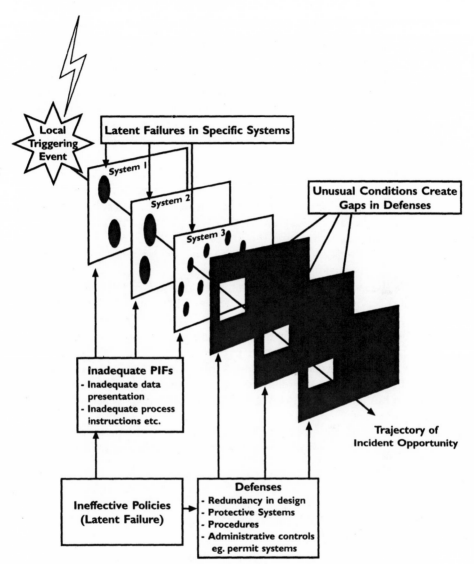

FIGURE 1.3   **The Dynamics of Incident Causation (adapted from Reason, 1990).**

phasize the importance of a philosophy that gets to the underlying causes of errors leading to quality lapses rather than attempting to control error by blame or punishment. Crosby (1984) explicitly advocates the use of error cause removal programs. Other experts such as Deming (1986), and Juran (1979) also emphasize the central importance of controlling the variability of human performance in order to achieve quality objectives. The practical techniques presented in this book could form an integral part of such programs. In Europe and the United States there has been increasing interest in the relationship between quality and safety (see, e.g., Whiston and Eddershaw, 1989; Dumas, 1987). Both quality and safety failures are usually due to the same types of human errors with the same underlying causes. Whether or not a particular error has a safety or quality consequence depends largely on when or where in a process that it occurs. This indicates that any investment in error reduction is likely to be highly cost effective, since it should produce simultaneous reductions in both the incidence of accidents and the likelihood of quality failures.

An additional reason for investing resources in error reduction measures is to improve the ability of the industry to conform to regulatory standards. It is likely that as the relationship between human error and safety becomes more widely recognized, regulatory authorities will place more emphasis on the reduction of error-inducing conditions in plants. It is therefore important that the Chemical Process Industries take the lead in developing a systematic approach and a defensible position in this area.

Despite the lack of interest in human factors issues in the CPI in the past, the situation is now changing. In 1985, Trevor Kletz published his landmark book on human error in the CPI: *An Engineer's View of Human Error* (revised in 1991). Several other books by the same author e.g., Kletz (1994b) have also addressed the issue of human factors in case studies. Two other publications have also been concerned specifically with human factors in the process industry: Lorenzo (1990) was commissioned by the Chemical Manufacturers Association in the USA, and Mill (1992), published by the U.K. Institution of Chemical Engineers. In 1992, CCPS and other organizations sponsored a conference on Human Factors and Human Reliability in Process Safety (CCPS, 1992c). This was further evidence of the growing interest in the topic within the CPI.

1.5. THE TRADITIONAL AND SYSTEM-INDUCED ERROR APPROACH

From the organizational view of accident causation presented in the previous section, it will be apparent that the traditional approach to human error, which assumes that errors are primarily the result of inadequate knowledge or motivation, is inadequate to represent the various levels of causation involved. These contrasting views of error and accident causation have major implications for the way in which human error is assessed and the preventative measures that are adopted.

The structure of this book is based on a model of human error, its causes, and its role in accidents that is represented by Figures 1.4 and 1.5. This perspective is called the *system-induced error approach.* Up to now, only certain aspects of this approach have been discussed in detail. These are the concept of performance-influencing factors (e.g., poor design, training, and procedures) as being the direct causes of errors, and the role of organizational and management factors in creating these causes. The other aspect of the model describes how performance-influencing factors interact with basic error tendencies to give rise to errors with significant consequences.

This aspect of the model is illustrated in Figure 1.5. The error tendencies circle represents the intrinsic characteristics of people that predispose them to error. These tendencies include a finite capability to process information, a reliance on rules (which may not be appropriate) to handle commonly occurring situations, and variability in performing unfamiliar actions. These error tendencies are discussed in detail in Chapter 2.

The error-inducing environment circle denotes the existence of conditions (negative performance-influencing factors) which, when combined with innate error tendencies, will give rise to certain predictable forms of error. For example, the finite information processing capabilities of the human means that overload is very likely if the worker is required to perform concurrent tasks. Another form of error, losing place in a sequence of operations, is likely if a high level of distractions are present. In terms of the management influences on these immediate causation factors, policies for planning workload would influence the number of tasks the worker is required to perform. Job design policies would influence the level of distractions.

The overlap between the error tendencies circle and the error-inducing environment circle represents the likelihood that an error would occur. However, given appropriate conditions, recovery from an error is highly likely. Recovery may arise either if the person making the error detects it before its consequences (accidents, product loss, degraded quality) occur, or if the system as a whole is made insensitive to individual human errors and supports error recovery. These aspects of the system-induced error approach are represented as the third circle in Figure 1.5. Thus, the dark area in the center of the model represents the likelihood of unrecovered errors with significant consequences. At least two major influences can be controlled by the organization to reduce the likelihood of error. The first of these is the design of the system to reduce the mismatch between the demands of the job and the capabilities of the worker to respond to these demands. This area can be addressed by modifying or improving performance-influencing factors that either reduce the levels of demand, or provide greater capability for the humans (e.g., through better job design, training, procedures, team organization). The other area that will have a major impact on error is that of organizational culture. This issue is discussed in Chapter 8.

The system-induced error approach can be restated in an alternative form as an accident causation model (see Figure 1.4). This shows how error-inducing conditions in the form of inadequate PIFs interact with error tendencies to

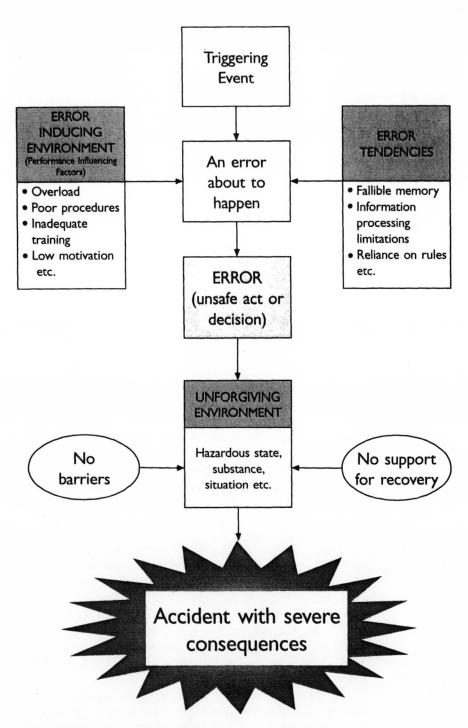

FIGURE 1.4 **Accident Causation Sequence.**

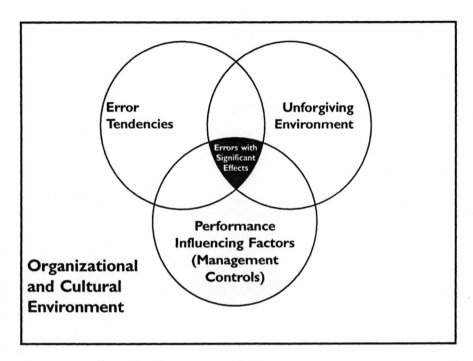

FIGURE 1.5 **System-Induced Error Approach.**

produce an unstable situation where there is a high probability of error. When a triggering event occurs, this gives rise to an error in the form of an unsafe act or decision. This in turn combines with an unforgiving environment that does not support recovery, to give rise to a severe accident. The ways in which the interaction between PIFs and error tendencies gives rise to error are discussed in Chapter 2. A comprehensive description of PIFs is given in Chapter 3.

1.6. A DEMAND–RESOURCE MISMATCH VIEW OF ERROR

A major cause of errors is a mismatch between the demands from a process system and the human capabilities to meet these demands. This is expressed in the model in Figure 1.6. One aspect of the demand side is the requirement for human capabilities that arises from the nature of the jobs in the process plant. Thus, physical capabilities such as craft skills (breaking flanges, welding pipe work, etc.) mental skills (diagnosing problems, interpreting trends) and sensory skills (e.g., being able to detect changes in process information) are all required to a lesser or greater extent by various jobs.

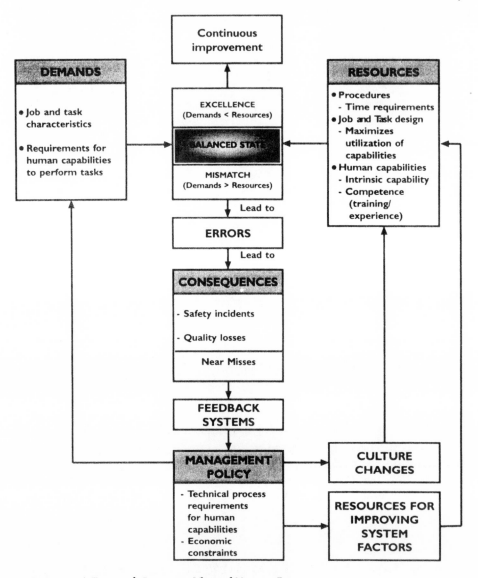

FIGURE 1.6 **A Demand–Resource View of Human Error.**

On the resources side, there are obviously upper limits on human capabilities in these areas. However, these capabilities will be considerably enhanced if the jobs and tasks are designed to utilize human capabilities effectively, if teams are constituted properly in terms of roles, and if personnel with sufficient capability (through training and selection) are available. In addition, these resources will be made more effective if an appropriate culture exists which releases the "discretionary energy" that is available if workers feel committed to and empowered by the organization.

In Figure 1.6, the relationship between demand and resources can produce three outcomes. Where demands and resources are in balance, errors will be at a low level. If resources exceed demands, the organization can be regarded as "excellent" using the terminology of Peters and Waterman (1982). The spare resources can be used to contribute to a continuous improvement process as defined by Total Quality Management. This means that errors can be progressively reduced over time. The existence of spare capacity also allows the system to cope more effectively when unusual or unpredictable demands occur. It should be emphasized that increasing resources does not necessarily equate to increasing numbers of personnel. The application of various design principles discussed in this book will often reduce errors in situations of high demand without necessarily increasing the size of the workforce. In fact, better designed jobs, equipment, and procedures may enable production and quality to be maintained in a downsizing situation. The third case, the mismatch state, is a major precondition for error, as discussed earlier.

The occurrence of errors gives rise to various consequences. The nature of the underlying causes needs to be fed back to policy makers so that remedial strategies can be implemented. A typical strategy will consist of applying existing resources to make changes that will improve human performance and therefore reduce error. This may involve interventions such as improved job design, procedures or training or changes in the organizational culture. These are shown by the arrows to the right of Figure 1.6. An additional (or alternative) strategy is to reduce the level of demands so that the nature of the job does not exceed the human capabilities and resources currently available to do it. An important aspect of optimizing demands is to ensure that appropriate allocation of function takes place such that functions in which humans excel (e.g., problem solving, diagnosis) are assigned to the human while those functions which are not performed well by people (e.g., long-term monitoring) are assigned to machines and/or computers.

1.7. A CASE STUDY ILLUSTRATING THE SYSTEM-INDUCED ERROR APPROACH

In a batch reaction plant, an exothermic reaction was cooled by water circulating in a jacket. The circulating pump failed and the reactor went out of control causing a violent explosion. A low flow alarm was present but was inoperable. A critical pump bearing had not been lubricated during maintenance, and the collapse of the bearing had led to the pump failure.

The incident report stated that the cause of the accident was human error. Although maintenance procedures were available, they had not been used. The maintenance technician was disciplined and a directive was issued that in the future more care should be exercised during maintenance and procedures should be used. This report was based on the traditional view of human error. The incident will now be analyzed from the systems-induced error perspective.

1.7.1. Error-Inducing Conditions

1.7.1.1. *Design and Culture Factors*
There were several reasons why the maintenance procedures, regarding pump bearing lubrication, were not used. They had been supplied by the original manufacturers of the pump and were written in highly technical language. The format of the procedures in terms of layout and typography made it difficult to find the appropriate section. The procedure was bound in a hard cover which made it physically unsuitable for workshop conditions. The nature of the maintenance operations had changed since the procedures were originally written, but these changes had not been incorporated. The general culture in the workshop was that only novices used procedures. Because the technicians had not participated in the development of the procedures there was no sense of ownership and no commitment to using procedures. Training was normally carried out "on the job" and there was no confirmation of competence.

1.7.1.2. *Organization and Policy Factors*
There were many distractions in the workshop from other jobs. The maintenance technicians were working under considerable pressure on a number of pumps. This situation had arisen because an effective scheduling policy was not in place. No policies existed for writing or updating procedures, or for training. In addition, pump bearing maintenance had been omitted on several occasions previously, but had been noticed before the pumps were put back into service. These occurrences had not been reported because of a lack of effective incident reporting systems for learning lessons from "near misses." The fact that the plant was being operated with an inoperable low flow alarm was also indicative of an additional deficiency in the technical risk management system.

1.7.2. Error Tendencies

The pump maintenance step that was omitted was in a long sequence of task steps carried out from memory. Memory limitations would mean that there was a high probability that the step would be omitted at some stage. The work was not normally checked, so the probability of recovery was low.

The steps for maintenance of the pump involved in the incident were very similar to those for other pumps that did not require bearing maintenance. These pumps were maintained much more frequently than the type requiring bearing lubrication. It is possible that in a distracting environment, the maintenance technician may have substituted the more frequently performed set of operations for those required. This is a basic error tendency called a **strong stereotype takeover** (see Chapter 2).

1.7.3. Unforgiving Environment

An opportunity for error recovery would have been to implement a checking stage by a supervisor or independent worker, since this was a critical maintenance operation. However, this had not been done. Another aspect of the unforgiving environment was the vulnerability of the system to a single human error. The fact that the critical water jacket flow was dependent upon a single pump was a poor design that would have been detected if a hazard identification technique such as a **hazard and operability study** (HAZOP) had been used to assess the design.

1.8 FROM THEORY TO PRACTICE: TURNING THE SYSTEMS APPROACH TO A PRACTICAL ERROR REDUCTION METHODOLOGY

This chapter has provided an overview of the book and has described its underlying philosophy, the system-induced error approach (abbreviated to the systems approach in subsequent chapters). The essence of the systems approach is to move away from the traditional blame and punishment approach to human error, to one which seeks to understand and remedy its underlying causes.

In subsequent chapters, the various theories, tools, and techniques required to turn the systems approach from a concept to a practical error reduction methodology will be described. The components of this methodology are described in Figure 1.7. Each of these components will now be described in turn, together with references to the appropriate sections of the book.

1.8.1. Performance Optimization

The first component of the systems approach to error reduction is the optimization of human performance by designing the system to support human strengths and minimize the effects of human limitations. The **human factors engineering and ergonomics** (HFE/E) approach described in Section 2.7 of Chapter 2 indicates some of the techniques available. Design data from the human factors literature for areas such as equipment, procedures, and the human–machine interface are available to support the designer in the optimization process. In addition the analytical techniques described in Chapter 4 (e.g., task analysis) can be used in the development of the design.

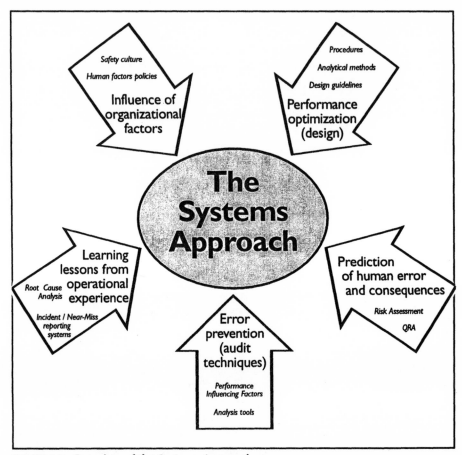

FIGURE 1.7 **Overview of the Systems Approach.**

1.8.2. Prediction of Human Error and Its Consequences

The application of human factors principles at the design stage can reduce the overall probability of errors occurring. However, beyond a certain point, the expenditure that will be required to reduce error rates in general to a very low level may become unacceptable. An approach is therefore required which specifies more accurately the nature of the errors that could occur and their significance compared with other sources of risk in the system. This is achieved by the techniques for the qualitative and quantitative prediction of errors that are described in Chapter 5. In particular, the System for Predictive Error Analysis and Reduction (SPEAR) methodology provides a comprehensive framework for predicting errors and their consequences. By using approaches such as SPEAR, it is possible to make rational decisions with regard to where resources should be most effectively spent in order to reduce the likelihood of errors that have the most severe implications for risk.

The importance of such risk assessment and risk management exercises is being increasingly recognized and can be highly cost-effective if it serves to prevent severe losses that could arise from unmanaged risk. In certain industry sectors, for example, offshore installations in the North Sea, safety cases are being required by the regulatory authorities in which formal risk assessments are documented.

1.8.3. Error Prevention (Audit Techniques)

Measures to reduce human error are often implemented at an existing plant, rather than during the design process. The decision to conduct an evaluation of the factors that can affect error potential at an existing plant may be taken for several reasons. If human errors are giving rise to unacceptable safety, quality or production problems, plant management, with the assistance of the workforce, may wish to carry out a general evaluation or audit of the plant in order to identify the direct causes of these problems.

The identification of the operational level deficiencies that contribute to increased error rates can be achieved by evaluations of PIFs as described in Chapter 3. Although the factors described in that chapter are not exhaustive in their coverage, they can provide a useful starting point for an evaluation exercise. Structured PIF evaluation systems are described in Chapter 2 which ensure that all the important factors that need to be evaluated are included in the exercise.

1.8.4. Learning Lessons from Operational Experience

The next component of the systems approach is the process of learning lessons from operational experience. In Chapter 6, and the case studies in Chapter 7, several techniques are described which can be used to increase the effectiveness of the feedback process. Incident and near-miss reporting systems are designed to extract information on the underlying causes of errors from large numbers of incidents. Chapter 6 provides guidelines for designing such systems. The main requirement is to achieve an acceptable compromise between collecting sufficient information to establish the underlying causes of errors without requiring an excessive expenditure of time and effort.

In addition to incident reporting systems, root cause analysis techniques can be used to evaluate the causes of serious incidents where resources are usually available for in-depth investigations. A practical example of root cause investigation methods is provided in Chapter 7.

1.8.5. Influence of Organizational Factors

The last area addressed by the systems approach is concerned with global issues involving the influence of organizational factors on human error. The major issues in this area are discussed in Chapter 2, Section 7. The two major perspectives that need to be considered as part of an error reduction program are the creation of an appropriate safety culture and the inclusion of human error reduction within safety management policies.

As discussed earlier in this chapter, the main requirements to ensure an appropriate safety culture are similar to those which are advocated in quality management systems. These include active participation by the workforce in error and safety management initiatives, a blame-free culture which fosters the free flow of information, and an explicit policy which ensures that safety considerations will always be primary. In addition both operations and management staff need feedback which indicates that participation in error reduction programs has a real impact on the way in which the plant is operated and systems are designed.

The other global dimension of the systems approach is the need for the existence of policies which address human factors issues at senior levels in the company. This implies that senior management realizes that resources spent on programs to reduce error will be as cost-effective as investments in engineered safety systems.

1.9. APPENDIX: CASE STUDIES OF HUMAN ERROR LEADING TO ACCIDENTS OR FINANCIAL LOSS

1.9.1. Introduction

The intention of this section is to provide a selection of case studies of varying complexity and from different stages of chemical process plant operation. The purpose of these case studies is to indicate that human error occurs at all stages of plant operation, and to emphasize the need to get at root causes. The case studies are grouped under a number of headings to illustrate some of the commonly recurring causal factors. Many of these factors will be discussed in later chapters.

In the shorter case studies, only the immediate causes of the errors are described. However, the more extended examples in the latter part of the appendix illustrate two important points about accident causation. First, the preconditions for errors are often created by incorrect policies in areas such as training, procedures, systems of work, communications, or design. These "root causes" underlie many of the direct causes of errors which are described in this section. Second, the more comprehensive examples illustrate the fact that incidents almost always involve more than one cause. These issues will be taken up in more detail in later chapters. In addition to the case studies in this chapter, further examples will be provided within each chapter to illustrate specific technical points.

1.9.2. Errors Occurring during Plant Changes and Stressful Situations

Insights into the human causes of accidents for a specific category of process plant installations are provided by the Oil Insurance Association report on boiler safety (Oil Insurance Association, 1971). This report provides a large number of case studies of human errors that have given rise to boiler explosions.

Plants are particularly vulnerable to human error during shutdowns for repair and maintenance. This is partly due to the higher level of direct human involvement with the plant, when errors are likely if procedures and supervisory systems are poor. Errors also occur during high stress situations such as emergency shutdowns. Workers need to be trained in how to handle these situations so that less stress is experienced (see Chapter 3).

Example 1.1

A boiler had been shut down for the repair of a forced draft fan. A blind was not installed in the fuel gas line, nor apparently was a double block and bleed in the fuel line utilized. Gas leaked into the firebox during the repair period and was not removed. A severe explosion occurred during the attempt to light of.

Example 1.2

Low water level had shut down a boiler. Flameout occurred on two attempts to refire the boiler. On the third attempt, a violent explosion occurred. The worker had not purged the firebox between each attempt to fire the boiler and this resulted in the accumulation of fuel–air mixture which exploded on the third attempt to ignite the pilot.

Example 1.3

A boiler house enclosed eight large boilers attended by two men. Failure of the combustion air supply shut down one of the boilers. This boiler shutdown created conditions beyond the control of just two men and lack of proper combustion control equipment finally caused seven of the eight boilers to shut down. Amid the confusion caused by low instrument air pressure, low steam pressure, constantly alarming boiler panels, the blocking-in of valves and attempts to get the boilers back on line, one boiler exploded. A purge interlock system was provided on the boilers but the individual burner valves were manually operated. The fuel gas header could not be charged until a timed purge period had been completed.

On the boiler that exploded the manual individual burner valves were not closed when the boiler shut down. After the purge period, fuel gas was admitted to the header from remote manual controls in the control room and into the firebox. Low fuel gas pressure tripped the master safety valve after each attempt to pressure the fuel header. Three attempts were made to purge the boiler and on each of these occasions fuel gas was dumped into the furnace through the open manual burner gas valves. On the third attempt a severe explosion occurred.

1.9.3. Inadequate Human–Machine Interface Design

The first set of case studies illustrates errors due to the inadequate design of the human–machine interface (HMI). The HMI is the boundary across which information is transmitted between the process and the plant worker. In the context of process control, the HMI may consist of analog displays such as chart records and dials, or modern video display unit (VDU) based control systems. Besides display elements, the HMI also includes controls such as buttons and switches, or devices such as trackballs in the case of computer controlled systems. The concept of the HMI can also be extended to include all means of conveying information to the worker, including the labeling of control equipment components and chemical containers. Further discussion regarding the HMI is provided in Chapter 2. This section contains examples of deficiencies in the display of process information, in various forms of labeling, and the use of inappropriate instrumentation scales.

1.9.3.1. *Inadequate Display of Process Information*

Example 1.4

The pump feeding an oil stream to the tubes of a furnace failed. The worker closed the oil valve and intended to open a steam valve to purge the furnace tubes free from oil. He opened the wrong valve, there was no flow to the furnace and as a result the tubes were overheated and collapsed. The error was not due to ignorance. The worker knew which was the right valve but nevertheless opened the wrong one.

This incident is typical of many that have been blamed on human failing. The usual conclusion is that the worker was at fault and there was nothing anyone could do. In fact, investigation showed that:

1. The access to the steam valve was poor and it was difficult to see which was the right valve.

2. There was no indication in the control room to show that there was no flow through the furnace coils.

3. There was no low-flow alarm or low-flow trip on the furnace.

This accident was therefore a typical example of "system-induced error." The poor design of the information display and the inaccessible steam valve created preconditions that were likely to contribute to the likelihood of an error at some time.

Example 1.5

A reactor was being started up. It was filled with the reaction mixture from another reactor which was already on line and the panel operator started to add fresh feed. He increased the flow gradually, at the same time watching the temperature on a recorder conveniently situated at eye level. He intended to start a flow of cooling water to the reaction cooler as soon as the temperature started to rise. Unfortunately, there was a fault in the temperature recorder and although the temperature actually rose, this was not recorded. As a result, a runaway reaction occurred.

The rise in temperature was indicated on a six-point temperature recorder at a lower level on the panel, but the worker did not notice this. The check instrument was about three feet above the floor and a change in one reading on a six-point recorder in that position was not obvious unless someone was actually looking for it.

Example 1.6

When a process disturbance occurred, the plant computer printed a long list of alarms. The operator did not know what had caused the upset and he did nothing. After a few minutes an explosion occurred. Afterwards, the designer admitted that he had overloaded the user with too much information.

1.9.3.2 Poor Labeling of Equipment and Components

Example 1.7

Small leaks from the glands of a carbon monoxide compressor were collected by a fan and discharged outside the building. A man working near the compressor was affected by carbon monoxide. It was then found that a damper in the fan delivery line was shut. There was no label or other indication to show whether the damper was closed or open. In a similar incident, a furnace damper was closed in error. It was operated pneumatically, and again there was no indication on the control knob to show which were the open and closed positions.

Example 1.8

Service lines are often not labeled. A mechanic was asked to fit a steam supply at a gauge pressure of 200 psi (13 bar) to a process line in order to clear a choke. By mistake, he connected up a steam supply at a gauge pressure of 40 psi (3 bar). Neither supply was labeled and the 40 psi supply was not fitted with a check valve. The process material flowed backwards into the steam supply line . Later the steam supply caught fire when it was used to disperse a small leak.

Example 1.9

Nitrogen was supplied in tank cars which were also used for oxygen. Before filling the tank cars with oxygen, the filling connections were changed and hinged boards on both sides of the tanker were folded down so that they read "oxygen" instead of "nitrogen." A tank car was fitted with nitrogen connections and labeled "nitrogen." Probably due to vibration, one of the hinged boards fell down, so that it read "oxygen." The filling station staff therefore changed the connections and put oxygen in it. The tank car was labeled "nitrogen" on the other side and so some nitrogen tank trucks were filled from it and supplied to a customer who wanted nitrogen. He off-loaded the oxygen into his plant, thinking it was nitrogen. Fortunately, the mistake was found before an accident occurred. The customer looked at his weigh scale figures and noticed that on arrival the tanker had weighed three tons more than usual. A check then showed that the plant nitrogen system contained 30% oxygen.

1.9.3.3. *Inappropriate Instrumentation Scales*

Example 1.10

A workman, who was pressure testing some pipe work with a hand operated hydraulic pump, told his foreman that he could not get the gauge reading above 200 psi. The foreman told him to pump harder. He did so, and burst the pipeline. The gauge he was using was calibrated in atmospheres and not psi. The abbreviation "atm." was in small letters, and in any case the workman did not know what it meant.

Example 1.11

A worker was told to control the temperature of a reactor at 60°C, so he adjusted the setpoint of the temperature controller at 60. The scale actually indicated 0–100% of a temperature range of 0–200°C, so the set point was really 120°C. This caused a runaway reaction which overpressured the vessel. Liquid was discharged and injured the worker.

1.9.3.4. *Inadequate Identification of Components*

Example 1.12

A joint that had to be broken was marked with chalk. The mechanic broke another joint that had an old chalk mark on it and was splashed with a corrosive chemical. The joint should have been marked with a numbered tag.

Example 1.13

An old pipeline, no longer used, was marked with chalk at the point at which it was to be cut. Before the mechanic could start work, heavy rain washed off the chalk mark. The mechanic "remembered" where the chalk mark had been and he was found cutting his way with a hacksaw through a line containing a hazardous chemical.

1.9.4. Failures Due to False Assumptions

In order to cope with a complex environment, people make extensive use of rules or assumptions. This rule based mode of operation is normally very efficient. However, errors will arise when the underlying assumptions required by the rules are not fulfilled. Chapter 2 discusses the causes of these rule based errors in detail.

Example 1.14

During the morning shift, a worker noticed that the level in a tank was falling faster than usual. He reported that the level gauge was out of order and asked an instrument mechanic to check it. It was afternoon before he could do so. He reported that it was correct. Only then did the worker find that there was a leaking drain valve. Ten tons of material had been lost. In this case an inappropriate rule of the form "If level in tank decreases rapidly then level gauge is faulty" had been used instead of the more general rule: "If level in tank decreases rapidly then investigate source of loss of material."

Example 1.15

Following some modifications to a pump, it was used to transfer liquid. When the movement was complete, the operator pressed the stop button on the control panel and saw that the "pump running" light went out. He also closed a remotely operated valve in the pump delivery line. Several hours later the high-temperature alarm on the pump sounded. Because the operator had stopped the pump and seen the running light go out, he assumed the alarm was faulty and ignored it. Soon afterward there was an explosion in the pump.

When the pump was modified, an error was introduced into the circuit. As a result, pressing the stop button did not stop the pump but merely switched off the running light. The pump continued running-dead-headed, overheated, and the material in it decomposed explosively.

Example 1.16

An ethylene oxide plant tripped and a light on the panel told the operator that the oxygen valve had closed. Because the plant was going to be restarted immediately, he did not close the hand-operated isolation valve as well, relying totally on the automatic valves. Before the plant could be restarted an explosion occurred. The oxygen valve had not closed and oxygen continued to enter the plant (Figure 1.8).

The oxygen valve was closed by venting the air supply to the valve diaphragm, by means of a solenoid valve. The light on the panel merely said that the solenoid had been deenergized not, as the operator assumed, that the oxygen valve had closed. Even though the solenoid is deenergized the oxygen flow could have continued because:

1. The solenoid valve did not open.
2. The air was not vented.
3 The trip valve did not close.

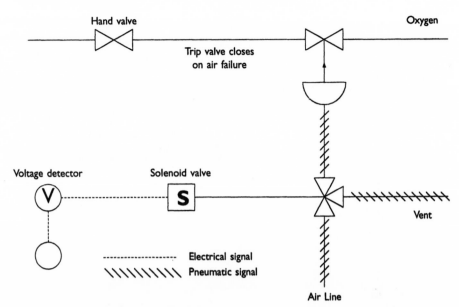

FIGURE 1.8 **The Light Shows That the Solenoid Is Deenergized, Not That the Oxygen Flow Has Stopped (Kletz, 1994b).**

In fact, the air was not vented. The 1-inch vent line on the air supply was choked by a wasp's nest. Although this example primarily illustrates a wrong assumption, a second factor was the inadequate indication of the state of the oxygen valve by the panel light. A similar error was a major contributor to the Three Mile Island nuclear accident.

Example 1.17

A permit was issued to remove a pump for overhaul. The pump was deenergized, removed, and the open ends blanked. Next morning the maintenance foreman signed the permit to show that the job—removing the pump—was complete. The morning shift lead operator glanced at the permit. Seeing that the job was complete, he asked the electrician to replace the fuses. The electrician replaced them and signed the permit to show that he had done so. By this time the afternoon shift lead operator had come on duty. He went out to check the pump and found that it was not there.

The job on the permit was to remove the pump for overhaul. Permits are sometimes issued to remove a pump, overhaul it, and replace it. But in this case the permit was for removal only. When the maintenance foreman signed the permit to show that the job was complete, he meant that the job of **removal** was complete. The lead operator, however, did not read the permit thoroughly. He assumed that the **overhaul** was complete.

When the maintenance foreman signed the permit to show that the job was complete, he meant he had completed the job **he thought he had to do.** In this case this was not the same as the job the lead operative expected him to do.

1.9.5. Poor Operating Procedures

This section gives an example of an error caused by poor operating procedures. In industries such as nuclear power, incident reporting systems indicate that inadequate or nonexistent operating instructions or procedures account for a high proportion of errors. Although there is little hard evidence, because of the incident reporting policies in the CPI (see Chapter 6), this cause probably contributes to many of the incidents discussed in this chapter. The effective design of procedures is discussed further in Chapter 7, Case Study 2.

Example 1.18

When the preparation of a batch went wrong the investigation showed that the worker had charged 104 kg of one constituent instead of 104 grams. The instructions to the worker were set out as shown below (originally the actual names of the chemicals were included).

Operating Instructions	
BLENDING INGREDIENTS	QUANTITY (TONS)
Chemical 1	3.75
Chemical 2	0.250
Chemical 3	0.104 kg
Chemical 4	0.020
Chemical 5	0.006
TOTAL	4.026

1.9.6. Routine Violations

This section is concerned with errors that are often classified as "violations," that is, situations where established operating procedures appear to have been deliberately disregarded. Such violations sometimes arise because the prescribed way of performing the task is extremely difficult or is incompatible with the demands of production. Another cause is lack of knowledge of the reasons why a particular activity is required. The case studies illustrate both of these causes.

Example 1.19

Experience shows that when autoclaves or other batch reactors are fitted with drain valves, they may be opened at the wrong time and the contents will then discharge on to the floor, often inside a building. To prevent this, the drain valves on a set of reactors were fitted with interlocks so that they could not be opened until the pressure was below a preset value. Nevertheless, a drain valve was opened when a reactor was up to pressure and a batch emptied on to the floor. The inquiry disclosed that the pressure measuring instruments were not very reliable. So the workers had developed the practice of defeating the interlock either by altering the indicated pressure with the zero adjustment screw or by isolating the instrument air supply. One day, having defeated the interlock, a worker opened a drain valve by mistake instead of a transfer valve.

Example 1.20

A small tank was filled every day with sufficient raw material to last until the following day. The worker watched the level in the tank and switched off the filling pump when the tank was 90% full. The system worked satisfactorily for several years before the inevitable happened and the worker allowed the tank to overfill. A high level trip was then installed to switch off the pump automatically if the level exceeded 90%. To the surprise of engineering staff the tank overflowed again after about a year. When the trip was installed it was assumed that:

1. The worker would occasionally forget to switch off the pump in time, and the trip would then operate.

2. The trip would fail occasionally (about once in two years).

3. The chance that both would occur at the time same time was negligible.

However, these assumptions were incorrect. The worker decided to rely on the trip and stopped watching the level. The supervisor and foreman knew this, but were pleased that the worker's time was being utilized more productively. A simple trip fails about once every two years so the tank was bound to overflow after a year or two. The trip was being used as a process controller and not as an emergency instrument. The operating and supervisory staff probably assumed a much higher level of reliability for the trip than was actually the case.

Example 1.21

A permit issued for work to be carried out on an acid line stated that goggles must be worn. Although the line had been drained, there might have been some trapped pressure. The man doing the job did not wear goggles and was splashed in the eye.

Further investigations showed that **all** permits issued asked for goggles to be worn, even for repairs to water lines in safe areas. The mechanics therefore frequently ignored this instruction and the supervisors and foremen tolerated this practice.

Example 1.22

Two men were told to wear breathing apparatus while repairing a compressor that handled gas containing hydrogen sulfide. The compressor had been purged but traces of gas might have been left in it. One of the men had difficulty in handling a heavy valve close to the floor and removed his mask. He was overcome by hydrogen sulfide or possibly nitrogen gas. It was easy to blame the man, but he had been asked to do a job which was difficult wearing breathing apparatus.

1.9.7. Ineffective Organization of Work

Error free operation and maintenance can only occur within an effective management system. At the level of the task itself, this is provided by operating instructions. However, at a more global level, separate tasks have to be organized in a systematic manner, particularly if hazardous operations are involved, and where several individuals need to coordinate to achieve an overall objective. This section illustrates some accidents due to poor organization of work or failure to carry out checks.

Example 1.23

A plumber foreman was given a work permit to modify a pipeline. At 4:00 PM. the plumbers went home, intending to complete the job on the following day.

During the evening the process foreman wanted to use the line the plumbers were working on. He checked that the line was safe to use and he asked the shift mechanic to sign off the permit. Next morning the plumbers, not knowing that their permit had been withdrawn, started work on the line while it was in use.

Example 1.24

A manhole cover was removed from a reactor so that some extra catalyst could be put in. After the cover had been removed, it was found that the necessary manpower would not be available until the next day. The supervisor therefore decided to replace the manhole cover and regenerate the catalyst overnight. By this time it was evening and the maintenance foreman had gone home and left the work permit in his office, which was locked. The reactor was therefore boxed up and catalyst regeneration carried out with the permit still in force. The next day a mechanic, armed with the work permit, proceeded to remove the manhole cover again, and while doing so was drenched with process liquid. Fortunately, the liquid was mostly water and he was not injured.

Example 1.25

A pump was being dismantled for repair. When the casing was removed, hot oil, above its autoignition temperature, came out and caught fire. Three men were killed and the plant was destroyed. Examination of the wreckage after the fire showed that the pump suction valve was open and the pump drain valve was shut.

The pump had been awaiting repair for several days when a work permit was issued at 8:00 AM. on the day of the fire. The foreman who issued the permit should have checked, before doing so, that the pump suction and delivery valves were shut and the drain valve open. He claimed that he did so. Either his recollection was incorrect or, after he inspected the valves and before work started, someone closed the drain valve and opened the suction valve. When the valves were closed, there was no indication on them of **why** they were closed. A worker might have opened the suction valve and shut the drain valve so that the pump could be put on line quickly if required. A complicating factor was that the maintenance team originally intended to work only on the pump bearings. When they found that they had to open up the pump they told the process team, but no further checks of the isolations were carried out.

Example 1.26

While a plant was on-line a worker noticed a blind in a tank vent. The blind had been fitted to isolate the tank from the blowdown system while the tank was being repaired. When the repairs were complete, the blind was overlooked. Fortunately, the tank, an old one, was stronger than it needed to be for the duty, or it would have burst. The omission of an isolated step at the end of a long sequence of operations is a common failure mode, which often occurs in the absence of formal checklists or operating procedures.

1.9.8. Failure to Explicitly Allocate Responsibility

Many errors have occurred due to failure to explicitly allocate responsibility between different individuals who need to coordinate their efforts. This is illustrated by the case study in this section.

Example 1.27

The following incident occurred because responsibility for plant equipment was not clearly defined, and workers in different teams, responsible to different supervisors, operated the same valves.

The flare stack shown in Figure 1.9 was used to dispose of surplus fuel gas, which was delivered from the gas holder by a booster through valves B and C. Valve C was normally left open because valve B was more accessible. One day the worker responsible for the gas holder saw that the gas pressure had started to fall. He therefore imported some gas from another unit. Nevertheless, a half hour later the gas holder was sucked in.

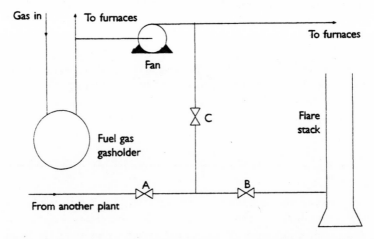

FIGURE 1.9 **Valve B was Operated by Different Workers (Kletz, 1994b).**

Another flare stack at a different plant had to be taken out of service for repair. A worker at this plant therefore locked open valves A and B so that he could use the "gas holder flare stack." He had done this before, though not recently, and some changes had been made since he last used the flare stack. He did not realize that this action would result in the gas holder emptying itself through valves C and B. He told three other men what he was going to do but he did not tell the gas holder worker as he did not know that this man needed to know.

1.9.9. Organizational Failures

This section illustrates some of the more global influences at the organizational level which create the preconditions for error. Inadequate policies in areas such as the design of the human–machine interface, procedures, training, and the organization of work will also have contributed implicitly to many of the other human errors considered in this chapter.

In a sense, all the incidents described so far have been management errors but this section describes two incidents which would not have occurred if the senior managers of the companies concerned had realized that they had a part to play in the prevention of accidents over and above exhortations to their employees to do better.

Example 1.28

A leak of ethylene from a badly made joint on a high pressure plant was ignited by an unknown cause and exploded, killing four men and causing extensive damage. After the explosion many changes were made to improve the standard of joint-making: better training, tools, and inspection.

Poor joint-making and the consequent leaks had been tolerated for a long time before the explosion as all sources of ignition had been eliminated and so leaks could not ignite, or so it was believed. The plant was part of a large corporation in which the individual divisions were allowed to be autonomous in technical matters. The other plants in the corporation had never believed that leaks of flammable gas could ignite. Experience had taught them that sources of ignition were liable to occur, even though everything was done to remove known sources, and therefore strenuous efforts had been made to prevent leaks. Unfortunately the managers of the ethylene plant had hardly any technical contact with the other plants, though they were not far away; handling flammable gases at high pressure was, they believed, a specialized technology and little could be learned from those who handled them at low pressure.

Example 1.29

Traces of water were removed from a flammable solvent in two vessels containing a drying agent. While one vessel was on-line, the other was emptied by blowing with nitrogen and then regenerated. The changeover valves were operated electrically. Their control gear was located in a Division 2 area and as it could not be obtained in a nonsparking form, it was housed in a metal cabinet which was purged with nitrogen to prevent any flammable gas in the surrounding atmosphere leaking in. If the nitrogen pressure fell below a preset value (about ½-inch water gauge) a switch isolated the power supply. Despite these precautions an explosion occurred in the metal cabinet, injuring the inexperienced engineer who was starting up the unit.

The nitrogen supply used to purge the metal cabinet was also used to blow out the dryers. When the nitrogen supply fell from time to time (due to excessive use elsewhere on the site), solvent from the dryers passed through leaking valves into the nitrogen supply line, and found its way into the metal cabinet. The nitrogen pressure then fell so low that some air diffused into the cabinet.

Because the nitrogen pressure was unreliable it was difficult to maintain a pressure of ½-inch water gauge in the metal cabinet. The workers complained that the safety switch kept isolating the electricity supply, so an electrician reduced the setpoint first to ¼ inch and then to zero, thus effectively bypassing the switch. The setpoint could not be seen unless the cover of the switch was removed and the electrician told no one what he had done. The workers thought he was a good electrician who had prevented spurious trips. Solvent and air leaked into the cabinet, as already described, and the next time the electricity supply was switched there was an explosion.

The immediate causes of the explosion were the contamination of the nitrogen, the leaky cabinet (made from thin steel sheet) and the lack of any procedure for authorizing, recording, and checking changes in trip settings. However, the designers were also at fault in not realizing that the nitrogen supply was unreliable and liable to be contaminated and that it is difficult to maintain a pressure in boxes made from thin sheet. If a hazard and operability study had been carried out on the service lines, with operating staff present, these facts, well known to the operating staff, would have been made known to the designers. It might also have brought out the fact that compressed air could have been used instead of nitrogen to prevent diffusion into the cabinet.

The control cabinet did not have to be in a Division 2 area. A convenient location was chosen and the electrical designers were asked to supply equipment suitable for the location. They did not ask if the cabinet had to be in a Division 2 area. This was not seen as their job. They perceived their job as being to provide equipment suitable for the classification which had already been agreed.

2

Understanding
Human Performance and Error

2.1. PURPOSE OF THE CHAPTER

The purpose of this chapter is to provide a comprehensive overview of the main approaches that have been applied to analyze, predict, and reduce human error in industrial systems. The practical application of specific techniques to achieve these goals must be built upon an understanding of the theory that led to the development of these techniques. Just as it would be inadvisable for an engineer to attempt to design a venting system without an underlying knowledge of the behavior of chemical reactions, it is recommended that the user of human factors techniques becomes acquainted with their underlying rationale.

This chapter is organized into four sections, which comprise four complementary approaches to human error in industrial systems:

- Traditional safety engineering
- Factors/ergonomics
- Cognitive systems engineering
- Sociotechnical systems

Prior to the sections that give a detailed description of these approaches, the following overview section provides a summary of the concepts and terminology used in the study of error. This is followed by an introduction to each of the approaches, which are then described in more detail in subsequent sections.

2.2. CONCEPTS OF HUMAN ERROR

A single, all-embracing definition of human error is difficult to achieve. For the engineer, the worker in a system such as a chemical process plant may be perceived as being there to perform a set of tasks to achieve specific operational objectives. There is therefore relatively little interest in the underlying mechanisms of failure. For the human reliability specialist, however, who is attempting to predict and optimize human performance, the underlying organizational and psychological causes of errors are of considerable importance.

The analysis of accidents and disasters in real systems makes it clear that it is not sufficient to consider error and its effects purely from the perspective of individual human failures. Major accidents are almost always the result of multiple errors or combinations of single errors with preexisting vulnerable conditions (Wagenaar et al., 1990). Another perspective from which to define errors is in terms of when in the system life cycle they occur. In the following discussion of the definitions of human error, the initial focus will be from the engineering and the accident analysis perspective. More detailed consideration of the definitions of error will be deferred to later sections in this chapter where the various error models will be described in detail (see Sections 5 and 6).

2.2.1. Engineering Concepts of Error

From a reliability engineering perspective, error can be defined by analogy with hardware reliability as "The likelihood that the human fails to provide a required system function when called upon to provide that function, within a required time period" (Meister, 1966). This definition does not contain any references to *why* the error occurred, but instead focuses on the consequences of the error for the system (loss or unavailability of a required function). The disadvantage of such a definition is that it fails to consider the wide range of other actions that the human might make, which may have other safety implications for the system, as well as not achieving the required function.

Meister (1977) classified errors into four major groupings:

- Performance of a required action incorrectly
- Failure to perform a required action (omission error)
- Performance of a required action out of sequence (combined commission/omission error)
- Performance of a nonrequired action (commission error)

This classification underscores the inadequacy of the approach common in reliability engineering of simply classifying errors into omission and commission categories.

An additional category related to the above was suggested by A. D. Swain:

- Failure to perform a required action within an allotted time

This is particularly relevant in situations where a human intervention is required in response to a potentially hazardous plant situation.

Although the above descriptions are, strictly speaking, **classifications** rather than **definitions** of error, they share the same characteristics as the first definition in that they describe *what* happened rather than *why* it happened. They are therefore much more easily related to the observable *consequences* of an error than to its causes.

2.2.2. Human Error in Accident Causation

Analysis of accidents and major losses in the CPI indicates that they rarely arise from a single human error or component failure. Often there is a combination of some triggering event (hardware or human) together with preexisting conditions such as design errors, maintenance failures or hardware deficiencies.

It is therefore useful to distinguish between active and latent errors or failures. An *active human error* has an immediate effect in that it either directly causes a hazardous state of the system or is the direct initiator of a chain of events which rapidly leads to the undesirable state.

Example 2.1: Active Human Error (Kletz, 1994b)

A plant worker opened the hatch of a reactor and manually charged it with caustic soda. However, he had failed to check the reactor prior to charging, and the caustic soda reacted with chemicals already present to release a toxic by-product. The worker was overcome, and only survived following emergency treatment.

In the case of a **latent human error** the consequences of the error may only become apparent after a period of time when the condition caused by the error combines with other errors or particular operational conditions. Two types of latent error can be distinguished. One category originates at the operational level and leads to some required system function being degraded or unavailable. Maintenance and inspection operations are a frequent source of this type of latent failure.

Example 2.2: A Latent Error Due to Misplaced Priorities

In an offshore oil production platform, a major accident occurred partly because pump seals failed and therefore an antifoaming agent was not delivered to a crude oil separator. The fact that the pump seals were defective should have been picked up during routine inspections, but the inspections were neglected because of production pressures. The failure to carry out the inspections was a latent error.

The other category of latent failures can occur at the level of engineering design or management policy. For example, the design of a scrubbing system may not be adequate to handle all credible releases. If an active human error initiates the production of an excessive volume of product the system may allow toxic materials to be released to the environment.

Example 2.3: A Latent Error Due to Lack of Design Knowledge (Kletz, 1994b)

In the Flixborough disaster, one of six reactors in series, through which hot cyclohexane was passed, was removed from service (see Figure 2.1). Each reactor was connected by a short pipe with a bellows at each end to allow for expansion. The fifth reactor was replaced by a temporary bypass pipe with two bends in it to allow for differences in height between reactors 4 and 6. Because the bypass was not properly supported and had a bellows at either end, it moved when there were pressure variations. This movement eventually caused the bellows to fail, releasing 50 tons of cyclohexane which exploded, killing 28 men.

Inadequate ergonomic design in areas such as control panels and the labeling and placement of valves on the plant can also be regarded as a latent failure because it will increase the probability of active errors. For example, a worker may misread process information from a poorly designed display. Poorly labeled and situated valves can cause the wrong valve to be selected, with possibly disastrous consequences.

Management policies are the source of many of the preconditions that give rise to systems failures. For example, if no explicit policy exists or if resources are not made available for safety critical areas such as procedures design, the effective presentation of process information, or for ensuring that effective communication systems exist, then human error leading to an accident is, at some stage, inevitable. Such policy failures can be regarded as another form of latent human error, and will be discussed in more detail in Section 2.7.

Because errors are frequently recoverable, it is also appropriate to define another category of errors, recovery failures. These are failures to recover a chain of events leading to a negative consequence (assuming that such a recovery was feasible) before the consequence occurs. This includes recovery from both active and latent failures.

For the sake of completeness, it is also useful to define at this stage the category of errors known as **violations**. Violations occur when a worker carries out actions that are either prohibited or are different from those which are prescribed by the organization and carry some associated risks. Since violations are deliberate acts, they are not, strictly speaking, errors. However, the violations category is useful when classifying human caused failures.

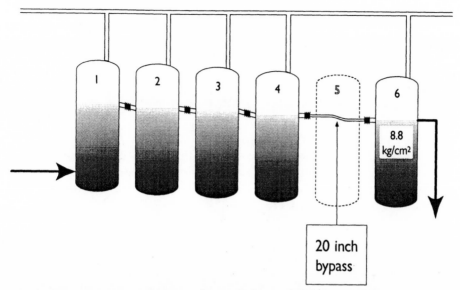

FIGURE 2.1 **Arrangement of Bypass Pipe at Flixborough (Kletz, 1994b).**

2.2.3. Summary of Definitions

Active Error/Failure: An active human error is an unintended action or an intended action based on a mistaken diagnosis, interpretation, or other failure, which is not recovered and which has significant negative consequences for the system.

Latent Human Error/Failure (operational level): A latent human error is similar to an active error, but the consequences of the error may only become apparent after a period of time or when combined with other errors or particular operational conditions.

Latent Human Error/Failure (management level): A management level human error is an inadequate or nonexistent management policy which creates the preconditions for active or latent human, hardware, or software failures.

Violation Error/Failure: A violation error occurs when an intended action is made which deliberately ignores known operational rules, restrictions, or procedures. However, this definition excludes actions that are deliberately intended to harm the system, which come within the category of sabotage.

Recovery Error/Failure: A recovery failure occurs if a potentially recoverable active or latent error is not detected or remedial action is not taken before the negative consequences of the error occur.

In the above definitions, the term "error" is used for the error event itself, and "failure" for the consequences of the error event.

2.3. AN OVERVIEW OF THE FOUR PERSPECTIVES ON HUMAN ERROR

The four perspectives to be discussed in detail later in this chapter are contrasted in Table 2.1 in terms of the error control strategies that are usually employed, their main areas of application and the frequency that the approaches are applied in the CPI.

2.3.1. Traditional Safety Engineering

The first perspective is the traditional safety engineering approach (Section 2.4). This stresses the individual factors that give rise to accidents and hence emphasizes selection, together with motivational and disciplinary approaches to accident and error reduction. The main emphasis here is on *behavior modification*, through persuasion (motivational campaigns) or punishment. The main area of application of this approach has been to occupational safety, which focuses on hazards that affect the individual worker, rather than process safety, which emphasizes major systems failures that could cause major plant losses and impact to the environment as well as individual injury.

2.3.2. Human Factors Engineering/Ergonomics

The second perspective to be considered in this chapter is the human factors engineering (or ergonomics) approach (HFE/E). This approach, described in Section 2.5, emphasizes the mismatch between human capabilities and system demands as being the main source of human error. From this perspective, the primary remedy is to ensure that the design of the system takes into account the physical and mental characteristics of the human. This includes consideration of factors such as:

- Workplace and job design to accommodate the job requirements of workers with differing physical and mental characteristics
- Design of the human–machine interface (HMI) such as control panels to ensure that process information can be readily accessed and interpreted and that appropriate control actions can be made
- Design of the physical environment (e.g., heat, noise, lighting), to minimize the negative physical and psychological effects of suboptimal conditions
- Optimizing the mental and physical workload on the worker

TABLE 2.1
Comparisons between Various Perspectives on Human Error

SOURCE OF ERROR AP-PROACH AND CONTROL STRATEGY	MAIN AREAS OF APPLICATION	TYPICAL APPROACHES	CURRENT USE BY THE CPI
Traditional Safety Engineering approach (control of error by motivational, behavioral, and attitude change)	• Occupational safety • Manual operations	• Selection • Behavior change via motivational campaigns • Rewards/punishment	Very common
Human Factors Engineering/Ergonomics approach (control of error by design, audit, and feedback of operational experience)	• Occupational/process safety • Manual/control operations • Routine operation	• Task analysis • Job design • Workplace design • Interface design • Physical environment evaluation • Workload analysis	Infrequent
Cognitive Engineering approach (control of error by design, audit, and feedback of operational experience, with particular reference to mental skills such as problem-solving and diagnosis)	• Process safety • Decision making/problem solving • Abnormal situations	• Cognitive task analysis • Decision support during emergencies • Incident analysis for human error root causes	Rare
Sociotechnical approach (control of error through changes in management policy and culture)	• Occupational/process safety • Effects of organizational factors on safety • Policy aspects • Culture	• Interviews • Surveys • Organizational redesign • Total Quality Management	More frequent in recent years

The emphasis on factors that can be manipulated during the design of a plant has led to the human factors engineering approach being described as "fitting the job to the person." This is in contrast to the approach of "fitting the person to the job," which focuses on training, selection, and behavior-modification approaches. The latter perspective is closer to the traditional safety approach. In fact, training is also usually considered by the human factors engineer, whereas occupational psychologists focus on the selection aspects. The HFE/E approach can be applied to both occupational and process safety and to manual and control room operations. The techniques and data available from the HFE/E approach have been largely developed and applied within the military, aerospace, and power generation sectors in the United States, although in Europe there has also been a long standing human factors research tradition in the process industries (see, e.g., Edwards and Lees, 1974; Goodstein et al., 1988). The practical application of these approaches to the CPI in both Europe and the United States has, however, been somewhat limited.

2.3.3. Cognitive Systems Engineering

The third approach, cognitive systems engineering (CSE) is described in Section 2.6. This is particularly useful in analyzing the higher level human functions involved in CPI operations, for example, problem solving, decision making, and diagnosis. It also provides an explanation of the underlying causes of errors in a wide range of CPI operations.

The approach developed from a general change in emphasis in applied psychology during the 1970s and 1980s, from viewing the human as a passive black box, analogous to an engineering component, to the view that individuals were purposeful in that their actions were influenced by future goals and objectives. The cognitive systems engineering approach is particularly applicable to activities such as planning and handling abnormal situations. Its methods include cognitive task analysis, which focuses on information processing failures, and the use of decision support systems of varying levels of sophistication to assist in the handling of abnormal situations. To date, the application of the approach has been limited in process plants, although the development of interest in the area by human factors specialists has stimulated research into the nature of the skills possessed by process workers. Nevertheless, this approach is the most comprehensive in terms of evaluating the underlying causes of errors. This means that it has particular relevance to analyzing the causes of recurrent errors and for predicting specific errors that may have serious consequences as part of safety analyses.

2.3.4. Sociotechnical Systems

The fourth approach, the sociotechnical systems perspective, is described in Section 2.7. This arose from a realization that human performance at the operational level cannot be considered in isolation from the culture, social factors and management policies that exist in an organization. For example, the availability of good operating procedures is well known as an important contributory factor in influencing the likelihood of errors leading to major disasters. The existence of good procedures requires a procedures design policy to be implemented by plant management. This should include elements such as participation by the eventual users of the procedures, design of the procedures based on analysis of operational tasks, their preparation in accordance with accepted human factors principles, and a system for modifying the procedures in light of operational experience. All of this requires resources to be allocated by managers at an appropriate level in the organization. The existence of good quality procedures does not guarantee that they will be used. If a culture exists that encourages workers to take shortcuts not specified in the procedures in order to achieve required production levels, then accidents and losses may still occur. These are typical issues that are considered by the sociotechnical systems approach.

The sociotechnical systems perspective is essentially top-down, in that it addresses the question of how the implications of management policies at all levels in the organization will affect the likelihood of errors with significant consequences. The sociotechnical systems perspective is therefore concerned with the implications of management and policy on system safety, quality, and productivity.

2.3.5. Conclusions

The approaches described in this chapter can be regarded as complementary rather than competing methodologies. They all have a part to play in an integrated approach to the management of human error to reduce accidents in the CPI. Having said this, we will place rather more emphasis on approaches other than the traditional safety approach in this book.

This is partly because the traditional approach is well known and documented in the industry, whereas the other approaches have received very little application to date. In addition, despite the successes of the traditional approach in the area of occupational safety, it may be less applicable in areas such as the prevention of major chemical accidents.

This is because many of the factors that have been shown to be the antecedents of major process accidents (e.g., poor procedures, inadequate training) are not usually under the control of the individual worker. The other approaches can also be applied to improving quality and productivity as well as process safety and can be readily integrated with engineering system safety techniques, as will be described in Chapters 4 and 5.

2.4. THE TRADITIONAL SAFETY ENGINEERING APPROACH TO ACCIDENTS AND HUMAN ERROR

The traditional safety engineering approach to accident causation focuses on the individual rather than the system causes of error. Errors are primarily seen as being due to causes such as lack of motivation to behave safely, lack of discipline or lack of knowledge of what constitutes safe behavior. These are assumed to give rise to "unsafe acts." These unsafe acts, in combination with "unsafe situations" (e.g., unguarded plant, toxic substances) are seen as the major causes of accidents.

One of the origins of this view of error and accident causation is the theory of accident proneness, which tried to show that a small number of individuals were responsible for the majority of accidents. Despite a number of studies that have shown that there is little statistical evidence for this idea (see, e.g., Shaw and Sichel, 1971); the belief remains, particularly in traditional industries, that a relatively small number of individuals account for the majority of accidents. Another element in the emphasis on individual responsibility has been the legal dimension in many major accident investigations, which has often been concerned with attributing blame to individuals from the point of view of determining compensation, rather than in identifying the possible system causes of error.

2.4.1. Accident Prevention from the Traditional Perspective

Based on this view of accident causation, certain strategies for prevention emerge. The control of unsafe conditions is achieved partly by methods such as eliminating the hazard at its source or by the use of guards or protective equipment. However, the majority of resources are directed at eliminating unsafe acts, either by motivating the worker to change his or her behavior or by retraining, on the assumption that much unsafe behavior is simply due to lack of knowledge, or because the correct way to do things has been forgotten. Retraining, in this context, usually refers to reinforcing existing work practices, or "more of the same."

The basic assumption is that the individual always has the choice of whether or not to behave in an unsafe manner. The implication of this assumption is that the responsibility for accident prevention ultimately rests with the individual worker. It also implies that as long as management has expended reasonable efforts to persuade an individual to behave responsibly, has provided training in safe methods of work, and has provided appropriate guarding of hazards or personal protection equipment, then it has discharged its responsibilities for accident prevention. If these remedies fail, the only recourse is disciplinary action and ultimately dismissal.

In some cases, more subtle approaches to behavior modification have been employed. Applications of behavior modification to safety are discussed in McKenna (1989), Hale and Glendon (1987), and Petersen (1984).

Modern behavior-modification programs rely on the identification and reinforcement of safe behaviors. Considerable improvements in measures of safety performance have been attributed to the introduction of these approaches (see McSween, 1993, for a petrochemical example). However, other studies have indicated that performance may return to its original level if the programs are withdrawn. It is therefore important to maintain a continuing program to ensure that the initial levels of improvements are maintained. Also, the benefits of behavior modification programs have mainly been demonstrated in the context of work activities where there is a high level of discretion with regard to how tasks are carried out. Thus, existing "unsafe behaviors" can be identified and alternative acceptable behaviors substituted in their place. In the case study cited in Marcombe et al. (1993) for example, the main unsafe behaviors that were cited as precursors to accidents were as follows: not checking out equipment, tools, and the work area; not using personnel protective equipment; and not using the proper body position required by the task. These behaviors were the focus of the program.

2.4.2. Disadvantages of the Traditional Approach

Despite its successes in some areas, the traditional approach suffers from a number of problems. Because it assumes that individuals are free to choose a safe form of behavior, it implies that all human error is therefore inherently blameworthy (given that training in the correct behavior has been given and that the individual therefore knows what is required). This has a number of consequences. It inhibits any consideration of alternative causes, such as inadequate procedures, training or equipment design, and does not support the investigation of root causes that may be common to many accidents. Because of the connotation of blame and culpability associated with error, there are strong incentives for workers to cover up incidents or near misses, even if these are due to conditions that are outside their control. This means that information on error-inducing conditions is rarely fed back to individuals such as engineers and managers who are in a position to develop and apply remedial measures such as the redesign of equipment, improved training, or redesigned procedures. There is, instead, an almost exclusive reliance on methods to manipulate behavior, to the exclusion of other approaches.

The traditional approach, because it sees the major causes of errors and accidents as being attributable to individual factors, does not encourage a consideration of the underlying causes or mechanisms of error. Thus, accident data-collection systems focus on the characteristics of the individual who has the accident rather than other potential contributory system causes such as inadequate procedures, inadequate task design, and communication failures.

The successes of the traditional approach have largely been obtained in the area of occupational safety, where statistical evidence is readily available concerning the incidence of injuries to individuals in areas such as tripping and falling accidents. Such accidents are amenable to behavior modification approaches because the behaviors that give rise to the accident are under the direct control of the individual and are easily predictable. In addition, the nature of the hazard is also usually predictable and hence the behavior required to avoid accidents can be specified explicitly. For example, entry to enclosed spaces, breaking-open process lines, and lifting heavy objects are known to be potentially hazardous activities for which safe methods of work can be readily prescribed and reinforced by training and motivational campaigns such as posters.

In the case of process safety, however, the situation is much less clear cut. The introduction of computer control increasingly changes the role of the worker to that of a problem solver and decision maker in the event of abnormalities and emergencies. In this role, it is not sufficient that the worker is trained and conditioned to avoid predictable accident inducing behaviors. It is also essential that he or she can respond flexibly to a wide range of situations that cannot necessarily be predicted in advance. This flexibility can only be achieved if the worker receives extensive support from the designers of the system in terms of good process information presentation, high-quality procedures, and comprehensive training.

Where errors occur that lead to process accidents, it is clearly not appropriate to hold the worker responsible for conditions that are outside his or her control and that induce errors. These considerations suggest that behavior-modification-based approaches will not in themselves eliminate many of the types of errors that can cause major process accidents.

Having described the underlying philosophy of the traditional approach to accident prevention, we shall now discuss some of the specific methods that are used to implement it, namely motivational campaigns and disciplinary action and consider the evidence for their success. We shall also discuss another frequently employed strategy, the use of safety audits.

2.4.3. Safety Campaigns

On the assumption that poor motivation or lack of safety awareness have a major contribution to accidents, most companies carry out safety campaigns. A safety campaign may be defined as "an operation or program aimed at influencing people to think or act in a safe manner." Such programs are designed to influence behavior using praise, punishment or fear. In addition, they may also provide specific information as a reinforcement for safety training.

There are at least three different forms of motivational campaigns: posters, films, and incentive schemes.

For posters, there are broadly four distinct types: (1) those appealing to a general awareness of safety issues; (2) those containing a warning or information on specific hazards; (3) pPosters providing general information on, for example, regulatory requirements; and (4) fear-inducing posters.

Films or videos cover the same broad areas as posters. They are typically fairly short (not more than 30 minutes) and are usually intended to be used during training. Instructor's notes are often supplied with the audiovisual material.

Many companies operate incentive schemes, ranging from competitions among departments or factories for an award (e.g., a certificate or trophy) to elaborate schemes involving inspection and auditing to check for the achievement of certain safety objectives, which are rewarded with prizes.

The question of the effectiveness of motivational campaigns is not easy to answer. The obvious method would be to look at accident rates. However, recorded accident rates vary widely according to the propensity to report or not report events.

A safety campaign may only reduce the willingness of the workforce to report an accident rather than significantly reducing the underlying accident occurrences and hazards.

This is a problem that is not unique to motivational campaigns but is common to all approaches involving the monitoring of accidents or human error, as will be discussed in Chapter 6.

An indirect way to evaluate the effectiveness of safety campaigns is to look at some other observable "performance indicator" such as the use of personal protection equipment (PPE). Many campaigns are targeted at increasing the use of different types of PPE. Monitoring the results of such campaigns is done by establishing a baseline level of use of the equipment prior to the campaign and then looking at the percentage change in this use by the same workforce shortly after the campaign and then after some months have passed. Table 2.2 gives some summary results from a study by Pirani and Reynolds (1976) showing the effects of different types of motivational schemes on the use of PPE for head, hands, eyes, and feet. The first column shows the change from the baseline measurement 2 weeks after the campaign. The second column records the change from the baseline 4 months after the campaign.

In Table 2.2 the results from the use of posters and films are shown in the first three rows. Two points should be noted. First, all three measures show only short term gains. After four months the change in the pattern of use of

TABLE 2.2
Effect of Different Motivational Schemes on Use of PPE (adapted from Pirani and Reynolds, 1976)

MEASURE	PERCENT CHANGE AFTER 2 WEEKS	PERCENT CHANGE AFTER 4 MONTHS
General safety posters	+51%	+11%
Appropriate films	+40%	+11%
Fear posters	+18%	− 2%
Disciplinary measures	+39%	− 7%
Discussion + opinion leaders	+ 9%	+ 2%
Role playing	+71%	+68%

PPE is very similar, if not lower, than the baseline level. This result has been verified by other researchers. Second, the use of fear-inducing posters was not as effective as the use of general safety posters. This is because unpleasant material aimed at producing high levels of fear often affects peoples' *attitudes* but has a varied effect on their *behavior*. Some studies have found that the people for whom the fearful message is least relevant—for example, nonsmokers in the case of anti-smoking propaganda—are often the ones whose attitudes are most affected. Some posters can be so unpleasant that the message itself is not remembered.

There are exceptions to these comments. In particular, it may be that horrific posters change the behavior of individuals if they can do something immediately to take control of the situation. For example, in one study, fear-inducing posters of falls from stairs, which were placed immediately next to a staircase, led to fewer falls because people could grab a handrail at once. In general, however, it is better to provide simple instructions about how to improve the behavior rather than trying to shock people into behaving more safely. Another option is to link competence and safe behavior together in people's minds. There has been some success in this type of linkage, for example in the oil industry where hard hats and safety boots are promoted as symbols of the professional.

Table 2.2 indicates that the most successful campaign to encourage the use of PPE involved the use of role playing. This is where people are asked to advocate differing views from their own or to act in ways which differed from their usual behavior. In this case, those workers who did not normally wear protective equipment could, for example, be asked to take part in a discussion supporting the wearing of PPE. Such role playing may be effective for two reasons. First, the person will gain greater familiarity with the opposing view. Second, and more importantly, people need to justify why they are doing something and, in this case, advocating the opposite position competently might only be explainable to themselves in terms of partly believing in that position.

Table 2.2 does not include any reference to the effectiveness of incentive schemes. The evidence in this regard is not conclusive. There have often been reports of quite spectacular improvements in accident rates. However, these do not form a controlled evaluation. The main difficulty in trying to establish the effectiveness of incentive schemes is that such campaigns are often only part of a "total safety climate" approach which includes changes in work procedures, job design, etc. In such cases it is difficult to separate out the effects of the incentive scheme alone. However, researchers suggest that simple competitions are not as effective as such "total safety climate" programs, especially when the latter include elaborate setting and monitoring of safety targets.

In summary, the following conclusions can be drawn with regard to motivational campaigns:

- Success is more likely if the appeal is direct and specific rather than diffuse and general. Similarly, the propaganda must be relevant for the workforce at their particular place of work or it will not be accepted.
- Posters on specific hazards are useful as short-term memory joggers if they are aimed at specific topics and are placed in appropriate positions. Fear or anxiety inducing posters must be used with caution. General safety awareness posters have not been shown to be effective
- The safety "campaign" must not be a one-shot exercise because then the effects will be short-lived (not more than 6 months). This makes the use of such campaigns costly in the long run despite the initial appearance of a cheap solution to the problem of human error.
- Motivational campaigns are one way of dealing with routine violations (see Section 2.5.1.1). They are not directly applicable to those human errors which are caused by design errors and mismatches between the human and the task. These categories of errors will be discussed in more detail in later sections.

2.4.4. Disciplinary Action

The approach of introducing punishment for accidents or unsafe acts is closely linked to the philosophy underlying the motivational approach to human error discussed earlier. From a practical perspective, the problem is how to make the chance of being caught and punished high enough to influence behavior. From a philosophical perspective, it appears unjust to blame a person for an accident that is due to factors outside his or her control. If a worker misunderstands badly written procedures, or if a piece of equipment is so badly designed that it is extremely difficult to operate without making mistakes, then punishing the individual will have little effect on influencing the recurrence of the failure.

In addition, investigations of many major disasters have shown that the preconditions for failure can often be traced back to policy failures on the part of the organization. Disciplinary action may be appropriate in situations where other causes have been eliminated, and where an individual has clearly disregarded regulations without good reason. However, the study by Pirani and Reynolds indicates that disciplinary measures were ineffective in the long term in increasing the use of personal protective equipment. In fact, four weeks after the use of disciplinary approaches, the use of the equipment had actually declined. The major argument against the use of disciplinary approaches, apart from their apparent lack of effectiveness, is that they create fear and inhibit the free flow of information about the underlying causes of accidents. As discussed earlier, there is every incentive for workers and line managers to cover up near accidents or minor mishaps if they believe punitive actions will be applied.

2.4.5. Safety Management System Audits

The form of safety audits discussed in this section are the self-contained commercially available generic audit systems such as the International Safety Rating System (ISRS). A different form of audit, designed to identify specific error inducing conditions, will be discussed in Section 2.7. Safety audits are clearly a useful concept and they have a high degree of perceived validity among occupational safety practitioners. They should be useful aids to identify obvious problem areas and hazards within a plant and to indicate where error reduction strategies are needed. They should also support regular monitoring of a workplace and may lead to a more open communication of problem areas to supervisors and managers. The use of safety audits could also indicate to the workforce a greater management commitment to safety.

Some of these factors are among those found by Cohen (1977) to be important indicators of a successful occupational safety program. He found that the two most important factors relating to the organizational climate were evidence of a strong management commitment to safety and frequent, close contacts among workers, supervisors, and management on safety factors. Other critical indicators were workforce stability, early safety training combined with follow-up instruction, special adaptation of conventional safety practices to make them applicable for each workplace, more orderly plant operations and more adequate environmental conditions.

Despite these potential benefits, there are possible problems associated with the use of generic safety audit systems. Questions that need to be considered in the case of such standardized audits include:

- How are the critical factors identified?
- What validation exists for such schemes?
- What does it really mean to do well on such audits i.e. what evaluation criteria are being used?
- What is the likelihood of missing an industry specific hazard when using a general scheme?

Such audits may therefore be useful as a method of increasing safety awareness and management commitment to safety as part of a more general attempt to reduce accidents. They should be treated as first steps and management must be prepared to do more than just carry out a safety audit. The authors of safety audits must be prepared to provide guidance on the next steps in error reduction once the problems have been identified.

Problems can also arise when the results of safety audits are used in a competitive manner, for example, to compare two plants. Such use is obviously closely linked to the operation of incentive schemes. However, as was pointed out earlier, there is no evidence that giving an award to the "best plant" produces any lasting improvement in safety. The problem here is that the competitive aspect may be a diversion from the aim of safety audits, which is to identify problems. There may also be a tendency to "cover-up" any problems in order to do well on the audit. Additionally, "doing well" in comparison with other plants may lead to unfounded complacency and reluctance to make any attempts to further improve safety.

2.4.6 Training

There is no question that training, particularly where it is task specific, is extremely important in the attempt to reduce human failures. Safety campaigns must always support, not replace safety training. However, all too often, organizations have attacked the problem of human error as simply a matter of training. Training departments have become the dumping grounds for problems created by factors such as bad design and poor management. It must be recognized that even the best-trained worker will experience difficulties if he or she is faced with a complex problem, a poorly designed human–machine interface, unrealistic task and workload demands and a "turbulent" environment with noise, interruptions, and stress. No amount of training can totally compensate for all of these adverse factors. Training should therefore consider the design of the task, equipment, job aids, and similar factors rather than be used instead of them. Training has to be directed at the underlying causes of an error and for this reason, reporting systems need to explicitly identify these root causes. Unfortunately, in many cases, the training approach adopted in response to errors is to provide "more of the same."

2.5. THE HUMAN FACTORS ENGINEERING AND ERGONOMICS APPROACH (HF/E)

Human factors engineering (or ergonomics), is a multidisciplinary subject that is concerned with optimizing the role of the individual in human–machine systems. It came into prominence during and soon after World War II as a result of experience with complex and rapidly evolving weapons systems. At one stage of the war, more planes were being lost through pilot error than through enemy action. It became apparent that the effectiveness of these systems, and subsequently other systems in civilian sectors such as air transportation, required the designer to consider the needs of the human as well as the hardware in order to avoid costly system failures.

The practical needs of military and aerospace systems tended to focus interest on human–machine interfaces (e.g., aircraft cockpits), with particular emphasis on information displays and the design of controls to minimize error. The predominant model of the human prevalent at that time (called *behaviorism*) concentrated exclusively on the inputs and outputs to an individual and ignored any consideration of thinking processes, volition, and other distinctively human characteristics. However, this model considerably influenced the early workers in HF/E. In fact many of the tasks that were studied in military systems were highly proceduralized and therefore involved little use of higher level skills such as decision making or problem solving. It was therefore possible for early HF/E practitioners to make a contribution to the design of more effective systems even though they only considered a limited subset of human skills and capabilities.

From the 1960s onward, there was a greater interest in psychological issues, dominated by the concept of the human as a single-channel processor of information. This stimulated research into a number of areas. Studies of mental workload were concerned with the ability of humans to cope with extremely high levels of information in situations such as air traffic control. Vigilance studies, which focused on the human's role in situations with very low levels of stimulation such as radar monitoring, represented the other extreme of human performance that was considered.

The conceptualization of the human as a single-channel processor of information was useful in emphasizing the need to design systems to take into account human capabilities and limitations. It did not, however, consider issues such as the meaning that people assign to their work, their intentions, and topics such as problem solving, decision making, and diagnosis. Despite these limitations, the traditional HF/E approach has been the source of many of the practical approaches and techniques which will be described in subsequent chapters. Some of the key concepts used in this approach will therefore be described in this section.

From the traditional HF/E perspective, error is seen as a consequence of a mismatch between the demands of a task and the physical and mental capabilities of an individual or an operating team. An extended version of this perspective was described in Chapter 1, Section 1.7. The basic approach of HF/E is to reduce the likelihood of error by the application of design principles and standards to match human capabilities and task demands. These encompass the physical environment (e.g., heat, lighting, vibration), and the design of the workplace together with display and control elements of the human–machine interface. Examples of the approach are given in Wilson and Corlett (1990) and Salvendy (1987).

2.5.1. The Human–Machine Interface

The human–machine interface (usually abbreviated to interface) is a major focus of interest for the HF/E approach to the reduction of human error. A representation of the interface in a CPI context is provided in Figure 2.2. The interface is the boundary across which information from the process is transduced by sensors and then displayed in a form that can be utilized by the human process controllers. It also allows control actions to be made to change the state of the system.

Figure 2.2 provides a more detailed description of the human side of the interface. This is based on the information processing model of Wickens (1984). It describes how the information presented at the interface (e.g., a control panel) goes through various stages of processing before a response is eventually made in the form of a control action (e.g., pressing a button to close a valve). The first stage, sensing and perception, involves the information being captured by a sensory channel, for example, vision, after which it will be stored in a limited-capacity store called working memory. The way in which information is acquired will be influenced by the knowledge and experience of the world, which is part of the observer's long-term memory. For example, an operator scanning a control panel for indications of problems will tend to focus on sources of information (e.g., alarms) that have proved to be particularly important in the past.

Interpretation of the information in working memory involves the use of knowledge and experience from long-term memory. For example, on the basis of experience, the panel operator may interpret a rapid rise in temperature as indicative of a dangerous situation. The process of diagnosis and then deciding on and selecting an appropriate response occurs at the next stage of processing (represented by the next box in Figure 2.2). Finally, an appropriate response is initiated (e.g., closing the valve), which will change the state of the system. This, in turn, will be displayed by the interface, thus completing the processing loop.

The Wickens model suggests that there are finite information-processing or attentional resources available, as represented by the box in Figure 2.2. These resources can be distributed in different ways but cannot be increased. Thus, interpretation of complex or unusual information displayed by the interface will leave fewer resources available for handling the response selection and decision making demands. This provides a theoretical basis for the view of human error described in Section 1.7, which described error as a mismatch between demands and capabilities.

A familiar example of limited attentional resources being distributed among different mental and physical processes occurs in car driving. A driver in a foreign country who is required to operate a manual gear change system, and at the same time drive on the opposite side of the road, may find that he or she has little capacity available to navigate or respond to a sudden stop by the car in front.

In the CPI, the most extensively studied human–machine interface is in the central control room in automated plants where plant information is displayed on visual display units (VDUs) and appropriate control actions are made by the operating team. In the case of a highly automated plant, the primary role of the human is to respond to unexpected contingencies such as plant states that have not been anticipated by the designers of the automatic

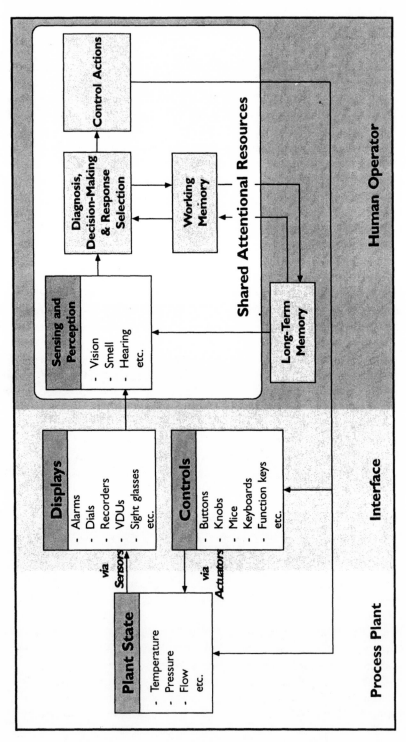

FIGURE 2.2. The Human–Machine Interface (adapted from Wickens, 1984).

control and protection systems. Even in the most automated systems such interventions are likely to occur occasionally, due to the difficulty of anticipating every process state in a complex plant. Only extremely simple processes can be completely automated. Although a large number of highly automated plants exist, it is probably true to say that the majority still require considerable human intervention from the control room, together with manual operations on the plant. This is particularly true of batch processes. The issue of whether or not automation is the solution to the problem of human error will be discussed further in Section 2.5.5.

Although most research on human factors in process control has focused on the control room interface, in fact the human–machine interface concept can be applied to all situations where the CPI worker has to take actions based on information acquired directly or indirectly concerning the state of the process. For example, a local indicator situated close to a reactor or a charging vessel is also a process interface. In these cases, the displays may consist of a sight glass or level indicator and the controls may be manually operated valves. It is important to emphasize that the principles of interface design are of equal importance to these situations as in a central control room. In fact, there is a tendency to focus interface design resources in the control room at the expense of equally critical but less prominent interfaces on the plant. Neglect of operability considerations by designers often leads to highly error inducing situations on the plant such as gauges displaying important information being placed in inaccessible positions, and valves or process lines being unlabeled.

Example 2.4: An Error Due to a Poorly Designed Interface

In a resin plant, solvents were directed from storage tanks to a blender by means of solvent charging manifold. Because of the poor panel layout and labeling of the charging manifold, a worker made connections that pumped solvent to blender 21A instead of 12A as directed by the instructions. An earlier error had left the valve open from the charging manifold to blender 21A and hence the misdirected solvent degraded a batch already in the blender (this example will be analyzed in more detail in Chapter 7).

The functions of the interface can be summarized as follows:

- To allow the presentation of process information consistent with the worker's needs and expectations (e.g., monitoring normal operations, responding to abnormalities in emergencies)
- To provide immediate feedback for control actions
- To support diagnosis, decision making and planning
- To facilitate the selection of the correct control actions and minimize accidental activation of controls

A number of design principles exist to achieve these aims, and these are described in detail in handbooks such as those described in the Bibliography at the end of the book. Examples of such principles are given below:

- *Representational layout of control panels.* Where the physical location of items is important, for example, area displays in fire control systems, the layout of the displays on a control panel should reflect the geographical layout of the plant. In other cases a functional arrangement of the elements of the process plant will be appropriate, for example, when monitoring the status of the system via an alarm panel.
- *Sequential design.* When a particular procedure is always executed in sequential order, for example, the start-up of a distillation column, a similar sequential arrangement of the controls will help to ensure that parts of the sequence are not omitted.
- *Design according to frequency of use or importance.* Controls and displays that are frequently used or are of special importance (e.g., critical alarms), should be placed in prominent positions, for example, near the center of the control panel.
- *Hierarchical organization of information.* Information should be provided at a range of different levels of detail from major systems such as reactors, to individual components such as valves, in order to satisfy a range of different process control requirements.

2.5.2. Human Error at the Human–Machine Interface

The following sections discuss how errors can arise at each of the stages of perception, decision-making and control actions. The account given below of how information is processed by the human perceptual system is highly simplified. More technical descriptions are provided in many textbooks, for example, Wickens (1984).

2.5.2.1. Perception
As described earlier, in the first stage of perception, information is acquired via the senses from a number of sources. These may include gauges and chart recorders, VDU screens in a control room, verbal communication with individuals on the plant, or direct observation of process variables. In the short term, this information provides feedback with regard to specific control actions.

For example, if a worker turns on a stirrer in a reactor, he or she may use a local or control room indicator to verify that current is flowing to the agitator motor. Errors may arise at several points in the input process. At the sensory stage, there may be so many sources of information that the worker may be unable to scan them all in the time available. This can be a particular problem when a large number of alarms occur following a major process disturbance.

The information may not be readily distinguishable either because it is too faint or because it may not be easily separated from other similar information. For example, a critical measurement on a multipoint temperature recorder may be lost in the surrounding clutter of irrelevant information. As discussed in the cognitive engineering approach described in Section 2.6, the worker may also ignore sources of information because of preconceptions that lead him or her to believe they are not significant.

The sensory input information is interpreted according to the worker's **mental model** of the process. The mental model is stored in long-term memory and is an internal representation of the process and its dynamics, which is used as a basis for decision making. This model is built up on the basis of the worker's experience in operating the plant and gaining an intuitive "feel" of the effects of various control actions. The model may be quite different from a chemical engineering model of the plant process but may be perfectly adequate as a basis for controlling the plant. However, if the model is based only on the worker's experience of the plant under normal operating conditions, errors could occur if actions are made in unusual situations for which the model does not apply.

This implies that plant controllers need frequent exposure to problem-solving training and evaluation to ensure that their mental model is kept up to date. A more detailed consideration of mental models is contained in Lucas (1987).

2.5.2.2. Decision Making

During the decision-making stage, evidence acquired from the system is used in an individual's working memory in conjunction with information from long-term memory to decide on an appropriate course of action. The long-term store contains the "mental model" mentioned earlier. The compatibility of the mental model to the actual state of the system and the process dynamics has an important bearing on the likelihood of an error being made. The translation between the actual state of the system and the mental model is facilitated by the use of displays such as schematic diagrams and the availability of hierarchically organized display systems. These have the effect of reducing the information-processing load involved in translating the plant display into an internal mental representation of the process.

Decision making may involve calculations, reference to procedures and past experience, and other demands on long-term memory. This contributes further to the overall mental workload. From the HF/E perspective, many errors are likely to arise from information processing overload, essentially from the mismatch between demands and capabilities. Information-processing demands can be reduced by the provision of information in the form of job aids such as flow charts or decision trees.

2.5.2.3. Control Actions

The final stage of the information-processing chain involves the selection and execution of a particular control action or response, on the basis of the decisions made in the preceding stage. The complexity of the selection process is influenced by the number of alternative control strategies the worker has to choose from, the physical characteristics of the control to be operated and the familiarity of the control action. For example, if the shutdown button for a distillation column is clearly and unambiguously marked, very little searching or information processing is necessary. If controls are ambiguous, or closely crowded together, the likelihood of accidental activation increases, as do the processing demands on the worker. Ergonomics textbooks, such as those described in the general bibliography, contain extensive guidelines for the use of different control types depending on the application.

2.5.3. Information Processing and Mental Workload

As discussed in the last section, attentional resources (see Figure 2.2) are taken up whenever the worker takes in data via the sensory channels such as vision or hearing, and when carrying out processes such as decision making, and making control actions. Since there are limitations on these resources, information processing overload is a common cause of errors. If, for example, a person is trying to perform a complex fault diagnosis, he or she may not have any spare capacity to deal effectively with an unexpected process deviation. The total information-processing load on the individual from inputs, central processing, and outputs, is known as the mental workload. Comprehensive accounts of research in this area are provided by Moray (1979, 1988). Techniques for measuring mental workload are reviewed in Hockey et al. (1989) and Wierwille and Eggemeier (1993).

In order to minimize errors, it is important that the mental workload is within a person's capabilities. If the workload exceeds these capabilities by a moderate amount, the individual may be able to utilize short-term "coping strategies," to maintain performance. However, in addition to the fact that such strategies may involve elements of risk taking, they often lead to some physical or psychological cost. Experiments have shown that experienced workers such as air traffic controllers can maintain their performance even if they are asked to cope with increasing numbers of flights. However, they often exhibit chronic stress symptoms if they are required to maintain this performance over long periods of time. If the worker is forced to use coping strategies as a regular part of his or her work, it is likely that feelings of physical or mental strain will ensue. This may lead to long term problems such as stress illnesses and absenteeism. At very high levels of mental workload, even coping strategies will be inadequate and errors will start to increase rapidly. There has been a considerable amount of research carried out in the area of mental workload, particularly in the aerospace industry. This has been partly driven by the desire to reduce staffing levels on flight decks.

In the case of the CPI, there are relatively few situations where control room workers are likely to face continuous periods of overload. However, when overload does occur it is likely to be associated with situations when the plant is in an unusual or abnormal state for which the workers may not have any rules or procedures available. In these situations, knowledge-based processing (see Section 2.6.2), which needs considerable mental resources, will be required and errors of diagnosis are likely to occur.

2.5.4. Automation and Allocation of Function

A commonly suggested solution to the problem of human error is to automate the plant process. The aim of automation is to replace human manual control, planning, and problem solving by automatic devices and computers. The topic of allocation of function is becoming increasingly important with the use of computer-based process control systems, which tend to change the role of the control process operator from that of a direct controller to a system monitor. Allocation of function is concerned with which functions to assign to human control and which to delegate to automatic systems such as computers. The engineering approach is normally to automate all functions for which it is technically feasible to develop an automatic system. In practice there are a number of problems associated with this approach. For example, operating conditions (e.g., the characteristics of feed stocks, the reliability of automatic controllers) are often more variable than the designer is able to take into account. This means that the automated system actually has to be controlled manually during a proportion of its operating range.

This form of unplanned manual operation is unsatisfactory on a number of counts. The fact that the operator may normally be insulated from the process by the automatic control systems means that he or she will probably not be able to develop the knowledge of process dynamics ("process feel") necessary to control the system manually, particularly in extreme conditions. Also, the fact that manual control was not "designed into" the systems at the outset may mean that the display of process information and the facilities for direct control are inadequate. A number of techniques are available to assist designers in the allocation of function process. Some of these are described in Meister (1985). In a paper entitled "Ironies of Automation" Bainbridge (1987) notes four areas where the changed role of the human in relation to an automated system can lead to potential problems. These will be discussed below.

2.5.4.1. The Deterioration of Skills
With automatic systems the worker is required to monitor and, if necessary, take over control. However, manual skills deteriorate when they are not used.

Previously competent workers may become inexperienced and therefore more subject to error when their skills are not kept up to date through regular practice. In addition, the automation may "capture" the thought processes of the worker to such an extent that the option of switching to manual control is not considered. This has occurred with cockpit automation where an alarming tendency was noted when crews tried to program their way out of trouble using the automatic devices rather than shutting them off and flying by traditional means.

Cognitive skills (i.e., the higher-level aspects of human performance such as problem solving and decision making), like manual skills, need regular practice to maintain the knowledge in memory. Such knowledge is also best learned through hands-on experience rather than classroom teaching methods. Relevant knowledge needs to be maintained such that, having detected a fault in the automatic system, the worker can diagnose it and take appropriate action. One approach is to design-in some capability for occasional hands-on operation.

2.5.4.2. The Need to Monitor the Automatic Process

An automatic control system is often introduced because it appears to do a job better than the human. However, the human is still asked to monitor its effectiveness. It is difficult to see how the worker can be expected to check in real time that the automatic control system is, for example, using the correct rules when making decisions. It is well known that humans are very poor at passive monitoring tasks where they are required to detect and respond to infrequent signals. These situations, called vigilance tasks, have been studied extensively by applied psychologists (see Warm, 1984). On the basis of this research, it is unlikely that people will be effective in the role of purely monitoring an automated system.

2.5.4.3. The Need to Hold an Accurate and Up-to-Date
Mental Model of the Plant Processes

As discussed earlier, the successful diagnosis of faults in automated control systems is highly dependent on the mental model the worker has built up of the current state of the plant processes. Such a model takes time to construct. An individual who has to act quickly may not be able to make the necessary diagnoses without time to build up and consult his or her mental model. Even in a highly automated plant, provision needs to be made to display major process deviations quickly.

Example 2.5: Failure of Automated Process System Because Critical Information Was Not Displayed

In a highly automated plant, a violent exothermic reaction occurred because of an unanticipated interaction between a chemical process and a by-product. The symptoms of the problem, a sudden temperature rise, went unnoticed because the process plant VDU display page for alarms was not being displayed, and there was no alarm algorithm to detect rapid temperature changes.

2.5.4.4. The Possibility of Introducing Errors

Automation may eliminate some human errors at the expense of introducing others. One authority, writing about increasing automation in aviation, concluded that "automated devices, while preventing many errors, seem to invite other errors. In fact, as a generalization, it appears that automation tunes out small errors and creates opportunities for large ones" (Wiener, 1985). In the aviation context, a considerable amount of concern has been expressed about the dangerous design concept of "Let's just add one more computer" and alternative approaches have been proposed where pilots are not always taken "out of the loop" but are instead allowed to exercise their considerable skills.

Example 2.6: An Error Due to Overreliance on Technology (Wiener, 1985)

Overreliance on technology was a feature of an accident involving a China Airlines B747-SP that occurred approximately 300 miles northwest of San Francisco in 1989. Toward the end of the flight, the aircraft suffered an in-flight disturbance at 41,000 feet following the loss of its number 4 engine. The aircraft, which was flying on autopilot at the time, rolled to the right during attempts by the crew to relight the engine, following which it entered into an uncontrolled descent. The crew were unable to restore stable flight until the aircraft reached 9500 feet, by which time the aircraft had exceeded its maximum operating speed and sustained considerable damage. In conducting its inquiry, the National Transportation Safety Board concluded that the major contributory factor underlying the incident occurrence was the crew's overdependence on the autopilot during attempts to relight the malfunctioning engine. The correctly functioning autopilot effectively masked the onset of loss of control of the aircraft.

2.5.5. System Reliability Assessment and Human Error

The main thrust of the HF/E approach is to provide the conditions that will optimize human performance and implicitly minimize human error. However, there is rarely any attempt to predict the nature and likelihood of *specific* human errors and their consequences. By contrast, the study of human error in the context of systems reliability is concerned almost exclusively with these latter issues. It is appropriate to introduce the systems reliability assessment approach to human error at this stage because, until recently, it was largely based on the mechanistic view of the human in traditional HF/E which was described at the beginning of Section 2.5.

Interest in human error in **system reliability** originated in work on military missile systems in the 1950s, when it became apparent that a large proportion of system failures could be traced to errors in design, manufacturing, and assembly. The application of formal techniques such as fault tree analysis in nuclear safety and the occurrence of the Three Mile Island accident also emphasized the need for predictive analyses of human error. Human reliability assessment originated from a very specific engineering requirement: the need to insert human error probabilities in fault trees for assessing the likelihood that predefined procedures involving human actions would be successfully carried out. For this reason human reliability assessment in the context of safety analysis is very mechanistic in its philosophy and is based on a simplified version of the HF/E approach described in earlier sections. In the CPI, human reliability assessment forms an integral part of chemical process quantitative risk analysis (CPQRA). A comprehensive description of this application is given in a companion volume in this series, *Guidelines for Chemical Process Quantitative Risk Analysis* (1989b) published by CCPS. More detailed descriptions of specific quantification techniques are provided in Chapter 5 of this book and in Swain (1989), Miller and Swain (1987), Kirwan, Embrey, and Rea (1988), and Kirwan (1990).

In this mechanistic approach to human reliability, the individual is modeled as being analogous to a hardware component that provides a function when required. A definition of human reliability from this perspective is as follows:

Human reliability is the probability that a job or task will be successfully completed by personnel at any required stage in system operation within a required minimum time (if a time requirement exists) (Meister, 1966).

This has close affinities with definitions of system reliability from a hardware perspective, for example, "the probability of performing a function under specified conditions for a specific period of time" (Zorger, 1966).

When performing human reliability assessment in CPQRA, a qualitative analysis to specify the various ways in which human error can occur in the situation of interest is necessary as the first stage of the procedure. A comprehensive and systematic method is essential for this. If, for example, an error with critical consequences for the system is not identified, then the analysis may produce a spurious impression that the level of risk is acceptably low. Errors with less serious consequences, but with greater likelihood of occurrence, may also not be considered if the modeling approach is inadequate. In the usual approach to human reliability assessment, there is little assistance for the analyst with regard to searching for potential errors. Often, only omissions of actions in proceduralized task steps are considered.

Since this approach to human reliability has its roots in the traditional HF/E perspective, it does not include any systematic means for identifying errors due to failures in higher level human functions such as diagnosis. Nevertheless, such diagnostic errors can give rise to particularly serious failures, where they lead to an erroneous series of actions being initiated based on the mistaken diagnosis. The Three Mile Island accident was a typical result of these types of errors. In order to address cognitive errors of this type, a comprehensive model of human error is required, as is discussed in detail in Section 2.6.5 of this chapter. Techniques for systematically identifying human error in safety analyses are described in Chapter 5.

2.5.6. Summary and Evaluation of the HF/E Perspective on Human Error in the CPI

The traditional HF/E approach provides techniques and data relevant to optimizing human performance and minimizing certain categories of error in chemical process industry operations. The main application of human factors and ergonomics methods is in the design of new systems. However, audit checklists are available for evaluating HF/E deficiencies that could give rise to errors in existing systems. These are considered in Chapters 3 and 4. As part of this design process, many of the performance-influencing factors described in Chapter 3 are taken into account. Some of the techniques described in Chapter 4—for example, task analysis—are also employed during the design process.

The disadvantages of the classical HF/E perspective as a basis for human error prediction have been reviewed earlier. The approach focuses mainly on the external aspects of human performance and does not provide any systematic methods for error identification or for addressing underlying causes of errors. In addition, the HF/E approach does not provide a systematic framework for addressing and eliminating cognitive errors in areas such as diagnosis and problem solving.

2.6. THE COGNITIVE ENGINEERING PERSPECTIVE

The classical human factors engineering/ergonomics approach to human error was essentially based on a "black box" model of human behavior that focused primarily on information inputs and control action outputs. In this section a more modern perspective, based on approaches from cognitive psychology, is introduced. At one level, the cognitive perspective is still concerned with information processing, in that it addresses how people acquire information, represent it internally and use it to guide their behavior. The key difference from the HF/E approach is that the cognitive approach emphasizes the role of intentions, goals, and meaning as a central aspect of human behavior. The term "cognitive" is based on the Latin *cognoscere* meaning 'to know.'

Instead of the human being conceptualized as a passive system element, to be treated in the same way as a pump or valve, the cognitive approach emphasizes the fact that people impose meaning on the information they receive, and their actions are almost always directed to achieving some explicit or implicit goal.

In the context of a process plant, this could be long-term goals such as producing a given amount of product over several days, or more short-term objectives such as maintaining a particular temperature profile or flow rate. Thus, the cognitive approach opens up the black box that had represented the higher-level reasoning processes in the HF/E model of the worker.

The cognitive approach has had a major influence in recent years on how human error is treated in systems such as chemical process plants and nuclear power generation. In the next section we shall describe some of the key concepts that have emerged from this work, and how they apply to the analysis of error in the CPI. Discussion of the cognitive view of human performance are contained in Reason (1990), Hollnagel (1993), Kantowitz and Fujita (1990), Hollnagel and Woods (1983), and Woods and Roth (1990).

2.6.1. Explaining and Classifying Errors from the Cognitive Perspective

A major advantage of the cognitive perspective is that it provides a basis for the prediction and classification of errors in CPI operations. An effective classification system for errors is essential from several points of view. If we wish to aggregate data on human errors from industrial situations for the purpose of discerning trends, identifying recurrent types of errors, or for developing a quantitative data base of error frequencies, we need a basis for grouping together errors of a similar type. Although there was considerable interest in classification systems from the HF/E perspective, almost all of these systems attempted to classify errors in terms of their external characteristics, for example, action omitted, action too late or action in the wrong order. This was because a model or theory of errors had not been developed which connected the external form of the error or external error mode with the underlying mental processes that gave rise to it. Until such a connection had been made, it was not possible to classify errors in a systematic way, because the same external error mode could be due to a number of entirely different underlying causes.

For example, consider the error of a worker closing valve B instead of the nearby valve A, which is the required action as set out in the procedures. There are at least five possible explanations for this error.

1. The valves were close together and badly labeled. The worker was not familiar with the valves and therefore chose the wrong one. Possible cause: wrong identification compounded by lack of familiarity leading to wrong intention (once the wrong identification had occurred the worker *intended* to close the wrong valve).
2. The worker may have misheard instructions issued by the supervisor and thought that valve B was the required valve. Possible cause: communications failure giving rise to a mistaken intention.
3. Because of the close proximity of the valves, even though he intended to close valve A, he inadvertently operated valve B when he reached for the valves (correct intention but wrong execution of action).
4. The worker closed valve B very frequently as part of his everyday job. The operation of A was embedded within a long sequence of other operations that were similar to those normally associated with valve B. The worker knew that he had to close A in this case, but he was distracted by a colleague and reverted back to the strong habit of operating B. Possible cause: intrusion of a strong habit due to external distraction (correct intention but wrong execution).
5. The worker knew that valve A had to be closed. However, it was believed by the workforce that despite the operating instructions, closing B had a similar effect to closing A and in fact produced less disruption to downstream production. Possible cause: violation as a result of mistaken information and an informal company culture to concentrate on production rather than safety goals (wrong intention).

These explanations do not exhaust the possibilities with regard to underlying causes, but they do illustrate an important point: the analysis of human error purely in terms of its external form is not sufficient. If the underlying causes of errors are to be addressed and suitable remedial strategies developed, then a much more comprehensive approach is required. This is also necessary from the predictive perspective. It is only by classifying errors on the basis of underlying causes that specific types of error can be predicted as a function of the specific conditions under review.

2.6.2. The Skill-, Rule-, and Knowledge-Based Classification

An influential classification of the different types of information processing involved in industrial tasks was developed by J. Rasmussen of the Risø Laboratory in Denmark. This scheme provides a useful framework for identifying the types of error likely to occur in different operational situations, or within different aspects of the same task where different types of information processing demands on the individual may occur. The classification system, known as the skill-, rule-, knowledge-based (SRK) approach is described in a number of publications (e.g., Rasmussen, 1979, 1982; Reason, 1990). An extensive discussion of Rasmussen's influential work in this area is contained in Goodstein et al. (1988), which also contains a comprehensive bibliography. This book contains a paper by Sanderson and Harwood that charts the development of the SRK concept.

The terms "skill-, rule-, and knowledge-based" information processing refer to the degree of conscious control exercised by the individual over his or her activities. Figure 2.3 contrasts two extreme cases. In the *knowledge*-based mode, the human carries out a task in an almost completely conscious manner. This would occur in a situation where a beginner was performing the task (e.g., a trainee process worker) or where an experienced individual was faced with a completely novel situation. In either of these cases, the worker would have to exert considerable mental effort to assess the situation, and his or her responses are likely to be slow. Also, after each control action, the worker would need to review its effect before taking further action, which would probably further slow down the responses to the situation.

The *skill*-based mode refers to the smooth execution of highly practiced, largely physical actions in which there is virtually no conscious monitoring. Skill-based responses are generally initiated by some specific event, for example, the requirement to operate a valve, which may arise from an alarm, a procedure, or another individual. The highly practiced operation of opening the valve will then be executed largely without conscious thought.

In Figure 2.4, another category of information processing is identified that involves the use of rules (*rule*-based mode). These rules may have been learned as a result of interacting with the plant, through formal training, or by working with experienced process workers. The level of conscious control is intermediate between that of the knowledge- and skill-based modes.

2.6.3. The Generic Error Modeling System (GEMS)

GEMS is an extension of the SRK approach and is described in detail in Reason (1990). GEMS is intended to describe how switching occurs between the different types of information processing (skill, rule, knowledge) in tasks such as those encountered in the CPI. GEMS is shown in Figure 2.5. The way in which GEMS is applied is illustrated most effectively by means of a specific example.

Consider a process worker monitoring a control panel in a batch processing plant. The worker is executing a series of routine operations such as opening and closing valves and turning on agitators and heaters. Since the worker is highly practiced, he or she will probably be carrying out the valve operations in an automatic skill-based manner only occasionally monitoring the situation at the points indicated by the "OK?" boxes at the skill-based level in Figure 2.5.

If one of these checks indicates that a problem has occurred, perhaps indicated by an alarm, the worker will then enter the rule-based level to

KNOWLEDGE-BASED MODE CONSCIOUS	SKILL-BASED MODE AUTOMATIC
Unskilled or occasional user	Skilled, regular user
Novel environment	Familiar environment
Slow	Fast
Effortful	Effortless
Requires considerable feedback	Requires little feedback
Causes of error: • Overload • Manual variability • Lack of knowledge of modes of use • Lack of awareness of consequences	Causes of error: • Strong habit intrusions • Frequently invoked rule used inappropriately • Changes in the situation do not trigger the need to change habits

FIGURE 2.3. **Modes of Interacting with the World (Reason, 1990).**

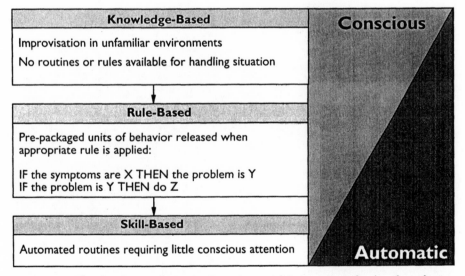

FIGURE 2.4. **The Continuum between Conscious and Automatic Behavior (based on Reason, 1990).**

determine the nature of the problem. This may involve gathering information from various sources such as dials, chart recorders and VDU screens, which is then used as input to a diagnostic rule of the following form:

<IF> symptoms are X <THEN> cause of the problem is Y

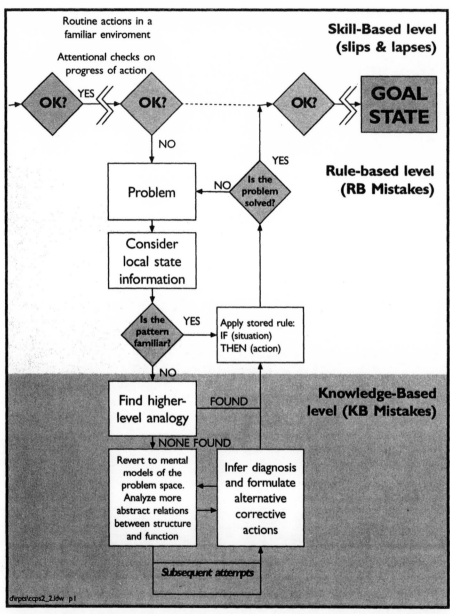

FIGURE 2.5. **Dynamics of Generic Error Modeling System (GEMS) (adapted from Reason, 1990).**

Having established a plausible cause of the problem on the basis of the pattern of indications, an action rule may then be invoked of the following form:

<IF> the cause of the problem is Y <THEN> do Z

If, as a result of applying the action rule, the problem is solved, the worker will then return to the original skill-based sequence. If the problem is not resolved, then further information may be gathered, in order to try to identify a pattern of symptoms corresponding to a known cause.

In the event that the cause of the problem cannot be established by applying any available rule, the worker may then have to revert to the knowledge-based level. The first strategy likely to be applied is to attempt to find an analogy between the unfamiliar situation and some of the patterns of events for which rules are available at the rule-based level. If such a diagnostic rule can be found that validly applies, the worker will revert back to the rule-based level and use the appropriate action rule. However, if a suitable analogy cannot be found, it may be necessary to utilize chemical or engineering knowledge to handle the situation. This process is illustrated in the following example:

Example 2.7: Moving among the Skill-, Rule-, and Knowledge-Based Levels in the GEMS Model

While scanning a control panel, a process worker notices that a pressure build-up is occurring during a routine transfer of reactant between the reactors (a skill-based check). He first checks if the appropriate valves have been opened. (Rule-based check: if pressure build-up, then transfer line may not have been opened.) Since the valve line-ups appear to be correct, he then moves to the knowledge-based level to draw upon other sources of information. The use of a data sheet of the chemical properties of the reactant and a piping diagram at the knowledge-based level identify the problem as solidification of the chemical in the line due to low ambient temperature. The formulation of corrective actions involves moving back up to the rule-based level to find an appropriate corrective action, for example turning on electric heat tracing at the point in the line where the blockage had occurred. If this action is successful, then the situation reverts to the skill-based level where the problem originally occurred.

This example illustrates the fact that several levels of processing may occur within the same task.

2.6.4. Classification of Errors from the Cognitive Perspective

2.6.4.1. Slips and Mistakes

The categorization set out in Figure 2.6 is a broad classification of the causes of human failures that can be related to the SRK concepts discussed in the last section. The issue of violations will be addressed later in Section 2.7.1.1. The distinction between slips and mistakes was first made by Norman (1981).

> Slips are defined as errors in which the intention is correct, but a failure occurring when carrying out the activities required. For example, a worker may know that a reactor needs to be filled but instead fills a similar reactor nearby. This may occur if the reactors are poorly labeled, or if the worker is confused with regard to the location of the correct reactor. Mistakes, by contrast, arise from an incorrect intention, which leads to an incorrect action sequence, although this may be quite consistent with the wrong intention. An example here would be if a worker wrongly assumed that a reaction was endothermic and applied heat to a reactor, thereby causing overheating. Incorrect intentions may arise from lack of knowledge or inappropriate diagnosis.

In Figure 2.6, the slips/mistakes distinction is further elaborated by relating it to the Rasmussen SRK classification of performance discussed earlier. Slips can be described as being due to misapplied competence because they are examples of the highly skilled, well practiced activities that are characteristic of the skill-based mode. Mistakes, on the other hand, are largely confined to the rule and knowledge-based domains.

In the skill-based mode, the individual is able to function very effectively by using "preprogrammed" sequences of behavior that do not require much conscious control. It is only occasionally necessary to check on progress at particular points when operating in this mode. The price to be paid for this economy of effort is that strong habits can take over when attention to checks is diverted by distractions, and when unfamiliar activities are embedded in a familiar context. This type of slip is called a "strong but wrong" error. The examples given in Section 2.6.1 can be classified as slips, mistakes, and violations using the categorization scheme in Figure 2.6.

2.6.4.2. Rule-Based Mistakes

With regard to mistakes, two separate mechanisms operate. In the rule-based mode, an error of intention can arise if an incorrect diagnostic rule is used. For example, a worker who has considerable experience in operating a batch reactor may have learned diagnostic rules that are inappropriate for continuous process operations. If he or she attempts to apply these rules to evaluate the cause of a continuous process disturbance, a misdiagnosis could result, which could then lead to an inappropriate action. In other situations, there is a tendency to overuse diagnostic rules that have been successful in the past.

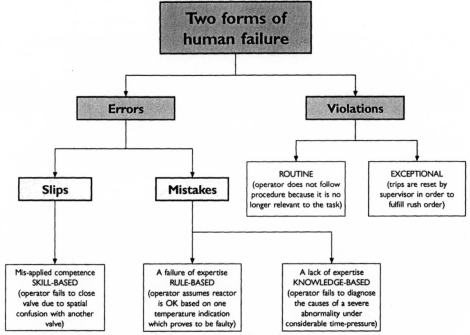

FIGURE 2.6. **Classification of Human Errors (adapted from Reason, 1990).**

Such "strong" rules are usually applied first, even if they are not necessarily appropriate.

There is a tendency to force the situation into the mold of previous events. Case study 1.15 was an example of this type of mistake. Following some modifications to a pump, it was used to transfer liquid. When movement was complete, the worker pressed the stop button on the control panel and saw that the "pump running" light went out. He also closed a remotely operated valve in the pump delivery line. Several hours later the high-temperature alarm on the pump sounded. Because the worker had stopped the pump and seen the running light go out, he assumed the alarm was faulty and ignored it. Soon afterward there was an explosion in the pump. When the pump was modified, an error was introduced into the circuit. As a result, pressing the stop button did not stop the pump but merely switched off the running light. The pump continued running, overheated, and the material in it decomposed explosively.

In this example, a major contributor to the accident was the worker's assumption that the pump running light being extinguished meant that the pump had stopped even though a high-temperature alarm occurred, which would usually be associated with an operating pump. The rule "If pump light is extinguished then pump is stopped" was so strong that it overcame the evidence from the temperature alarm that the pump was still running. By analogy with the "strong but wrong" action sequences that can precipitate skill-based slips, the inappropriate use of usually successful rules can be described as "strong but wrong" rule failures. Other types of failure can occur at the rule-based level and these are described extensively by Reason (1990).

2.6.4.3. Knowledge-Based Mistakes

In the case of knowledge-based mistakes, other factors are important. Most of these factors arise from the considerable demands on the information processing capabilities of the individual that are necessary when a situation has to be evaluated from first principles. Given these demands it is not surprising that humans do not perform very well in high stress, unfamiliar situations where they are required to "think on their feet" in the absence of rules, routines, and procedures to handle the situation. Kontogiannis and Embrey (1990) and Reason (1990) describe a wide range of failure modes under these conditions. For example, the "out of sight, out of mind" syndrome means that only information that is readily available will be used to evaluate the situation. The "I know I'm right" effect occurs because problem solvers become overconfident of the correctness of their knowledge. A characteristic behavior that occurs during knowledge-based problem solving is "encystment" where the individual or the operating team become enmeshed in one aspect of the problem to the exclusion of all other considerations (the Three Mile Island accident is a notable example). The opposite form of behavior, "vagabonding" is also observed, where the overloaded worker pays attention superficially to one problem after another, without solving any of them. Janis (1972) provides detailed examples of the effects of stress on performance.

2.6.4.4. Error Recovery

In the skill-based mode, recovery is usually rapid and efficient, because the individual will be aware of the expected outcome of his or her actions and will therefore get early feedback with regard to any slips that have occurred that may have prevented this outcome being achieved. This emphasizes the role of feedback as a critical aspect of error recovery. In the case of mistakes, the mistaken intention tends to be very resistant to disconfirming evidence. People tend to ignore feedback information that does not support their expectations of the situation, which is illustrated by case study 1.14. This is the basis of the commonly observed "mindset" syndrome.

2.6.5. The Stepladder Model

The GEMS model is based on a more detailed model of human performance known as the **stepladder model** developed by Rasmussen (see Rasmussen 1986) and illustrated in Figure 2.7. In this model, Rasmussen depicted the various stages that a worker could go through when handling a process disturbance.

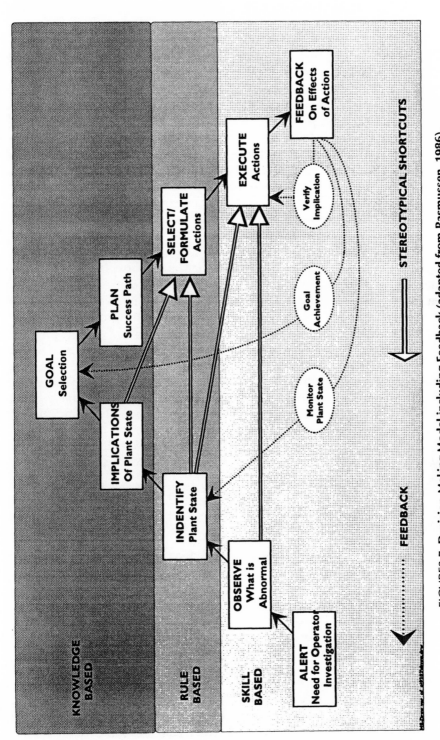

FIGURE 2.7. Decision-Making Model including Feedback (adapted from Rasmussen, 1986).

77

Only if the worker has to utilize the knowledge-based mode will he or she traverse every information processing stage represented by the boxes connected by the black arrows. As in the GEMS model (Section 2.6.3), if the situation is immediately recognized, then a preprogrammed physical response will be executed in the skill-based mode (e.g., by moving the process on to the next stage by pressing a button).

If the nature of the problem is not readily apparent, then it might be necessary to go to the rule-based level. In this case a diagnostic rule will be applied to identify the state of the plant and an action rule used to select an appropriate response. Control will revert to the skill-based level to actually execute the required actions. More abstract functions such as situation evaluation and planning will only be required at the knowledge-based level if the problem cannot not be resolved at the rule-based level.

The lighter arrows represent typical shortcuts, which omit particular stages in the information-processing chain. These shortcuts may be "legitimate," and would only lead to errors in certain cases. For example, the worker may erroneously believe that he or she recognizes a pattern of indicators and may immediately execute a skill-based response, instead of moving to the rule-based level to apply an explicit diagnostic rule.

The dotted lines in the diagram indicate the various feedback paths that exist to enable the individual to identify if a particular stage of the processing chain was executed correctly. Thus, if the operating team had planned a strategy to handle a complex plant problem, they would eventually obtain feedback with regard to whether or not the plan was successful. Similar feedback loops exist at the rule and skill-based levels, and indicate opportunities for error correction. The application of the stepladder model to a process industry example is given in Appendix 2A at the end of this chapter.

2.6.6. How Can the Cognitive Approach Be Applied to Process Safety in the CPI?

Up to this point, various models have been described that provide a comprehensive description of the mental functions that underlie the whole range of activities performed by a process plant worker, from simple skill-based physical actions, to rule-based diagnosis, and more complex knowledge-based problem solving. Although these models are certainly not the only explanations of process control behavior available (see, e.g., Edwards and Lees, 1974, and papers in Goodstein et al., 1988) they have proved valuable in providing a link between the work of cognitive psychologists and the practical concerns of engineers in the process industries. A number of practical applications of these concepts will now be described. These applications include the development of error-reduction design strategies, error prediction for safety analysis, and identification of the root causes of errors in accident analysis.

Many of these applications require tasks or parts of a task to be categorized according to the SRK scheme. Although this is difficult in some cases, a simple flowchart may assist in this process. This is given in Figure 2.8. This assumes that the tasks will be performed by a worker of average competence. This assumption is necessary, since the actual mode that the task will be performed in (skill, rule, or knowledge) obviously depends on the characteristics of the individual (how well trained, how capable) as well as the task.

2.6.6.1. Error Reduction

If we can classify a task or a part of a task as being, for example, predominantly skill- rather than rule-based (given that no task falls exactly into each category), this has a number of implications for various approaches to error reduction. From a training perspective, this means that extensive practice of the largely physical and manipulative aspects of the task, together with frequent feedback, will be required in order to ensure that the required actions can be smoothly executed and coordinated without conscious thought. From the standpoint of procedures, there is no point in developing extensive step-by-step written procedures, since skill-based actions will be largely executed automatically when the appropriate cue for action is received. Thus, the most appropriate form of job aid is likely to be a simple checklist which specifies the starting point of each sequence of actions with perhaps specific checks to verify that each activity has been correctly performed.

2.6.6.2. Error Prediction

As implied in the diagram representing the GEMS model (Figure 2.5) and discussed in Section 2.6.3, certain characteristic error forms occur at each of the three levels of performance. This information can be used by the human-reliability analyst for making predictions about the forms of error expected in the various scenarios that may be considered as part of a predictive safety analysis. Once a task or portion of a task is assigned to an appropriate classification, then predictions can be made. A comprehensive set of techniques for error prediction is described in Chapter 5.

The SRK model can also be used as part of a approach for the elimination of errors that have serious consequences proactive for the plant. Once specific errors have been identified, based on the SRK model, interventions such as improved procedures, training or equipment design can be implemented to reduce their likelihood of occurrence to acceptable levels. This strategy will be discussed in more detail in Chapter 4.

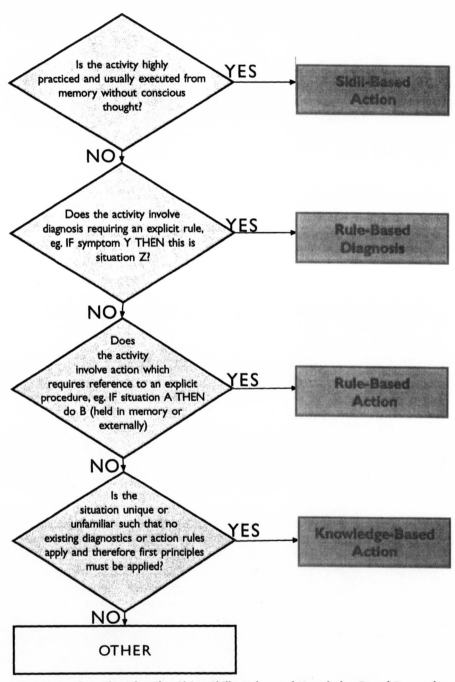

FIGURE 2.8. **Flow Chart for Classifying Skill-, Rule-, and Knowledge-Based Processing.**

80

2.6.6.3. Analysis of Incident Root Causes Using the Sequential Error Model

In addition to the proactive uses of the SRK model described in the two previous sections, it can also be employed retrospectively as a means of identifying the underlying causes of incidents attributed to human error. This is a particularly useful application, since causal analyses can be used to identify recurrent underlying problems which may be responsible for errors which at a surface level are very different. It has already been indicated in Section 2.4.1 that the same observable error can arise from a variety of alternative causes. In this section it will be shown how several of the concepts discussed up to this point can be combined to provide a powerful analytical framework that can be used to identify the root causes of incidents.

The block diagram shown in Figure 2.9 was developed by Rasmussen (see Rasmussen 1981, 1986) as a sequential model of the causal chain leading to an error. Basically, the model identifies the various processes that intervene between the initiating or triggering event, and the external observable form of the error, referred as the external error mode. This external error mode may or may not lead to an accident, depending on the exact conditions that apply. The internal error mechanisms have been discussed in earlier sections (e.g., the strong stereotype takeovers discussed in Section 2.6.4.2). They are intrinsic error tendencies. The "internal error mode" represents the point in the various stages of handling a situation (e.g., failed to detect problem, failed to act) where the failure occurred.

For each of the stages of the model, Petersen (1985) provided a series of flow diagrams to assist analysts in using the model for incident analysis. These are given in Appendix 2B. The use of the model and the flow charts for detailed psychological analysis of incidents is illustrated by a case study in Appendix 2C.

2.6.7. Summary of the Use of Cognitive Models in CPI Safety

The applications of the SRK, GEMS, stepladder and sequential block diagram models to human error in process safety can be summarized as follows:

Error Reduction by Design

This is a proactive process which involves the following stages:
1. Perform task analysis (see Chapter 4) and identify skill, rule or knowledge-based tasks or aspects of tasks (the flow diagram in Figure 2.7 may be used to assist in this classification).
2. Depending on the results of the classification select an appropriate error reduction strategy in areas such as training, procedures or equipment design, as illustrated in Table 2.3.

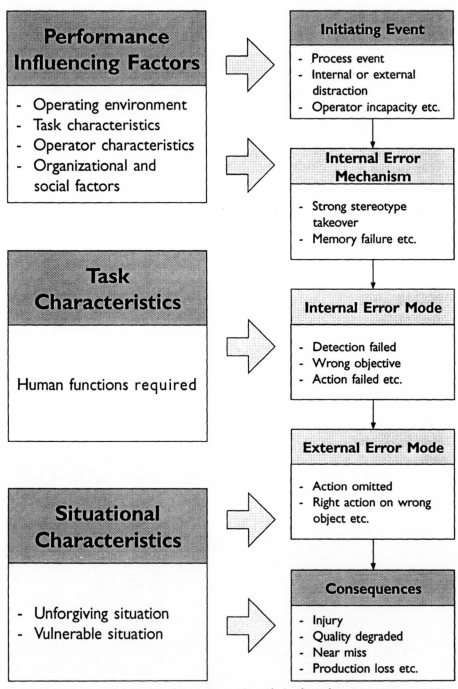

FIGURE 2.9. **Sequential Model of Error Causation Chain (based on Rasmussen, 1982).**

3. Evaluate the effectiveness of the strategy by reviewing operational experience when the task has been performed for some time, and identifying the error root causes by the process set out below.

TABLE 2.3
Example Error Reduction Recommendations Arising from the SRK Model

TYPICAL ERRORS ASSOCIATED WITH DIFFERENT INFORMATION PROCESSING LEVELS	EXAMPLES OF ERROR REDUCTION STRATEGIES		
	TRAINING	PROCEDURES/JOB AIDS	EQUIPMENT DESIGN
Skill-based Errors • manual variability • strong but wrong action sequences	Train for physical and manipulative skills (repeated practice and feedback)	Checklists setting out starting and finishing activities and checks	Layout and labeling of controls and process lines Distinguish between plant areas with similar appearance but different functions Provide feedback
Rule-based Errors • incorrect diagnosis due to strong but wrong rule • incorrect action chosen due to incorrect or inappropriate rule	Identify diagnostic and action rules required to perform job. Ensure worker is given extensive practice in using rules. Explain exceptions and possible errors due to confusing symptoms and strong rules	For complex or infrequently used rules, provide job aids, for example, fault/symptom matrices to facilitate correct diagnosis and to support selection of appropriate actions	Ensure information displays designed so that workers do not use inappropriate rules based on similar symptoms with differing causes Provide feedback
Knowledge-based Errors • information processing • perceptual tunnel vision	Where possible provide simulations of complex events to encourage development of strategies in forgiving environment. Provide training in principles of process dynamics	Provide data on plant (P & I diagrams, plant configuration) in readily accessible form. Provide problem-solving schematics to ensure all information taken into account	As above

Error Prediction for Safety Analysis and Proactive Error Reduction

This procedure is performed when error modes are being identified (e.g., critical action omitted, alternative unsafe action carried out) as part of a predictive safety analysis (e.g., CPQRA) or as part of a proactive error reduction process (see Chapter 4).

1. Perform task analysis and classify skill, rule or knowledge-based behaviors involved in the scenario being evaluated.
2. Perform a preliminary screening analysis to identify aspects of human
performance where failures can have serious consequences.
3. For these tasks identify likely internal and external error modes using
flow charts and methods described in Chapter 6.
4. Quantify error probabilities for these error modes using methods described in Chapter 5.
5. For errors with serious consequences and/or high likelihood of occurrence, develop appropriate error reduction strategies.

Analysis of Operational Experience

Detailed methods for incident analysis are described in Chapter 6. The methods described in this chapter provide the basis for a psychological analysis of
incident causes.

1. Taking the observed error or near miss as a starting point, perform task
analysis (see Chapter 4) to describe overall context of the error.
2. Use methods such as STEP (see Chapter 6) to evaluate the event
sequence.
3. Use the flow charts as a basis for asking questions relating to each stage
of the sequential causal block diagram. Work backward from the
observable error to the initiating event. A careful analysis of the performance-influencing factors (Chapter 3) will form part of this analysis.

These various aspects of evaluating, predicting, and reducing human
error form part of a general strategy for managing error which will be described in Chapter 5.

2.6.8 Conclusions Regarding Application of the Cognitive
Modeling Perspective to Errors in the CPI

The previous sections have presented an extensive description of some of the
central concepts from the cognitive modeling perspective. These topics have
been dealt with in some depth because they provide a comprehensive basis
for the reduction of human error in the CPI.

Several examples have already been provided of the use of cognitive
models of error to evaluate the possible causes of accidents that have already
occurred. This form of retrospective analysis performs a vital role in providing
information on the recurring underlying causes of accidents in which human
error is implicated. The advantage of an analytical framework driven by a
model of human error is that it specifies the nature of the questions that need

to be asked and the contextual information that should be collected in order to establish root causes and therefore develop effective remedial strategies. In the longer term, it also provides the basis for the evaluation of the effectiveness of these strategies by indicating if the same underlying causes recur even after error reduction measures are implemented (see Chapter 6).

The use of a model of human error allows a systematic approach to be adopted to the prediction of human failures in CPI operations. Although there are difficulties associated with predicting the precise forms of mistakes, as opposed to slips, the cognitive approach provides a framework which can be used as part of a comprehensive qualitative assessment of failure modes. This can be used during design to eliminate potential error inducing conditions. It also has applications in the context of CPQRA methods, where a comprehensive qualitative analysis is an essential precursor of quantification. The links between these approaches and CPQRA will be discussed in Chapter 5.

2.7. THE SOCIOTECHNICAL PERSPECTIVE

The approaches described so far tackle the problem of error in three ways. First, by trying to encourage safe behavior (the traditional safety approach), second by designing the system to ensure that there is a match between human capabilities and systems demands (the human factors engineering approach) and third by understanding the underlying causes of errors, so that error inducing conditions can be eliminated at their source (the cognitive modeling approach). These strategies provide a technical basis for the control of human error at the level of the individual worker or operating team.

The control of human error at the most fundamental level also needs to consider the impact of management policy and organizational culture. The concepts introduced in Chapter 1, particularly the systems-induced error approach, have emphasized the need to go beyond the direct causes of errors, for example, overload, poor procedures, poor workplace design, to consider the underlying organizational policies that give rise to these conditions. Failures at the policy level which give rise to negative performance-influencing factors at the operational level are examples of the latent management failures discussed in Chapter 1 and in Section 2.2.2.

Another way in which management policies affect the likelihood of error is through their influence on organizational culture. For example, a culture may arise at the operational level where the achievement of production objectives is given greater emphasis than safe practices. Of course, no responsible company would sanction such a situation if they knew it existed. However, without effective communications or incident feedback systems, management may never realize that safety is being compromised by an inappropriate culture and the working practices it produces.

Studies of major accidents have shown that they almost always arise from a combination of active errors, latent failures and inappropriate culture. Examples of such analyses from the sociotechnical perspective are available from a number of sources, for example, Reason (1990), Rasmussen (1990), Wagenaar and Groenweg (1987), and Kletz (1994a). These analyses have considered accidents as diverse as Three Mile Island, Chernobyl, the *Challenger* Space Shuttle, Bhopal, Flixborough, and Piper Alpha. Although these accidents may appear to be far removed from the day-to-day concerns of a plant manager in the CPI, they indicate the need to look beyond the immediate precursors of accidents to underlying systemic causes. Methods for addressing these issues during the retrospective analysis of incidents are discussed in Chapter 6.

2.7.1. The TRIPOD Approach

From the point of view of accident prevention, approaches have been developed which seek to operate at the level of organizational factors affecting error and accident causation. One of the most extensive efforts has been the development of the TRIPOD system with the support of the Shell International Petroleum Company. In this system, the direct causes of errors leading to accidents are called "tokens" and the generic management level factors that create latent failure conditions are called "general failure types." (See Wagenaar et al., 1990, and Wagenaar, 1992, for a more detailed description.)

These general failure types are used to produce profiles which indicate the accident potential of a facility on a number of dimensions. An example of these profiles is shown in Figure 2.10 (from Wagenaar, 1992). Scores on these factors are derived from checklists which comprise a series of yes/no questions concerning relevant "indicators." For example, whether or not people have worked 24 hours continuously is taken as an indicator of increased error likelihood. Such a question would be one component of the general failure type "error enforcing conditions." There is a list of questions corresponding to each of the general failure types, which varies depending on the nature of the activity, country or ethnic culture. In the terminology of this book, TRIPOD provides an auditing method which can be used to identify negative performanc-e influencing factors. Those factors which score poorly are used to guide subsequent corrective actions. Wagenaar (1992) states that analyses of accident data show that situations where accidents occur correlate highly with poor scores on the general failure type profiles.

The benefits claimed for the TRIPOD approach are that it provides a consistent method for auditing a situation to identify deficiencies in the factors that are likely to give rise to errors. These deficiencies can then be corrected to reduce the likelihood of accidents occurring in the future.

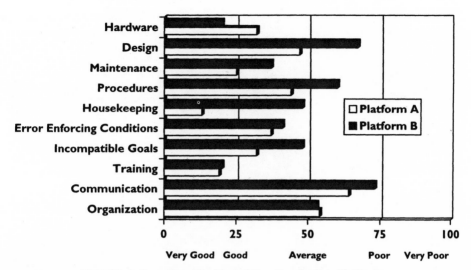

FIGURE 2.10 **TRIPOD Failure-State Profiles of Two Production Platforms** (Wagenaar, 1992).

2.7.2. Human Factors Analysis Methodology

Another strategic initiative in this area is the development of a human factors analysis methodology (HFAM) by a U.S.-based multinational chemical processing company. Preliminary descriptions of this approach are available in Pennycook and Embrey (1993). This methodology is based on the systems-induced error philosophy set out in this book. This states that control of error can be most effectively achieved by attacking the environmental or system causes of error which are under the control of management rather than trying to change behavior directly. HFAM has a similar philosophy to TRIPOD in that it defines a comprehensive set of factors which together address the primary system causes of error. These factors in turn are broken down into a series of diagnostic questions which can be used to make numerical assessments of the dimensions which make up the higher level factors. The current set of factors that make up the HFAM tool are given in Figure 2.11. It can be seen that the factors can be divided into three groups, management level, generic, and job specific.

HFAM has 20 groups of factors instead of the 10 general failure types of the TRIPOD approach. The reason for this is that all of the 10 TRIPOD GFTs would be applied in all situations, even though the actual questions that make up the factors may vary. In the case of HFAM, it would be rare to apply all of the factors unless an entire plant was being evaluated. HFAM uses a screening process to first identify the major areas vulnerable to human error. The generic factors and appropriate job specific factors are then applied to these areas. For example, control room questions would not be applied to maintenance jobs.

MANAGEMENT-LEVEL FACTORS	OPERATIONAL-LEVEL GENERIC FACTORS	OPERATIONAL-LEVEL JOB-SPECIFIC FACTORS
• Safety priorities • Degree of participation • Effectiveness of communications • Effectiveness of incident investigation • Effectiveness of procedures development system • Effectiveness of training system • Effectiveness of design policies	• Process management • Job design and work planning • Safe systems of work • Emergency response plan • Training • Work group factors • Work patterns • Stress factors • Individual factors • Job aids and procedures	• Computer-based systems • Control panel design • Field workplaces • Maintenance

FIGURE 2.11   **Factors in Human Factors Assessment Methodology.**

The components of each factor can be evaluated at two levels of detail. An example of these levels for the factor "Procedures and Job Aids" is provided in Figure 2.12. If the question indicates that the first level (e.g., content and reliability) is deemed to be inadequate then more questions are available at the next level of detail (the topic level) to provide additional information on the nature of the problem. For each topic, further questions are provided at a greater level of detail. These detailed questions (diagnostics) are intended to pinpoint the precise nature of a deficiency and also to provide insights for remedial action.

Problems identified at the operational level by the generic and job specific factors are regarded as being indicative of a failure of management level controls of that factor. The corresponding management level factor would then be evaluated to identify the nature of this latent failure. Although specific human factors design deficiencies might be identified at the operational level (e.g., inadequacies in control panel design, poor procedures), inadequacies within the higher level management factor, for example, "Effectiveness of design policies affecting human error" would affect a number of the operational level situations. Thus, the process of remedying the problem would not be confined to addressing the specific operational deficiencies identified but would also consider the changes in management policies needed to address these deficiencies across the whole site (or even the company). Figure 2.12 provides an example of how the system can be applied.

In addition to the management level factors which can be specifically linked to operational level factors (procedures, training, and design), the HFAM tool also provides an assessment of other management level factors which will impact upon error likelihood in a less direct way. Some of these factors, for example, "safety priorities" and "degree of participation," are

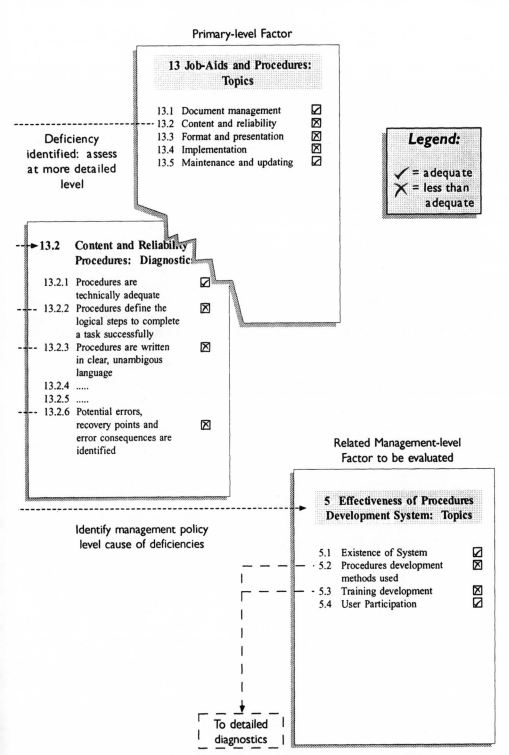

GURE 2.12 **Example of use of HFAM tool for evaluation (Pennycook et al., 1993).**

89

intended to address conditions that have been found to be good indicators of the quality of the safety culture. The remaining factors, "communications" and "incident investigation" are intended to provide an indication of how effectively information is transmitted vertically and horizontally in the organization, and the capability of the organization to learn lessons from operational experience.

2.7.3 The UK Health & Safety Executive Research Program on Sociotechnical Systems

A program of research has been supported for several years by the United Kingdom Health & Safety Executive (HSE) to address the effects of sociotechnical factors on risk in the CPI. The initial emphasis of this work was to develop a methodology so that chemical process quantitative risk analysis (CPQRA) would take into account the effects of the quality of the management factors of plant being assessed. This work has been described in a series of publications (e.g., Bellamy et al., 1990; Hurst et al., 1991; Geyer et al., 1990; and Hurst et al., 1992).

The project began with an extensive evaluation of 900 reported incidents involving failures of fixed pipework on chemical and major hazard plant. As part of the analysis a failure classification scheme was developed which considered the chief causes of failures, the possible prevention or recovery mechanism that could have prevented the failure and the underlying cause. The classification scheme is summarized in Figure 2.13. A typical event classification would be

> **Corrosion** *(direct cause)* due to **Design error** *(basic or root cause)*
> not recovered by **Inspection** *(failure of recovery)*

These results, together with other research on reactor failures assisted in the development of an audit tool called MANAGER, based on the model shown in Figure 2.14. This allowed an assessment to be made of the different levels of engineering and Sociotechnical factors contributing to the overall risk for a particular plant. The results of this audit process are used to generate a Management Factor for the facility. This is then used to modify the overall risk estimates calculated by traditional CPQRA approaches (e.g., the fault tree analysis) by a factor varying between 10^{-1} and 10^{-3}.

Although the main thrust of the HSE work is directed to providing inputs to the CPQRA process, the audit procedure generates valuable qualitative information regarding both the quality of the overall plant management and also the specific human factors dimensions which affect risk.

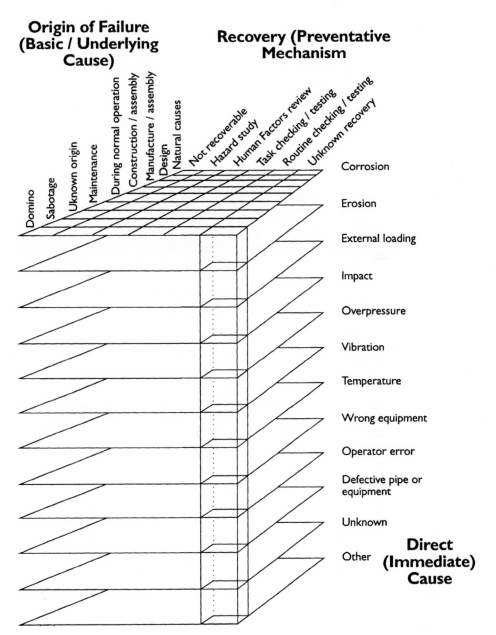

FIGURE 2.13. **Classification of Causal Factors (from Hurst et al., 1992).**

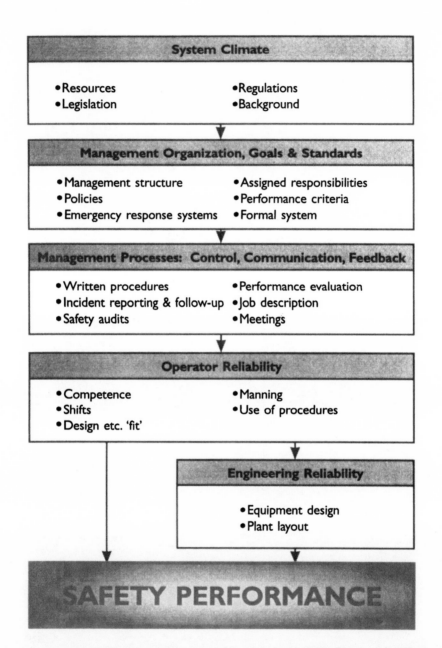

FIGURE 2.14 **Sociotechnical Model Underlying Audit Tool (from Hurst et al., 1992).**

2.7.4. Comparisons among the Sociotechnical Approaches

The similarities between TRIPOD and HFAM are considerable in that they are both based on a systems view of error, and the importance of policy in influencing the immediate causes of error. They have also both been developed iteratively, using extensive field trials. Although both systems are ultimately directed at the reduction of human error leading to accidents, there are differences as how they are applied. It appears that TRIPOD is mainly intended as a proactive evaluation tool which will be used by auditors to evaluate sites and recommend improvement strategies. By contrast, the initial focus of HFAM is to encourage operations staff to evaluate their own environments to identify error potential and develop appropriate remedial strategies. By this means, it is hoped to encourage active participation by individuals with a strong stake in accident prevention as part of the process of continuous improvement. Although both systems are primarily directed at error prevention, they can also be applied as part of the retrospective analysis of accidents that have already occurred.

The focus of MANAGER is somewhat different, in that it was primarily developed to provide a numerical output for use in risk assessment. Nevertheless, the qualitative dimensions included in the audit trail will undoubtedly provide information which can be used as part of an error prevention program.

The fact that these systems exist and have been given considerable support by companies and regulators in the CPI, must be taken as a positive indication of an increasing realization of the importance of human performance in ensuring safe and profitable operation of chemical facilities.

2.8. SUMMARY

The intention of this chapter has been to provide an overview of the wide range of strategies available to the CPI for the management of error. The traditional safety approach described in Section 2.4 concentrates on modifying individual behavior, and has been successful in many areas of occupational safety. Section 2.4 provided a review of some of the methods used in this approach and assessed their effectiveness. Section 2.5 considered some of the major technical issues within the human factors engineering approach. Detailed description of the various design approaches and techniques for the optimization of human performance that have emerged from this perspective, will be considered in Chapter 4. The cognitive modeling perspective reviewed in Section 2.6 provides an approach to modeling human errors that can be applied both at the design stage and for deriving the root causes of errors. Both of these applications will be developed in later chapters. Section 2.7 reviewed the organizational perspective, and emphasized the need for error reduction techniques to be supported by a consideration of the role of management policies in influencing the immediate causes of errors—a description was provided of three approaches that have been developed by chemical companies and regulators to provide comprehensive systems for managing error in the CPI.

2.9. APPENDIX 2A: PROCESS PLANT EXAMPLE OF THE STEPLADDER MODEL

In order to explain each box in the stepladder model shown in Figure 2.7 (reprinted on the facing page), we shall use the same batch processing example as in Section 2.6.3.

Consider a process worker monitoring a control panel in a batch processing plant. The worker is executing a series of routine operations such as opening and closing valves and turning on agitators and heaters.

Alert (need for investigation)
An alarm sounds which indicates a problem.

Observe (what is abnormal?)
Scan information sources (dials, chart recorders, etc.). If the pattern of indicators is very familiar, the worker will probably immediately branch to the Execute Actions box (via the thin arrow) and make the usual response to this situation (e.g., pressing the alarm accept button if the indications suggest a nonsignificant event).

Identify Plant State
If the pattern does not fit into an immediately identifiable pattern, the process worker may then consciously apply more explicit "if–then" rules to link the various symptoms with likely causes. Three alternative outcomes are possible from this process. If the diagnosis and the required actions are very closely linked (because this situation arises frequently) then a branch to the Execute Actions box will occur. If the required action is less obvious, then the branch to the Select/Formulate Actions box will be likely, where specific action rules of the form: "if situation is X then do Y" will be applied. A third possibility is that the operating team are unable or unwilling to respond immediately to the situation because they are uncertain about its implications for safety and/or production. They will then move to the Implications of plant state box.

Implications of Plant State
At this stage the implications of the situation will be explored, using the operating team's general functional knowledge of the process. This explicit

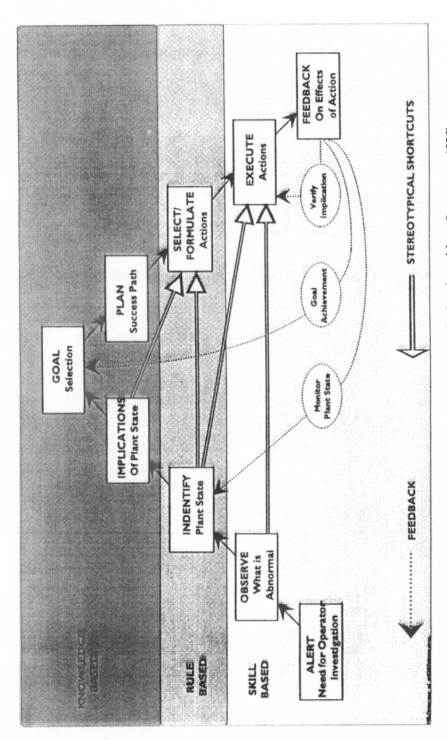

FIGURE2.7. Decision-Making Model including Feedback (adapted from Rasmussen, 1986).

evaluation procedure is classified as occurring in the knowledge-based domain, whereas the previous stage was rule based. If the required response to the situation is obvious, that is, there are no alternative goals, then the sequence branches to the Select/Formulate Actions box, where the required actions to achieve the objective are formulated and then acted upon in the Execute Actions box.

Goal Selection
During the goal selection stage, the operating team consider alternative objectives which they might wish to achieve. For example, if their assessment of the situation suggested that there was a major potential explosion hazard, then their objective would probably be to shut down the system as quickly as possible. If, on the other hand, the batch was simply off-specification as a result of the abnormal conditions, the strategy of mixing the batch with other batches in a blender might be considered.

Plan Success Path
Having decided on an appropriate objective, the next stage is to plan how to get from the current plant state to the required objective. This could involve deciding whether or not the batch requires cooling, how this would be achieved, what cross couplings are available to connect the reactor to a blender and so on.

Select/Formulate Actions
This step involves the formulation of a specific procedure or action sequence to achieve the plan decided upon at the previous stage. This may involve the linking together of an existing set of generic procedures which are employed in a variety of situations (e.g., executing a blowdown sequence). This phase uses action rules of the form "if Y then do Z" as opposed to the diagnostic rules of the "Identify Plant State" box, which is the other component of rule-based processing.

Execute Actions
This box, which is self-explanatory, involves highly practiced actions in the skill-based domain.

2.10. APPENDIX 2B: FLOWCHARTS FOR USING THE RASMUSSEN SEQUENTIAL MODEL FOR INCIDENT ANALYSIS (Petersen, 1985)

Start

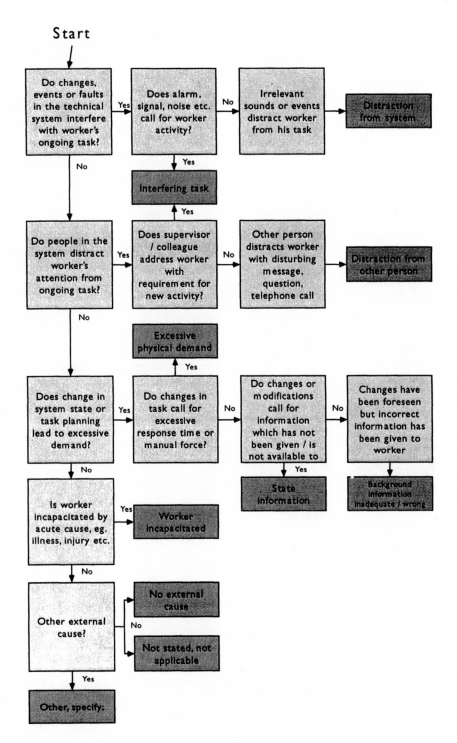

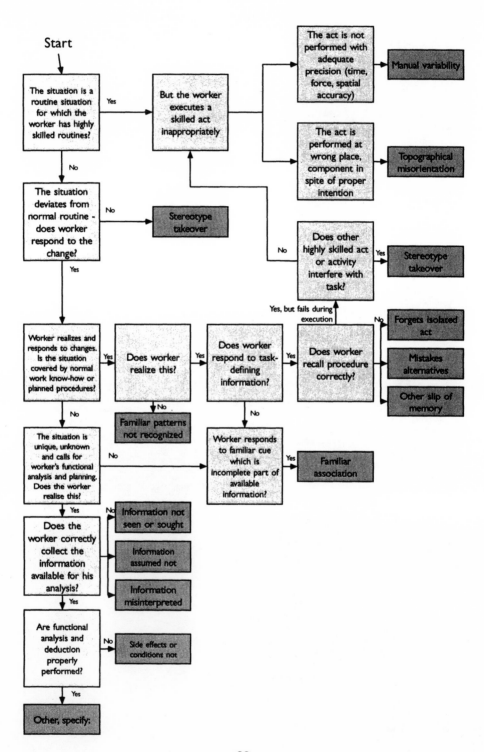

Start

The situation is a routine situation for which the worker has highly skilled routines? — Yes → But the worker executes a skilled act inappropriately

The act is not performed with adequate precision (time, force, spatial accuracy) → Manual variability

The act is performed at wrong place, component in spite of proper intention → Topographical misorientation

No

The situation deviates from normal routine - does worker respond to the change? — No → Stereotype takeover

Yes

Does other highly skilled act or activity interfere with task? — No ... Yes → Stereotype takeover

Worker realizes and responds to changes. Is the situation covered by normal work know-how or planned procedures? — Yes → Does worker realize this? — Yes → Does worker respond to task-defining information? — Yes → Does worker recall procedure correctly?

Yes, but fails during execution

No → Forgets isolated act

Mistakes alternatives

Other slip of memory

No → Familiar patterns not recognized

No

Worker responds to familiar cue which is incomplete part of available information? — Yes → Familiar association

No

The situation is unique, unknown and calls for worker's functional analysis and planning. Does the worker realise this? — No

Yes

Does the worker correctly collect the information available for his analysis? — No → Information not seen or sought

Information assumed not

Information misinterpreted

Yes

Are functional analysis and deduction properly performed? — No → Side effects or conditions not

Yes

Other, specify:

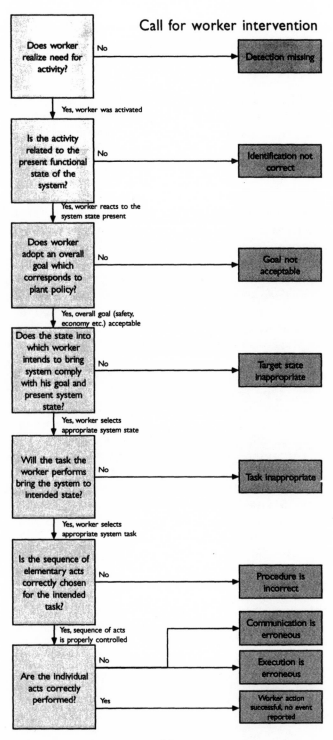

Call for worker intervention

Does worker realize need for activity?	No → Detection missing

Yes, worker was activated

Is the activity related to the present functional state of the system?	No → Identification not correct

Yes, worker reacts to the system state present

Does worker adopt an overall goal which corresponds to plant policy?	No → Goal not acceptable

Yes, overall goal (safety, economy etc.) acceptable

Does the state into which worker intends to bring system comply with his goal and present system state?	No → Target state inappropriate

Yes, worker selects appropriate system state

Will the task the worker performs bring the system to intended state?	No → Task inappropriate

Yes, worker selects appropriate system task

Is the sequence of elementary acts correctly chosen for the intended task?	No → Procedure is incorrect

Yes, sequence of acts is properly controlled

Communication is erroneous

Are the individual acts correctly performed?	No → Execution is erroneous
	Yes → Worker action successful, no event reported

2.11. APPENDIX 2C: CASE STUDY ILLUSTRATING THE USE OF THE SEQUENTIAL MODEL OF ERROR IN INCIDENT ANALYSIS

A process worker is monitoring the rise in temperature in reactor A. An exothermic reaction occurs producing an alarm requiring the opening of a valve on a circuit which provides cooling water to the reactor. Instead of opening the correct valve, he operates another valve for reactor B, which is the reactor which he monitors on most shifts. Reactor A is destroyed by a runaway reaction.

Initiating Event
At the time that the alarm occurred, the worker was helping a colleague to fix a problem on an adjacent panel. The initiating event was therefore a distraction from another person (see Figure 2.15).

Internal Error Mechanism
Internal error mechanisms can be regarded as intrinsic human error tendencies. The particular error mechanisms that will be triggered depend on the performance-influencing factors (PIFs) in the situation (see Chapter 3). However, use of Figure 2.16 allows certain preliminary conclusions to be drawn.

The fact that the worker normally operated reactor B, and he reverted to this operating mode when distracted, indicates that the internal error mechanism was a **Stereotype Takeover**.

It can be seen that the various boxes in the flowchart can be associated with different stages of the stepladder model. For example, the first box on the left corresponds to skill-based behavior and its associated internal failure mechanisms. The second box illustrates the situation (Stereotype Fixation) where the worker erroneously does not change to a rule-based mode when encountering an unusual situation in the skill-based mode (see also the discussion of the GEMS model in Section 2.6.3).

Performance-Influencing Factors
Performance-influencing factors are general conditions which increase or decrease the likelihood of specific forms of error. They can be broadly grouped into the following categories:

- Operating environment (e.g., physical work environment, work patterns)
- Task characteristics (e.g., equipment design, control panel design, job aids)
- Operator characteristics (e.g., experience, personality, age)
- Organizational and social factors (e.g., teamwork, communications)

All of these factors can influence both the likelihood of various internal error mechanisms, and also the occurrence of specific initiating events. (See Chapter 3 for a comprehensive description of PIFs.)

The PIFs increased the likelihood of the strong stereotype takeover in the case study were the fact that the worker was more used to operating the valve for reactor B than reactor A, together with the distracting environment. In addition, the panel was badly designed ergonomically, and valves A and B were poorly labeled and quite close physically. On the basis of the evaluation of the PIFs in the situation, the internal error mechanisms could be stereotype takeover or spatial misorientation.

Internal Error Mode

This is the actual mental function required by the task that failed (see Figure 2.17). In the case study under consideration the failure was at the **Execute Action** stage of the stepladder model, since the worker intended to operate the valve for reactor A, so there was no question of failure in the selection of actions. The connection with the task characteristics box indicates the fact that action is a function required by the task.

External Error Mode

The external error mode is the observable form of the error. This can often be classified in several ways. In the current example the external error modes were "right action on wrong object" (wrong valve closed) and "action omitted" (the correct valve was not closed). The exact form of the external error mode will obviously depend on the nature of the task. A comprehensive classification of external error modes is provided in Chapter 4.

Consequences

The consequences of an external error mode will depend on the context in which it occurs. Consequences for the same error may be trivial (near misses) or catastrophic, depending on the design of the plant and the recoverability of the error. In the example under consideration, a serious accident occurred.

3

Factors Affecting
Human Performance
in the Chemical Industry

3.1. INTRODUCTION

In the previous chapter, a comprehensive description was provided, from four complementary perspectives, of the process of how human errors arise during the tasks typically carried out in the chemical process industry (CPI). In other words, the primary concern was with the process of error causation. In this chapter the emphasis will be on the why of error causation. In terms of the system-induced error model presented in Chapter 1, errors can be seen as arising from the conjunction of an error inducing environment, the intrinsic error tendencies of the human and some initiating event which triggers the error sequence from this unstable situation (see Figure 1.5, Chapter 1). This error sequence may then go on to lead to an accident if no barrier or recovery process intervenes. Chapter 2 describes in detail the characteristics of the basic human error tendencies. Chapter 3 describes factors which combine with these tendencies to create the error-likely situation. These factors are called **performance-influencing factors** or PIFs.

In the nuclear power industry the term **performance-shaping factors** (PSF) has been used to describe a similar concept to PIFs (Swain and Guttmann, 1983). The decision to use the alternative term was taken for the following reasons. First, the concept of PSFs has largely been applied in the context of quantifying human error probabilities. In this book, and other applications, PIFs have been used primarily in a qualitative sense, particularly with respect to designing and auditing systems to minimizing the likelihood of error. When used in quantitative risk assesment (QRA). applications (see Chapter 5), the two terms are more or less synonymous. Another reason for using an alternative term is to emphasise the fact that the factors which influence performance in the CPI may be different from those which affect human error in nuclear power systems. This is a similar distinction to probabilistic safety analysis (PSA) used in nuclear power, and quantitative risk assessment (QRA) used in the CPI. Although broadly the same methods are used, there are differences in emphasis.

In general terms, PIFs can be defined as those factors which determine the likelihood of error or effective human performance. It should be noted that PIFs are not automatically associated with human error. PIFs such as quality of procedures, level of time stress, and effectiveness of training, will vary on a continuum from the best practicable (e.g., an ideally designed training program based on a proper training needs analysis) to worst possible (corresponding to no training program at all). When all the PIFs relevant to a particular situation are optimal then performance will also be optimal and error likelihood will be minimized.

It should be noted that, even with optimal PIFs, errors are still possible. There are two reasons for this. Even in the optimal case, some random variability in performance will remain. These random variations correspond to the "common causes" of process variability considered in statistical process control. Variations in PIFs correspond to the "special causes" of variability considered within the same framework.

Taking the hardware analogy further, PIFs can be seen as corresponding to the design, operational, and maintenance factors which affect the reliability of hardware equipment. The reliability of a pump, for instance, will be influenced by a number of factors such as:

- Type and temperature of liquid processed
- Presence of safety devices (e.g., nonreturn valves, remotely operated isolation valves)
- Any power supply problems
- Effectiveness of maintenance
- Environmental conditions (e.g., presence of corrosive vapors)
- Operational problems (e.g., allowing the pump to run against a closed delivery valve)

It is important, however, not to take this analogy too far. In general, the performance of a piece of hardware, such as a valve, will be much more predictable as a function of its operating conditions than will human performance as a function of the PIFs in a situation. This is partly because human performance is dependent on a considerably larger number of parameters than hardware, and only a subset of these will be accessible to an analyst. In some ways the job of the human reliability specialist can be seen as identifying which PIFs are the major determinants of human reliability in the situation of interest, and which can be manipulated in the most cost-effective manner to minimize error.

Further links exist between the PIF concept and topics considered in previous chapters. In Chapter 2 the sequential model developed by Rasmussen to represent the error process from its initiator to its consequences was described (Figure 2.9). In this process, the PIFs were shown as being involved in both the initiating event and the internal error mechanisms. In the application example of the model in Appendix 2C, the PIF which constituted the initiating event was the distracting environment, and poor ergonomics of the panel was a PIF which influenced the internal error mechanism.

Another link exists between the PIF concept and the sociotechnical assessment methods described in Section 2.7 The checklists used in the TRIPOD methodology are essentially binary questions which evaluate whether the sets of PIFs making up each of the general failure types are adequate or not. The hierarchical sets of factors in HRAM are essentially PIFs which are expressed at increasingly finer levels of definition, as required by the analyst. The audit tool which forms MANAGER also comprises items which can be regarded as PIFs which assess both management level and direct PIFs such as procedures.

In the next section of this chapter, some application areas for PIF analyses will be described. This will be followed by a classification scheme for PIFs based on the demand–resource mismatch model of error described in Chapter 1, Section 1.6. Subsequent sections will describe each of the PIF categories in turn, followed by examples where appropriate. These sections are followed by a discussion of the effects of interactions between PIFs and the implications of high levels of stress in emergencies for human performance.

3.2. APPLICATIONS OF PERFORMANCE INFLUENCING FACTORS

In subsequent sections the application of PIFs to various aspects of error reduction will be described. One of the most important of these applications is the use of comprehensive lists of PIFs as a means of auditing an existing plant to identify problem areas that will give rise to increased error potential. This is one aspect of the proactive approach to error reduction that forms a major theme of this book. This application of PIFs can be used by process workers as part of a participative error reduction program. This is an important feature of the human factors assessment methodology (HFAM) approach discussed in Section 2.7.

Another application area is the use of PIFs as part of the process of incident investigation. Any investigation which seeks to establish the underlying causes of minor or major incidents will benefit from a systematic framework for evaluating the factors which can contribute to the human contribution to such incidents. This topic will also be discussed in Chapter 6.

Performance-influencing factors analysis is an important part of the human reliability aspects of risk assessment. It can be applied in two areas. The first of these is the qualitative prediction of possible errors that could have a major impact on plant or personnel safety. The second is the evaluation of the operational conditions under which tasks are performed. These conditions will have a major impact in determining the probability that a particular error will be committed, and hence need to be systematically assessed as part of the quantification process. This application of PIFs will be described in Chapters 4 and 5.

The PIF concept is also useful during the process of design. Design guidelines to maximize the usability of plant and to minimize the potential for error can be based upon comprehensive descriptions of PIFs such as the factors which determine the most effective presentation of information in control rooms, or the characteristics of usable and clear operating instructions.

For some applications, for example, human reliability analysis, a situation needs to be rated on a numerical scale. In these cases, values such as those shown in the left-hand column of Table 3.1 can be generated by comparing the situation being evaluated with the descriptions in the second, third, and subsequent columns which represent other PIFs relevant to the situation being assessed. These represent the worst, average, and best conditions that are likely to occur in chemical plants in general and correspond to ratings of 1, 5, and 9 on the numerical scale in the left hand column of Table 3.1. Obviously,

TABLE 3.1

Examples of PIF Scales

PIF EVALUATION SCALE (QUALITATIVE AND QUANTITATIVE)	PROCEDURES	PHYSICAL WORK ENVIRONMENT
WORST 1	• No written procedures, or standard way of performing tasks • Not integrated with training	• High levels of noise • Poor lighting • High or very low temperatures and high humidity or wind chill factors
AVERAGE 5	• Written procedures available, but not always used • Standardized method for performing task	• Moderate noise levels • Temperature and humidity range
BEST 9	• Detailed procedures and checklists available • Procedures developed using task analysis • Integrated with training	• Noise levels at ideal levels • Lighting design based on analysis of task requirements • Temperature and humidity at ideal levels

it is possible to interpolate among these values for situations that are intermediate between the descriptions provided.

Unlike the hardware component in a system, human performance is much more variable and difficult to predict. The same combination of input conditions will produce nearly similar effects on hardware. This is not the case for humans who will process the inputs in the light of their intentions and biases in a unique manner.

3.3. A CLASSIFICATION STRUCTURE FOR PERFORMANCE-INFLUENCING FACTORS

The classification structure for PIFs used in this chapter is based on the model of human error as arising from a mismatch between demands and resources which was described in Chapter 1, Section 1.6 (Figure 1.6). In this model demands were seen as requirements for human performance which arise from the characteristics of the process environment (e.g., the need to monitor a panel or to be able to fix a seal in a flange) and the nature of the human capabilities to satisfy these demands (e.g., skills of perception, thinking, and physical action). These demands are met by the individual and group resources of personnel and the extent to which the design of the task allows these resources to be effectively deployed. Where demands exceeded resources, errors could be expected to occur.

In terms of the model, both demands and resources could be influenced by management policy. Demands can be set to fall within the range of human capabilities by ensuring that correct allocations of function are made between humans and machines (including computers). Resources can be maximized by optimizing the PIFs in the situation. This model provides a useful basis for classifying PIFs, since it implies that at least three categories of PIFs need to be considered, those relating to demands, resources, and policies.

As can be seen from Table 3.2, the classification scheme divides PIFs into four major groups and twelve intermediate categories (the numbering system corresponds to the sections in this chapter). The first group addresses those factors related to the chemical and work environment within which the task is carried out such as process hazards, novelty of events, time shortage, lighting, noise, work hours, shift rotation and others. The second group comprises those associated with the workers and their interaction with their tasks including design of control panels and equipment, job aids, procedures, and training. The third group concerns individual characteristics of the workers such as operating experience, personality traits, health, and age. The final group comprises the organizational and social environment and includes topics such as teamwork and communications, safety policies, design policies, systems of work, and others that affect human performance in an indirect manner.

TABLE 3.2

A Classification Structure of Performance Influencing Factors

3.4 OPERATING ENVIRONMENT	3.5 TASK CHARACTERISTICS

3.4.1 Chemical Process Environment
3.4.1.1 Frequency of Personnel Involvement
3.4.1.2 Complexity of Process Events
3.4.1.3 Perceived Danger
3.4.1.4 Time Dependency
3.4.1.5 Suddenness of Onset of Events
 	
3.4.2 Physical Work Environment
3.4.2.1 Noise
3.4.2.2 Lighting
3.4.2.3 Thermal Conditions
3.4.2.4 Atmospheric Conditions
 	
3.4.3 Work Pattern
3.4.3.1 Work Hours and Rest Pauses
3.4.3.2 Shift Rotation and Night Work
	3.5.1 Equipment Design
	3.5.1.1 Location/Access
	3.5.1.2 Labeling
	3.5.1.3 Personal Protective Equipment
	
	3.5.2 Control Panel Design
	3.5.2.1 Content and Relevance of Information
	3.5.2.2 Identification of Displays and Controls
	3.5.2.3 Compatibility with User Expectations
	3.5.2.4 Grouping of Information
	3.5.2.5 Overview of Critical Information and
	    Alarms
	
	3.5.3 Job Aids and Procedures
	3.5.3.1 Clarity of Instruction
	3.5.3.2 Level of Description
	3.5.3.3 Specification of Entry/Exit Conditions
	3.5.3.4 Quality of Checks and Warnings
	3.5.3.5 Degree of Fault Diagnostic Support
	3.5.3.6 Compatibility with Operational
	    Experience
	3.5.3.7 Frequency of Updating
	3.5.4 Training
	3.5.4.1 Conflicts between Safety and
	    Production Requirements
	3.5.4.2 Training in Using New Equipment
	3.5.4.3 Practice with Unfamiliar Situations
	3.5.4.4 Training in Using Emergency Procedures
	3.5.4.5 Training in Working with Automatic
	    Systems

3.6 OPERATOR CHARACTERISTICS	3.7 ORGANIZATION AND SOCIAL FACTORS
3.6.1 Experience 3.6.1.1 Degree of Skill 3.6.1.2 Experience with Stressful Process Events **3.6.2 Personality Factors** 3.6.2.1 Motivation 3.6.2.2 Risk-Taking 3.6.2.3 Risk Homeostasis Theory 3.6.2.4 Locus of Control 3.6.2.6 Emotional Control 3.6.2.6 Type "A" versus Type "B" **3.6.3 Physical Condition and Age**	3.7.1 Teamwork and Communications 3.7.1.1 Distribution of Workload 3.7.1.2 Clarity of Responsibilities 3.7.1.3 Communications 3.7.1.4 Authority and Leadership 3.7.1.5 Group Planning and Orientation **3.7.2 Management Policies** 3.7.2.1 Management Commitment 3.7.2.2 Dangers of a "rule book" Culture 3.7.2.3 Overreliance on Technical Safety Methods 3.7.2.4 Organizational Learning

It should be emphasized that the PIFs considered in this chapter, although generally considered important by human reliability specialists, are not meant to be exhaustive in their coverage. Other selections, such as those considered by the methods such as TRIPOD and HFAM (Chapter 2), are possible. It is recommended that the advice of an experienced human reliability or human factors specialist is sought when deciding which PIFs should be covered in a specific situation.

3.4. OPERATING ENVIRONMENT

There are three elements of the operating environment which play a crucial role in human reliability, namely:

- **The chemical process environment** which refers to the complexity and novelty of the process events, their perceived danger, and the imposed time constraints on the workers
- **The physical work environment** which includes conditions of lighting, thermal conditions, atmospheric conditions and noise levels
- **The patterns of work** such as work hours and pauses, night work and shift rotation

3.4.1. Chemical Process Environment

All of these factors determine the stress experienced by the workers and the extent to which operational errors will be recovered before disastrous consequences have ensued. In this context, hazard identification techniques, such as hazard and operability studies (HAZOP), failure modes and effects and criticality analysis (FMECA), fault trees, and others are useful in making the process environment more forgiving.

Throughout these guidelines it is argued that when engineering techniques for the design and assessment of process equipment and control systems are supplemented with human reliability techniques, then performance of both the hardware and humans will be optimized.

A characteristic of the PIFs that follow is that although they may not affect performance in a direct manner, they may interact with other factors such as inadequate procedures, training, and worker experience to give rise to errors.

3.4.1.1. *Frequency of Personnel Involvement*
The frequency with which a task is performed or a process event has been dealt with in the past, affects the likelihood of success. Process skills that are not frequently practiced (e.g., for tasks that are only required on an irregular

basis), may not be retained adequately and performance can deteriorate. Whether or not this deterioration will give rise to a significant error will depend on other factors which will be described in later sections such as refresher training, detailed procedures and so on.

Performance problems may be exacerbated during unfamiliar or novel process events, for example, situations not covered in the emergency procedures or in refresher training. These events require knowledge-based information processing for which people are not very reliable. The types of errors associated with knowledge-based performance have been discussed in Chapter 2.

3.4.1.2. Complexity of Process Events
Apart from the degree of novelty of a process event, its complexity (e.g., the range of operations to be carried out), the interrelationships of the process variables involved and the required accuracy, will affect performance. Start-up and shutdown operations are examples of tasks which, although are not entirely unfamiliar, involve a high degree of complexity.

3.4.1.3. Perceived Danger
One of the most serious stressors to personnel working in many chemical processes is the perception of danger by the workers arising from ineffective control and supervision of these systems. Despite the fact that modern plants are equipped with automated protection systems, there is always some perception of potential risk in their operation. Serious threats can be posed not only for those within the plant, but also for the neighboring public. An environment that is perceived as being highly dangerous will increase the stress experienced by the workers and may have a detrimental effect on their performance.

3.4.1.4. Time Dependency
Time dependency refers to the time available to cope with a process event. Time pressure is a well-known stress factor which affects human performance. Here, the time response of plant equipment and chemical processes will determine the time available to respond to an incident.

3.4.1.5. Suddenness of Onset of Events
In a process disturbance, the suddenness of the onset of the event will also play a significant role in human performance. This category refers to the time required for the process symptom to develop to the extent that it becomes detectable by the workers. If the symptom develops gradually, this leaves some scope for the workers to switch to a high mode of alertness. This allows them to develop an adequate mental model (see Chapter 2) of the process state. If an adverse condition develops extremely slowly it may not be detected by workers, particularly if its development spans more than one shift.

3.4.2. Physical Work Environment

The next four PIFs (noise, lighting, thermal conditions, atmospheric conditions) refer to the quality of the worker's environment. In general, if the quality of these factors is poor, they may cause anxiety and fatigue which may result in errors. Some of these stressors, such as noise and heat, produce psychological as well as physiological effects on performance. Even moderate levels of such stressors in the control room can interfere with task performance because workers can be distracted, lose concentration, and become irritated. Working under these stressors means that more work and more attentional and memory resources will have to be expended on each individual action. The existence of such stressors can also indicate a lack of management concern for the well being of the workers which can increase unsafe behavior.

Most of the research on the effects of these stressors on human performance has been done on simple laboratory tasks rather than actual work situations. As a result, the extent that such findings can carry over to tasks in the CPI is debatable. In addition, most of these studies have examined the effect of a single stressor (e.g., noise or heat) only, rather than the combined effect. Nevertheless, some useful guidelines about optimal levels of these stressors are available in the ergonomics literature (e.g., McCormick and Sanders, 1983; Salvendy, 1987).

3.4.2.1. Noise
The effects of noise on performance depend, among other things, on the characteristics of the noise itself and the nature of the task being performed. The intensity and frequency of the noise will determine the extent of "masking" of various acoustic cues, i.e. audible alarms, verbal messages and so on. Duration of exposure to noise will affect the degree of fatigue experienced. On the other hand, the effects of noise can vary on different types of tasks. Performance of simple, routine tasks may show no effects of noise and often may even show an improvement as a result of increasing worker alertness.

However, performance of difficult tasks that require high levels of information processing capacity may deteriorate. For tasks that involve a large working memory component, noise can have detrimental effects. To explain such effects, Poulton (1976, 1977) has suggested that "inner speech" is masked by noise: "you cannot hear yourself think in noise." In tasks such as following unfamiliar procedures, making mental calculations, etc., noise can mask the worker's internal verbal rehearsal loop, causing work to be slower and more error prone.

Another effect of noise on tasks involving monitoring and interpretation of a large number of information sources is the "narrowing of the span of attention." In a noisy environment, personnel monitoring the control panel would tend to concentrate on the most obvious aspects of the situation which seem to be familiar to them and fail to incorporate any novel features of the situation. Apart from causing distractions and communication difficulties, permanent exposure to a noisy environment may reduce any opportunities for social interaction and thus make the job more boring.

3.4.2.2. Lighting

Apart from physical discomfort and irritation, poor lighting can induce errors in reading valve labels or instruments on the control panel. Direct or reflected glare can be another problem in many work situations. Having to avoid the glare may constitute another task the worker has to perform, which can divert him or her from the primary job responsibility.

Example 3.1. Effects of Glare

Swain and Guttmann (1983) cite an incident in which the problem of glare had been so severe that the workers disconnected many of the lamps, with the result that a flashlight was considered a standard accessory for reading certain displays.

3.4.2.3. Thermal Conditions

The effect of high or low environmental temperature on skilled performance is important for industrial or service personnel. Operators often have to work in extreme thermal conditions, such as in furnaces or when they need to operate a pump in cold weather at night. Errors of omission are quite often due to the workers trying to minimize the time period they have to be exposed to high or low temperatures. Particular emphasis has been placed on the effects of cold on manual performance. Cold can affect muscular control, reducing such abilities as dexterity and strength.

Experience and familiarity with the task will affect the relationship between temperature and performance. Experience and practice will make performance largely skill based, and therefore, more resistant to impairments due to high temperatures. This explains why unskilled workers are affected more adversely when they have to work in extreme heat.

3.4.2.4. Atmospheric Conditions

Many operations may expose the workers to dust, fumes, gases, etc., and apart from causing personnel injuries these may lead to human errors. This is because protective clothing and apparatus is usually uncomfortable. Attempts to get the job finished quickly may therefore result in errors.

3.4.3. Work Pattern

Two important work pattern PIFs are the duration of work hours and rest pauses, and the type of shift rotation.

3.4.3.1. Work Hours and Rest Pauses

On many occasions, long hours of work are required because the worker may have to stay on duty at the end of the shift to fill in for someone on the next shift or because there are plant start-up or shutdown operations.

In some plants, workers voluntarily request 12-hour shifts in order to benefit from the long periods away from the job that this regime brings. However, on the basis of everyday experience one would expect the fatigue arising from prolonged work to give rise to performance decrements and errors. In this section we will review the evidence on this question from two perspectives: sleep loss and sleep disturbance and prolonged working hours.

Sleep Loss and Sleep Disturbance
The effects of acute sleep deprivation where subjects are deprived of sleep over successive days have been studied extensively. Research findings have demonstrated clear decrements in psychological performance and resulting behavioral impairments (see Froberg, 1985 for an overview). In particular, tasks of 30 minutes or more in duration; low in novelty, interest, or incentive; or high in complexity have been shown to deteriorate in a situation of prolonged work duty and no sleep. Memory has also been found to be affected in people who are required to stay awake (Wilkinson, 1964). However, such effects are reversed with only 1 to 2 nights of recovery sleep even in the longest deprivation studies.

The effects of chronic sleep deprivation or cumulative minor sleep losses have been relatively under investigated. Little is known about the relationships among the size of the sleep deficit, its rate of accumulation, the amount and timing of optimum recovery sleep, and their effect on human performance and productivity.

Those studies of partial sleep deprivation that have been carried out show that people can tolerate a degree of sleep loss and are able to keep up their performance level with sleep periods shorter than normal. The limit of tolerance for prolonged spells of reduced sleep seems to be around 4–5 hours of sleep per day. This seems to represent an obligatory quota. Providing this quota can mostly be reclaimed or retained, it is possible for psychological performance and day time tiredness to be maintained at normal or near normal levels. However, this depends on subjects maintaining a regular sleep schedule. People who are forced to take less sleep but who cannot maintain sleep regularity will have increased difficulty because of insufficient time to adapt.

In conditions of acute sleep deprivation, "microsleeps" will occur more and more often. These very short sleeps do not have the recuperative value of normal sleep, and the sleep-deprived person still feels sleepy and performance still degrades even though there may be a large number of microsleep periods.

The Effects of Prolonged Working Hours
Extended working weeks of 60 hours or more were common in the nineteenth century, but for much of the twentieth century the norm has been a 5-day/40-

hour working week. Inevitably much of the work related to the productivity and performance implications of extended working hours stems from studies carried out in atypical periods such as during and immediately following both world wars. Allusi and Morgan (1982), in their review of temporal factors in human performance and productivity, summarized the result of many of these early studies. The overall conclusion reached was that improvements in industrial productivity have generally been found following reductions in the total hours of work in both the work day and work week.

This increase in productivity is accounted for partly by a decrease in absenteeism and accidents as well as a general increase in working efficiency. For example, Vernon (1918), found that when women in a munitions factory worked a 12-hour day they incurred 2.5 times more accidents than when they worked a 10-hour day. One of the more comprehensive studies of the effects of total hours of work was carried out after World War II by the U.S. Department of Labor (Kossoris and Kohler, 1947). This covered over 3500 men and women in 78 work units. Data were collected on accidents and absenteeism as well as productivity. The overall findings were that exceeding the 8-hour work day, 5 days/40 hours work week resulted in lower productivity and higher absenteeism and accident rates.

Such historical evidence is always vulnerable to methodological criticisms. However, following the fundamental shift in working practice which subsequently occurred, such studies represent the only significant body of field studies which have assessed the repercussions of prolonged working hours in an industrial setting. Some more recent studies looking at the work of, for example, hospital doctors have reported on sleep loss and the effects of long hours of work. Studies such as that of Folkard and Monk (1985) which examined self-reports of work impairment show that a considerable percentage of junior doctors who responded (over one-third) felt that their ability to work with adequate efficiency was impaired by the long hours of duty. Objective tests of performance such as those used by Poulton (1978), again looking at hospital doctors, show less conclusive results. It is also difficult to generalize findings from such a highly specific work situation to other types of working environments.

The desirability of the standard 8-hour work day and 5-day/40-hour work week is currently being questioned in response to both economic and commercial pressures and worker preference for greater flexibility and leisure time. A variety of alternative schedules are now available and their introduction has led to a renewed interest in the effects of extended working times. In particular, these more recent studies have provided some further insights into the performance effects of long work days. For example, recent work by Rosa et al. (1986) has examined the effects of the introduction of the 12-hour day compressed work week. The principle underlying this schedule is to shorten the work week to 3 or 4 days by increasing the length of the work shift to 12 hours. However, there are persistent concerns about feelings of increased fatigue associated with long work days. Moreover, such concerns are supported by laboratory and worksite (Volle et al., 1979) comparisons of 8-hour and 12-hour days.

Rosa et al. (1986) evaluated changes in a range of variables associated with a switch from an 8-hour shift schedule with three rotations to a 12-hour shift schedule with two rotations. The workers involved were control room operators at a continuous processing plant. The authors report that, after 7 month's adaptation to the new schedules there were decrements in the tests of performance and alertness (National Institute of Occupational Safety and Health, NIOSH Fatigue Test Battery) attributable to the extra 4 hours of work per day. There were also reductions in sleep and disruption of other personal activity during 12-hour work days. In summary however, the study concludes that there have only been a few direct evaluations of the effects of long work days on individual functioning. Those that do exist have provided some suggestions of accumulated fatigue across a number of long work days. The overall conclusion is nevertheless that substantive further work is needed to clarify the performance effects of long work days.

Effects of Fatigue on Skilled Activity

"Fatigue" has been cited as an important causal factor for some everyday slips of action (Reason and Mycielska, 1982). However, the mechanisms by which fatigue produces a higher frequency of errors in skilled performance have been known since the 1940s. The Cambridge cockpit study (see Bartlett, 1943) used pilots in a fully instrumented static airplane cockpit to investigate the changes in pilots" behavior as a result of 2 hours of prolonged performance. It was found that, with increasing fatigue, pilots tended to exhibit "tunnel vision." This resulted in the pilot's attention being focused on fewer, unconnected instruments rather than on the display as a whole. Peripheral signs tended to be missed. In addition, pilots increasingly thought that their performance was more efficient when the reverse was true. Timing of actions and the ability to anticipate situations was particularly affected. It has been argued that the effects of fatigue on skilled activity are to regress to an earlier stage of learning. This implies that the tired person will behave very much like the unskilled operator in that he has to do more work, and to concentrate on each individual action.

Conclusions on Work Hours and Rest Pauses

In interpreting the above research findings it is important to consider a number of additional points.

- Most sleep deprivation experiments have used mentally and physically healthy young adults. For other types of individuals, particularly older people for whom the sleep function deteriorates in general, and also for "real world" conditions, sleep deprivation may be more significant.

- Fatigue effects associated with a long working day has been identified in the context of a 4 day week.
- Little is known about the cumulative effects of factors such as prolonged working hours and extra mural demands on workers and how such demands interact with workers performance reserves and productivity.

It is likely that a person experiencing fatigue over a long period of time would develop strategies to cope with the effects on his or her performance. Such coping strategies could include:

- Working more slowly
- Checking the work more thoroughly
- Using more memory "reminders"
- Relying on fellow workers
- Choosing to carry out less critical tasks

However, such strategies are vulnerable to additional factors such as increased time pressure, and working alone. The combined influences of such factors may be more important than each negative factor in isolation.

3.4.3.2 Shift Rotation and Night Work

There are two concerns about the effects of shift rotation and night work: disruption of "circadian rhythms" and sociological costs, that is, effects on the worker's family life.

The term *circadian rhythms* refers to variations in certain physiological variables (e.g., body temperature) over the 24-hour cycle. Individuals who are "day adjusted," that is, who are active and asleep during the normal periods of day and night, exhibit the characteristic variations of body temperature shown by the dark graph in Figure 3.2. Similar variations occur in psychological functions such as activation (see Chapter 2, Section 2.3.3) and self-estimates of alertness (shown by the light graph in Figure 3.2). These estimates are generated by asking subjects to rate how alert they feel by marking a scale between the range "almost asleep" and "fully alert." Figure 3.2 indicates the close relationship between body temperature and alertness. When individuals work on continuous night shifts for a protracted period, the circadian cycle gradually changes so that the peaks of body temperature and alertness tend to occur at night when the worker is active.

With regard to the effect of circadian cycles on performance, most studies have been carried out using individuals such as nurses or airline pilots, whose work involves shifts or the crossing of time zones. One study that specifically addressed process workers was carried out by Monk and Embrey (1981). In this study the body temperatures of six "day adjusted" workers working on a batch chemical plant were recorded over a one-month period of plant operations. The average temperature variations are shown in Figure 3.1, together with the workers" self-ratings of their alertness.

Research on circadian rhythms has generally indicated that performance on mental tasks broadly follows the same pattern of variations as body temperature and alertness. However, other work suggests that in fact this is only the case for mental tasks requiring little information processing capacity. For more complex "cognitive" tasks where working memory is more important, variations in performance are in the opposite phase to body temperature; that is, best performance, occurs when the body temperature is low (i.e., at night). This hypothesis was tested by asking the workers to perform two types of memory-based test every 2 hours. One test, the 2-MAST (memory and search test) involved a low memory load whereas the other (6-MAST) required a much greater memory load. These tests both involve the mental manipulation of numbers. The larger the sets of numbers, the greater will be the memory load. To some extent the tests mimic the mental demands of process control tasks of differing complexity. Performance on the tests is measured by the length of time they take to perform, better performance being indicated by a shorter time. Figure 3.3 confirms the predictions by indicating that performance on the 6-MAST (high memory load) was in opposite phase to the circadian body temperature cycle, whereas the performance on the low memory load task closely followed variations in body temperature.

The applicability of these findings for actual operational tasks was evaluated by considering the incidence of data entry errors recorded by the on-line plant computer system over the 24-hour shift cycles. It was judged that the data entry task, which involved evaluating the set point changes needed for

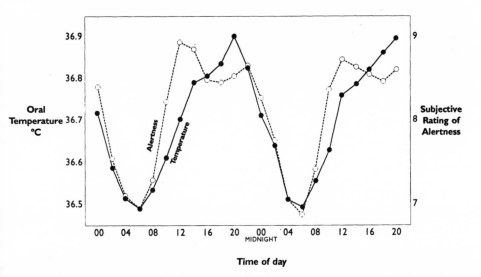

FIGURE 3.1.   **Circadian Variations in Oral Temperatures and Alertness for Six Process Workers (Monk and Embrey, 1981).**

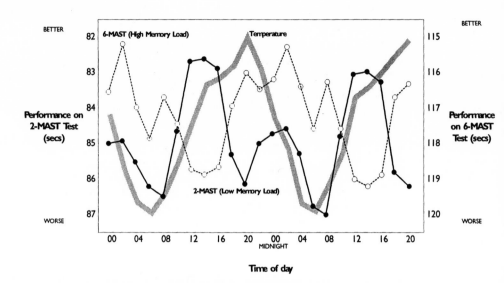

FIGURE 3.2. **Circadian Variations in Performance on High- and Low- Memory Load Tasks (adapted from Monk and Embrey, 1981).**

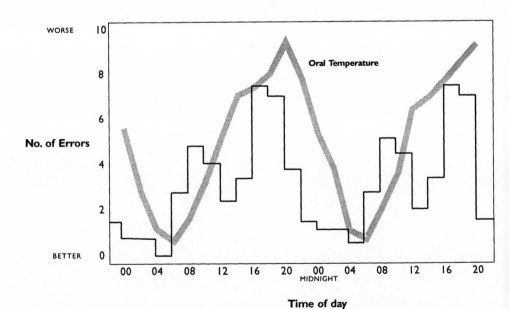

FIGURE 3.3. **Circadian Variations in Errors Made by Process Workers Compared with Body Temperature Changes (adapted from Monk and Embrey, 1981).**

the process and then entering these set points into the database, was a high memory load "cognitive" task. The results of the error evaluation, plotted in Figure 3.3, show that the variation in error rates follows the same temporal pattern as performance on the 6-MAST test shown in Figure 3.3 (this correlation was statistically significant) and is an opposite phase to variations in body temperature. (Note that the scales in Figure 3.3 are in the opposite direction to those in Figure 3.2, since more errors indicate worse performance). This appears to confirm the prediction that high memory load cognitive tasks would have lower error rates at night (for day adjusted workers).

The practical implications of this experiment are that when evaluating the effects of shift work due to circadian effects, the type of task being carried out by the worker must be taken into account. For example, skill-based tasks would be expected to exhibit the performance changes characteristic of low memory load tasks, whereas performance variations in knowledge-based tasks would be expected to follow the pattern of high memory load tasks. Performance on rule-based tasks may depend on the degree of frequency of use of the rules, which in turn may determine the memory load. If these results were confirmed by further process plant studies, it would have implications for when different types of operation (involving different levels of memory load) should be scheduled to reduce circadian rhythm effects and minimize errors.

Studies by Smith et al. (1982), Folkard et al. (1979), and Colquhoun et al. (1969), have investigated the disruption of circadian rhythms caused by having to be awake and work at unusual hours and by having to sleep during daytime. With respect to the sociological effects, studies by Kasl (1974) and Kahn (1974) concluded that fixed afternoon and night shifts lead to lower levels of social satisfaction because it becomes difficult to participate in family activities.

With regard to the scheduling of shift work, the general recommendation (putting aside social and lifestyle considerations) is that shifts should allow workers to either remain day or night adjusted. This is because it is the constant readjustment of circadian cycles which appears to produce the most acute feelings of fatigue and disorientation. This implies that permanent night or day shifts will be the most effective (In a union environment, where seniority provisions could lead to inexperienced operators being concentrated on the afternoon or evening shifts, there could be an offsetting problem of fixed shifts to rotating shifts.) Failing this, shifts should be operated over a sufficiently short cycle that they allow the operating team to remain day adjusted. However, it must be emphasized that determining optimal shift work regimes is a highly complex and controversial area of research. Comprehensive reviews of the state of the art are available in Folkard and Monk (1985) and Monk and Folkard (1991).

3.5. TASK CHARACTERISTICS

This section addresses aspects of the task which influence human reliability in both control-room and field operation situations. It includes the physical equipment to be used, control panels, job-aids and procedures, and the type of training provided.

3.5.1. Equipment Design

Plant equipment should have good access and controls and instruments should be clearly labeled. Under this category protective clothing and other equipment to enable safe operation is also included.

3.5.1.1. Location/Access
Process workers often complain that valves are inaccessible. Emergency valves should always be readily accessible but other valves, if they are operated, say, once a year or less often, can be out of reach. It is reasonable to expect workers to get a ladder or scramble into a pipe trench at this frequency. Designers should remember that if a valve is just within reach of an average person then half of the population cannot reach it. Equipment should be placed such that at least 95% of the population can reach it. Guidance on specific measurements to achieve this objective is available in a number of standard human factors textbooks (see Bibliography).

3.5.1.2. Labeling
Many incidents have occurred because equipment was not clearly labeled. Some have already been described in Section 1.2. Ensuring that equipment is clearly and adequately labeled and checking from time to time to make sure that the labels are still there is a dull job, providing no opportunity to exercise many technical and intellectual skills. Nevertheless, it is as important as more demanding tasks.

3.5.1.3. Personal Protective Equipment
The design and the enforcement of personal protective equipment can play the key role of protecting the worker from exposure to hazardous conditions. Such equipment includes goggles, gloves, breathing apparatus, helmets, earplugs, safety shoes, safety belts, and so forth. Standards for the design and performance of protective equipment are specified by regulatory agencies such as the National Institute for Occupational Safety and Health (NIOSH). A comprehensive overview of the human factors aspects of personal protection equipment is provided by Moran and Ronk (1987).

Many incidents are caused by the process workers not bothering to wear safety equipment or removing it during an operation. In these cases, blaming the worker is not necessarily appropriate, since a number of other factors may be held responsible, such as the difficulty of carrying out certain jobs when wearing safety equipment, or a work permit approach which emphasizes the use of equipment even for safe areas (see Examples 1.21 and 1.22)

3.5.2. Control Panel Design

The term control panel refers to the instrumentation console in a central control room through which process information is communicated to the process worker and via which the worker changes the state of the process. This category includes display elements such as chart recorders, bar indicators, dials, and modern VDU-based systems together with control elements such as buttons, switches, track balls and mice. The control panel is the human–machine interface (see Chapter 2) that has traditionally received the most attention from human factors specialists.

The content and organization of the displayed information are of critical importance in inferring the state of the process and subsequently evaluating the effects of alternative courses of action. The following factors will determine the demands of the control panel on the attentional and memory resources of the workers. For detailed data on the design of the control panel, the reader is referred to standard ergonomics textbooks (e.g., Salvendy, 1987).

3.5.2.1. Content and Relevance of Information
The first questions to be considered when designing a control panel are what information is required and how much of it will be appropriate. Too little information may increase the amount of inference that the worker is required to make to predict the state of process parameters that are not directly displayed. This is especially important for emergency situations where the human information processing system is taxed heavily with many tasks. On the other hand, too much redundant information can overload the worker. It is essential, therefore, that the information needs of the worker are identified through some form of task analysis and worker interviews.

The relevance of the information to the process worker is another factor in design. This principle is often violated with the introduction of new VDU-based computer systems where information needed to assist computer scientists or production managers is mixed with information relevant for the safe operation of the plant. Clearly, some kind of structuring and prioritization will be necessary for the different users of the system.

3.5.2.2. Identification of Displays and Controls
The issue of how controls and displays are identified on a control panel is usually referred to as coding. In the case of controls this can be achieved by techniques such as labeling, color, shape, location, or size. The relationship

between displays and controls needs to be carefully considered. Comprehensive recommendations for displays and controls are available in Salvendy (1987).A recurring problem in many process plants concerns the lack of demarcation lines for the tolerance limits of various critical parameters. Workers need to know how rapidly a parameter is moving toward its tolerance limits in order to understand the urgency of the situation.

3.5.2.3. Compatibility with Personnel Expectations

Compatibility refers to the degree of similarity between the direction of physical movement of a control or an instrument indicator and the worker's expectations. Many errors are due to the fact that the operation of the controls or the layout of the displays is incompatible with population stereotypes. For instance, on a control panel it is customary to increase the value of a parameter by turning the appropriate switch clockwise and reduce its value by turning it counterclockwise. (Note that this stereotype is the opposite for controls which control flow directly, e.g., valves.) If such a stereotype is violated, errors may occur. Although such errors may be recoverable in the short run, under the stress of a process transient they may lead to serious consequences.

Example 3.2. Design Fault Leading to Inappropriate Worker Expectations

In the Three Mile Island power plant, the light of the pilot operated relief valve (PORV) status indicator was designed to come on when an electrical signal was transmitted to the valve to open, and go out when a signal was transmitted for the valve to close. When the worker pushed the button to close the valve, the signal was transmitted but it was not received by the valve due to an electrical fault. As a result the light went out, but the valve remained open. For two hours, the workers were under the impression that this valve was closed, which resulted in radioactive coolant discharging from the reactor circuit. This design violated the worker's expectation that the light would indicate the status of the valve and not that of the signal. Similar incidents have been described in Examples 1.16 and 1.17

3.5.2.4. Grouping of Information

This factor refers to the spatial organization of the information displays. In general, instruments displaying process parameters that are functionally related should also be physically close. In this way, it is likely that a given fault will lead to a symptom pattern that is easier to interpret than a random distribution of information. Although violation of this principle may not induce errors in a direct manner, it may hinder human performance. The following example illustrates this point.

Example 3.3. Poor Control Panel Design Causes Lack of Diagnosis

In a power plant a failure of the steam regulator in the turbine gave rise to a high pressure profile in the three condensers downstream. Previously, one of the three cooling water pumps had failed, activating a high pressure alarm in the affected condenser. The crew did not notice the pattern of pressure rise in all three condensers (which was rapid, large, and of a similar amplitude) and thus failed to diagnose the latent failure in the steam regulator. A careful examination of the displays showed that two 2-channel recorders were used instead of one 3-channel recorder, making it difficult to perceive the dynamics of the pressure rise. Second, the steam regulator display was positioned in a different section of the panel to that showing the condenser system. This made it less likely that any deviation would be detected through the normal strategy of checking related subsystems.

3.5.2.5. *Overview of Critical Information and Alarms*
With the increasing complexity of plants, overview displays of critical process information and alarms can be very useful particularly for plant disturbances. In this regard, several investigators (Goodstein, 1982; Woods et al., 1981) have advocated the concept of the integrated or polar display which can be implemented on modern computer-based systems. The different radial scales are adjusted so that normal operation is represented by a normal geometric shape, while departures indicate distortions. This type of display capitalizes on human "pattern recognition" capabilities and can support early detection of abnormal process states.

3.5.3. Job Aids and Procedures

3.5.3.1. *Introduction*
As process plants become more complex, it becomes apparent that it is not possible to rely exclusively on the process worker's skills and memory required to perform the task. Job aids and procedures are devices which aim to reduce the need for human retention of procedures and references as well as the amount of decision making required. Job aids assume a variety of formats including flowcharts, checklists, decision tables, etc., while procedures refer to other systems of documentation such as standard operating instructions and emergency procedures.

3.5.3.2. Common Problems with Procedures

Which often lead to violations. The following deficiencies may occur in any applications of procedures, from operating instructions to permit to work systems:

Procedures Do Not Correspond to the Way the Job Is Actually Done.

Procedures are often developed when a system is first commissioned and are seldom revised to take into account changes in the hardware or the operating regime. In addition, procedures are often not written on the basis of a systematic analysis of the task as perceived by the workers or other personnel who have to use them. The remedy for this is to make sure that individuals who are going to use procedures are actively involved in their development. In addition, effective updating and auditing systems need to be in place to ensure that procedures are correct, and available to the persons who need them.

The Information Contained in Procedures Is Correct, but It Is Not Cast in a Form Usable by the Individual at His or Her Workplace.

Very often, voluminous procedures gather dust in cabinets where they have lain since the system was commissioned. For simple skill-based tasks carried out by experienced workers, no procedural support will be necessary. Other activities such as trouble shooting or diagnosis may, as discussed in Chapter 2, involve the use of formal or informal rules which are used infrequently. In these cases some form of job aid or checklist is the most effective type of procedure.

Detailed procedures will only be required in unusual situations where the usual rules of thumb do not apply and the worker is likely to be in the knowledge-based mode. In Chapter 4, and case study 3 in Chapter 7, a systematic framework for developing procedures, in which their format and content is based on a detailed analysis of the tasks to be performed and the normal skill level of the person who will perform the tasks, will be described.

Only task elements which are particularly critical (from the point of view of the consequences of failure) or where errors are particularly likely, are included in the job aid. The development of procedures obviously has to be closely integrated with the content of training, since the design of procedures has to assume that the individual has received appropriate training for certain aspects of the task.

The Distinction between Procedures as Regulatory Standards and as Instructions to Perform a Task Is Not Adequately Made.

In many industries, rule books have a tendency to become enshrined as policy statements, either for internal or external regulatory purposes. Unfortunately, the format that is appropriate for a regulatory or standards document is unlikely to fulfill the requirements of an effective operating instruction or procedure to provide assistance in carrying out a task effectively.

Procedures Are Not Updated on the Basis of Operational Experience.
If procedures are obviously out of date or do not take into account lessons learned throughout a system, they rapidly lose their credibility and are likely to fall into disuse.

Rules and Procedures Are Not Seen to Apply to the Individuals or the Situation in Question.
If there are situations where ordinary procedures may be suspended for specific purposes, these need to be carefully defined and controlled by the proactive development of "rules" which explicitly state the boundary conditions for such interventions.

The User of the Procedures Does Not Understand the Underlying Reasoning behind Them and Therefore Carries Out Alternative Actions That Appear to Achieve the Same Purpose but Are Easier to Perform.
This type of failure underscores the earlier comment that individuals should, if possible, be actively involved in the development of procedures that they are required to use, so that they understand the underlying purpose behind them.

3.5.3.3. Criteria for Selecting Job Aids

To select the most appropriate method to support the process worker, one needs to consider the characteristics of the task and the type of support to be provided. Flowcharts and decision tables, for instance, offer a concise organization of the information and the job criteria required to perform fault diagnosis and planning tasks. Checklists are more suitable for tasks which involve remembering sequences of steps. Procedures, on the other hand, provide step-by-step directions with regard to how and when to perform various tasks which involve stringent memory requirements, calculation, accuracy, and difficult decisions. Standard operating instructions are usually provided for critical tasks involving changes in the plant operating conditions such as plant start-up or shutdown or changes of fuel firing in a refinery furnace. Emergency procedures are provided for tasks which involve diagnosing plant or instrumentation failures and stabilizing and recovering abnormal plant conditions.

An important issue is how much of the job requirements should be supported by job aids and procedures as opposed to training. If job aids are developed at the expense of adequate training, the worker may become tied to the aid and thus vulnerable to situations where the aid contains errors or unforeseen plant conditions occur. On the other hand, overloading the worker with too much information and skills to be learned during training may result in performance decrements in the long run. To determine the extent of job aid provision versus training, the investment required to generate and validate the aids as well as develop and carry out extensive training programs should be considered. Joyce et al. (1973) and Smillie (1985) provide a thorough discussion of the criteria to be taken into account when examining these trade-offs.

In general, job aids and procedures are useful for tasks which are performed rarely or require complex logic, for example, diagnostic aids. They are also applicable for situations which involve following long and complex action sequences, and where reference to printed instructions is not disruptive. Training should be emphasized for tasks which are performed frequently, require complex manual skills, depend strongly on team efforts or involve unforeseen plant conditions. These considerations can be seen to be directly related to the skill-, rule-, and knowledge-based classification discussed in Chapter 2.

In order to judge the extent that the job aids and procedures provided will facilitate process worker performance or engage him or her in a time-consuming search for information, we need to look closer at a number of factors .

3.5.3.4. Clarity of Instruction

- This refers to the clarity of the meaning of instructions and the ease with which they can be understood. This is a catch-all category which includes both language and format considerations. Wright (1977) discusses four ways of improving the comprehensibility of technical prose.
- Avoid the use of more than one action in each step of the procedure.
- Use language which is terse but comprehensible to the users.
- Use the active voice (e.g., "rotate switch 12A" rather than "switch 12A should be rotated").
- Avoid complex sentences containing more than one negative.

The following example highlights how lack of clarity of instructions can lead to errors of misinterpretation.

Example 3.4. Error Due to Lack of Clarity of Instructions

In one plant, the operating procedures required that valve A should be placed into the "manual closed position." The process worker misinterpreted this information and instead of placing the valve controller in the manual position, he closed the block valve manually and deprived the plant of an essential feed.

The format of the procedure is also important in this respect. There may be situations where alternatives to prose are more efficient and acceptable. A flow diagram or a decision table may help the process worker to concentrate more easily on what indications are presented, and what decisions and control actions he or she has to make (see Wright. 1977).

3.5.3.5. Level of Description

An important issue in the writing of procedures is how much information is necessary for the process worker in order to minimize the likelihood of error. Too little may be inappropriate for an inexperienced process worker while too much may encourage a highly experienced worker not to use the procedure. It is obvious that the level of worker expertise and the criticality of the task will determine the level of description. This example shows how lack of detailed information can lead to errors of omission.

Example 3.5. Error Due to Lack of Detail of Instructions (Kletz, 1994b)

A day foreman left instructions for the night shift to clean the reactor. He wrote "agitate with 150 liters nitric acid solution for 4 hours at 80°C." He did not actually tell them to fill the reactor with water first, as he assumed that this was obvious since the reactor had been cleared this way in the past. The night shift did not fill the reactor with water. They added the nitric acid to the empty reactor via the normal filling pump and line which contained isopropyl alcohol. The nitric acid displaced the isopropyl alcohol into the reactor, and reacted violently with it, producing nitric fumes. As a result the reactor, which was designed for a gauge pressure of 3 bar, burst. Although this accident can also be said to be due to failure of the night shift to use their knowledge of chemistry, it clearly demonstrates the importance of the appropriate level of detail in the instructions

---

3.5.3.6. Specification of Entry/Exit Conditions

Many of the difficulties in using operating procedures stem from the fact that the conditions for applying a given section or branch and the conditions for completing or transferring to another section are not clearly specified. This is particularly important in emergency situations where a choice must be made under time pressure and excessive workload.

3.5.3.7. Quality of Checks and Warnings

Checks of critical process parameters and warnings about hazardous conditions that can cause injury or equipment damage are important factors which determine the occurrence and recovery of human error. The purpose of these checks is to emphasize critical process information. Because of the critical nature of this information, checks and warning should be highlighted in a way that distinguishes them from other notes, and should be located where process workers will not overlook them.

3.5.3.8. Degree of Fault Diagnostic Support

Emergency procedures usually require the process worker to make the correct diagnosis in order to select the right compensatory actions, a task which is often performed poorly under the duress of an abnormal situation. To overcome this problem, some procedures provide fault diagnostic support such as fault-symptom tables or other graphical aids relating to each plant failure for which recovery actions are specified. The degree of fault diagnostic support and their particular format will influence the likelihood of a correct human intervention in an emergency situation.

3.5.3.9. Compatibility with Operational Experience

It is common practice that procedures and job-aids are often developed either by plant manufacturing companies or process designers with minimal participation by the end-users, usually plant workers. This has led to situations where the indicated sequence of actions was incompatible with the way the job is done in practice. This presents great problems for the workers who will have to reconcile a potential violation of procedures with a well established method of operation.

Although manufacturing companies and process designers may have a thorough knowledge of plant equipment, factors such as subsequent modifications, age, and working hours of the equipment, changes in the product specifications, and maintenance problems, may not be foreseen. In addition, experience with the dynamic response of the plant provides workers with insights into its detailed operating characteristics which need to be factored into the procedures. These considerations emphasize the importance of the active participation of the operating team in the design and maintenance of procedural aids.

3.5.3.10. Frequency of Updating

The above factors also highlight the importance of updating the procedures frequently. There are many occasions where control loops are introduced in the plant without proper modification of the procedures, which means that the process worker will not be able to explain the behavior of the plant or understand the required intervention on his part.

3.5.4. Training

Control panel design, equipment design, and job-aids and procedures are factors which change the demands of the task to be performed. Training is a factor which determines the capability of the worker to cope with a task by providing the required knowledge and skills. Process worker training can fulfill various requirements, for example, the ability to perform a job, to use new equipment, job aids and procedures, to respond to emergency situations, to maintain process skills with the introduction of automation, and finally, to make teamwork effective. These types of training will be considered in detail below, in order to examine how deficiencies in their design may dispose the worker toward error.

A distinction can be made between the previous forms of training and the methods to provide the required skills. In process control, we may consider training people off-the-job, on the plant itsefl—but not actually carrying out the job, and while they are carrying out the job. Off-the-job training is best seen as a means of preparing trainees to benefit from real experience and not as a sole training method. Diagrams of the flow of the product, decision trees, and other job-aids are all very useful for off-the-job training.

For training which is done "on-the-job," the actual plant can be used as a context of training. Operations can be taught by "walking through" with the trainee, possibly using an operating manual. When it is safe, an experienced process worker or the supervisor can demonstrate some operations on the plant and subsequently let trainees operate the plant under close supervision and guidance.

A combination of on-the-job and off-the-job methods is usually the best solution in most types of training. The following factors should be examined in order to analyze the role of training in preventing human error. Team training will be considered in the social and organizational factors which follow in other sections.

3.5.4.1. Conflicts between Safety and Production Requirements
One of the most important aspects of training is to highlight those steps during an operation at which production and safety requirements may potentially conflict. The following incident illustrates the importance of addressing such conflicts explicitly during training.

Example 3.6. Conflicts between Production Pressures and Safe Practices

In a refinery furnace, the panel man observed that the burner fuel flow and the smoke meter were oscillating. A process worker arrived and checked the conditions of the two oil burners from underneath the furnace. Burner "A" appeared to be extinguished and burner "B" unstable. On similar occasions, there were two alternative strategies to be considered: (i) maintain or reduce production by shutting the oil cock of burner "B" and improving stability of burner "A"; or (ii) shut down furnace by closing the oil cocks of both burners and purge furnace with air. Training must emphasize these production–safety conflicts and specify how one can cope with them.

Unfortunately, this was not the case for the plant in this example, and the worker wrongly chose to maintain production. By the time he arrived at the furnace, some of the fuel oil from burner "A" was deposited on the furnace tubes. Due to the heat from burner "B," the oil had vaporized and had been carried into the furnace stack. An explosion occurred when the mixture of air and unburned fuel came into the flammable range.

3.5.4.2. Training in Using New Equipment

On many occasions, new equipment is installed or process workers have to work in other similar plant units in order to substitute for one of their colleagues. Despite the overall similarity of the new equipment, there might be some differences in their operation which may sometimes become very critical. We cannot always rely on the operator to discover these potentially critical differences in equipment design, especially under time pressure and excessive workload. If multiskill training in a range of plant equipment is not feasible, then training should be provided for the specific new equipment. The incident below was due to lack of training for a canned pump.

Example 3.7. Lack of Knowledge of Safety Prerequisites before Carrying out Work on a Pump (Kletz, 1994b)

In canned pumps the rotor (the moving part of the electric motor) is immersed in the process liquid; the stator (the fixed part of the electric motor) is separated from the rotor by a stainless steel can. If there is a hole in the can, process liquid can get into the stator compartment. A pressure relief plug is therefore fitted to the compartment and should be used before the compartment is opened for work on the stator. One day, an operator opened the pump without using the pressure relief plug. There was a hole in the can which had caused a pressure build-up in the stator compartment. When the cover was unbolted, it was blown off and hit a scaffold 6 feet above. On the way up it hit a man on the knee and the escaping vapor caused eye irritation. The worker was not familiar with canned pumps and did not realize that the pressure relief plug should be used before opening the compartment.

3.5.4.3. Practice with Unfamiliar Situations

It is not possible to predict all the potential situations which the process worker will have to deal with. Unfamiliar situations sometimes arise whose recovery is entirely dependent upon the operating team. When this is the case, the likelihood of success will depend upon the problem solving skills of the process workers. These skills can be trained in refresher training exercises where the team will have to respond to unfamiliar situations. Training simulators can be particularly useful for such scenarios. Techniques for training the diagnostic skills of process operators are described in Embrey (1986).

One of the classical responses to unfamiliar situations is that people revert to previously learned well established habits and strategies which bear some sort of similarity with the new situation yet they are totally inappropriate. These strategies may have worked effectively in the past or have been emphasized in the emergency procedures or during training. People have to learn how to remain vigilant to changing plant conditions and reevaluate their initial hypotheses. Other types of human errors during emergency conditions are discussed in Section 6.

3.5.4.4. Training in Using Emergency Procedures
Another aspect of the response to plant transients is the effective use of the emergency procedures. The process worker needs training in order to be able to apply these procedures correctly under time pressure. Conditions of entry or transfer to other procedures, profitability–safety requirements, and the response of the automatic protection systems need to be learned extensively in training exercises.

3.5.4.5. Training in Working with Automatic Control and Protection Systems
Although training in using emergency procedures may refer to the operation of the various automatic control and protection systems, this factor needs to be considered in its own right due to its significant effect on performance. Any training course should consider the potential risk which may arise where the automatic systems are defeated (see Example 1.19). It should also consider any cases where workers tend either to overrely on the good operation of the automatic systems or to mistrust them without appropriate checking. An example of overreliance on automation was described in Example 1.20, while Examples 1.15 and 1.16 illustrate the tendency of some workers to blame the instrumentation for any abnormal readings. A useful strategy to overcome these problems is the "cross-checking" of instruments measuring the identical or functionally related parameters, for example, temperature and pressure.

3.5.4.6. Developing a Training Program
In general, little use is made in the process industry of more sophisticated approaches such as job and task analysis (see Chapter 4) to define the mental and physical skills required for specific types of work, and to tailor the training program accordingly. Instead, informal on-the-job training is common, even in more complex types of work such as control room tasks. Although the necessary skills will eventually be acquired by this process, its inefficiency

leads to the need for extended periods of training. In addition, there is the problem that inappropriate or even dangerous practices may become the norm as they are passed from one generation of workers to the next. It is therefore essential that training programs are based upon a comprehensive and systematic procedure which involves the following stages:

- **Job and task analysis.** This involves applying techniques such as hierarchical task analysis (see Chapter 4) to provide a comprehensive description of the work for which training is required. The task analysis provides essential information on the **content** of training.
- **Skills analysis.** This stage of the training development process involves identifying the nature of the skills required to perform the job. For example, a control room job may involve perceptual skills such as being able to identify out of limit parameters on a visual display screen, and decision making skills in order to choose an appropriate course of action. By contrast, an electrical maintenance job may require training in fine manipulative skills. As discussed in Chapter 2, the classification of a task as being predominantly skill-, rule-, or knowledge-based can provide insights with which is the most appropriate form of training
- **Specification of training content.** The content of training, in terms of skills and knowledge required to do the job, is derived from the previous two steps. At this stage it is important to define the information that will be obtained from procedures (in the case of infrequency performed tasks) and generic knowledge that will be required for a wide range of different tasks and which the operator would be expected to know as part of the skill of the job.
- **Specification of training methods.** This stage of the design of the training system will specify the appropriate training methods to provide the skills and knowledge identified by the earlier stages. A wide variety of sophisticated training techniques exist, such as interactive videos, which can be used to impart the knowledge aspect of training. More complex mental skills such as those required for control room tasks benefit from the use of various types of simulation. In order to be effective as a training method, simulators do not have to be highly similar to the actual plant control room. Inexpensive personal computer-based simulators can be used to teach control, problem solving and decision making skills. Applications of simulations to training in the CPI are given in Shepherd et al. (1977), Patternotte and Verhaegen (1979), and West and Clark (1974). Craft-based mechanical skills are usually taught by experienced trainers, together with guided on the job training.
- **Definition of competence assessment methods.** The definition of formal methods of assessing competence is a neglected area in many training programs. It is obviously necessary to ensure that trainees possess the necessary skills to do the job at the end of the training program.

Competence assessment is also required if workers are assigned to new areas of work. In the offshore industry, considerable importance is being attached to the issue of demonstrating competence, following the recommendations of the inquiry that followed the Piper Alpha disaster.

- **Validation of training effectiveness.** The effectiveness of the training system in terms of its capability to equip people with the skills necessary to carry out a job safely and efficiently, can only be determined by long term feedback from operations. The types of feedback that are important in evaluating a training program include incident reports, which should explicitly identify the role of lack of knowledge and skills in accidents, and reports from line managers and supervisors.
- **Definition of skill maintenance training.** All skills decline with time and it is therefore important to specify the needs for skill maintenance training by means of refresher courses or other methods.

3.6. OPERATOR CHARACTERISTICS

This group of PIFs concerns the operator characteristics of personnel such as operating experience, personality, physical condition and age. Considerable emphasis is placed on individual differences by many managers. There is a strong belief that all problems can be solved by better motivation or more intrinsically capable people. However, although many of the individual factors discussed in this section might reasonably be expected to have an effect on human error, in practice there are few controlled studies that have actually established such a link. Nevertheless, it is important that engineers are aware of the wide range of factors that could impact on error.

3.6.1. Experience

Although training can provide workers with adequate practice in process control, some elements of expertise develop primarily with operational experience. The degree of skill and experience with stressful process events are two separate PIFs which will be discussed thoroughly in this section.

3.6.1.1. Degree of Skill
The amount of the "on-the-plant" experience of personnel determines the extent that well-known knowledge can be applied to real-life problems, particularly under time pressure and high workload. Although engineering schools make an effort to provide all the required theoretical knowledge to young graduates and process workers, many people find it difficult to apply such knowledge to the plant, especially in the beginning of their employment period.

As has been discussed in Chapter 2, people go through three stages in the acquisition of skills. An educational course usually gets people to the cognitive or knowledge-based stage, where principles of physics and chemistry are well learned. With further practice, possibly on the plant, people "compile" their knowledge into practical "know-how" in the form of rules which can solve applied problems. The transition to the rule-based stage is analogous to software source code being translated into an executable form of code. After considerable experience people can reach the skill-based stage, which requires the least attentional and memory resources for the performance of a task, as discussed in Chapter 2. It is only at the rule- and skill-based stage that people will be able to apply their theoretical knowledge effectively to real-life problems. The following two examples (Kletz, 1994b), illustrate failures to apply well-known knowledge.

Example 3.8 Failure to Apply Well Known Knowledge (Kletz, 1994b)

Scaffolding was erected around a 225-foot distillation column so that it could be painted. The scaffolding was erected when the column was hot and then everyone was surprised that the scaffolding became distorted when the column cooled down.

Example 3.9. Failure to Realize that Changed Physical Conditions Would Render Safety Systems Ineffective (Kletz, 1994b)

A tank burst when exposed to fire for 90 minutes. During this time the Fire Department had, on advice of the refinery staff, used the available water for cooling surrounding tanks to prevent the fire spreading. The relief valve, it was believed, would prevent the tank bursting. They failed to realize that the tank could burst because the metal could get too hot and lose its strength. Below the liquid level the boiling liquid kept the metal cool, but above the liquid level the metal softened and burst at a pressure below that at which the relief valve would operate.

3.6.1.2. Experience with Stressful Process Events

Experience with stressful process events can be obtained both through simulator training and "on-the-job" practice. Both types of practice have their pros and cons. In simulator training, greater control can be exercised over the course of the process transient and the operating team can benefit fully from well designed instructional methods. What can be missing however, is the stress factor arising from potentially disastrous plant consequences. "On-the-job" experience of stressful events can present process workers with many aspects of their work which cannot be represented faithfully in an artificial environment. However, it is questionable whether people can learn effectively under stress and there is little control over any sort of misunderstanding that process workers may develop. It is a combination of "controlled" and "real-life" stressful process events which will benefit the workers.

Studies by Berkun (1964), Abe (1978), and Gertman et al. (1985) have found that people who have coped successfully with many previous stressful experiences perform better under stress than those who have not had these experiences. What is not evident from these studies is the kind of attitudes and skills that experience equips people with in order to perform effectively in future stressful situations. One can postulate that such beneficial experiences may help people develop generic problem solving strategies, remain vigilant to changing system conditions, and continually evaluate their working assumptions. With regard to their work attitudes, they may become more confident that they can cope with the unexpected, and may therefore exert greater emotional control and maintain good working relationships with their colleagues.

3.6.2. Personality Factors

This category includes a number of personality factors which can have an influence of human performance, particularly under stress. Although it is desirable to devise personality assessment tests to select the most suitable individuals for a job, the usefulness of these tests is questionable for CPI operations. A recent review of the state of knowledge of current practices in selecting workers for process control jobs was carried out by Astley et al. (1990). A finding of this study was that the basis of the choice of various psychological tests and selection devices was often superficial. There were rarely any measures of performance that could be used as a basis for deciding on which tests are likely to be valid predictors of performance. This is an important point, because process control tasks may vary considerably from plant to plant according to the different levels of complexity and different control philosophies. It may therefore be inappropriate to use the same general selection procedures in all cases. The methodologies of task analysis which are described in Chapter 4, aim to identify the necessary types of skills for specific process worker tasks and to ensure that test items are matched to the real needs of the workers.

It is worth noting that personnel managers who were interviewed as part of the above study had few expectations that selection would enable them to overcome inadequate training, job, or work design. Selection was seen as something that had to be done completely and conscientiously in order to make the best decisions possible. There was no expectation that, on its own, selection would solve operating problems.

The following section will address six personality traits that may affect human reliability, namely, motivation, risk taking, risk homeostasis, locus of control, emotional control, and type "A" versus type "B" personality.

3.6.2.1. Motivation

Considerable attention has been focused on the kind of motives which drive the decisions and choices of individuals in a work setting. An influential model of motivation was the "scientific management" movement of F. W. Taylor (1911) which viewed motivation largely in terms of rational individual decisions to maximize financial gain. This theory claimed that workers only wanted to make as much as possible for as little effort as possible, and that they were neither interested in, nor capable of planning and decision-making.

Later theories by Maslow (1954) showed the narrowness of that view, and the importance of factors such as social, esteem, achievement, and other needs. Maslow has put forward a hierarchy of five types of needs in descending order of priority:

- *Existence needs:* food, drink, air, sex
- *Security needs:* shelter, secure sources of the existence needs, freedom from fear, need for structure in life
- *Social needs:* affection, belonging to a group
- *Esteem needs:* need to be valued by self and others, competence, independence, recognition
- *Self-actualization needs:* self-fulfillment, achievement

Maslow postulated that the most basic level of need which is not yet satisfied is the one that controls behavior at any moment in time. Hence, people will not be very concerned with pursuing needs for esteem if they are threatened with the loss of their job, and therefore their security. While there is evidence that the first two levels do need to be satisfied in most people, before much concern is shown with the remaining levels, there does not appear to be any clear progression among those higher levels.

Another influential theory of motivation was proposed by Herzberg et al. (1959). This theory postulates only two levels of motivation. Herzberg contrasted wages, working conditions, interpersonal relations and supervisory behavior which he called "hygiene" factors, with recognition, achievement, responsibility, and advancement which he called "motivators."

Although the theories of both Maslow and Herzberg seem to be conceptually simple, they were probably among the first to recognize the role that various "system factors," such as equipment design, procedures, training, organizational culture and so on, play in the motivation of workers. When management has applied sound human factors principles to CPI tasks, training has provided the required skills to cope with all contingencies, and workers are actively involved in their job through participation schemes, then it is likely that motivation will be high.

Recent research on motivation theories has provided more elaborate models of the factors which drive human behavior and has taken into account issues of individual differences and the influence of the social and cultural background of the process workers. More extensive discussion on motivation theories is provided in Warr (1978) and Hale and Glendon (1987).

3.6.2.2. Risk-Taking

The concepts of accident proneness and risk taking as a personal trait predisposing the individual to a relatively high accident rate was first suggested by three statisticians, Greenwood, Woods, and Yule in 1919. They published an account of accidents sustained by workers in a munitions factory during the First World War and showed that a small minority of workers had more accidents than they would have done if chance factors alone were operating. Despite these early findings, attempts to explain them in terms of personality characteristics have met with little success. Either these characteristics explained only a maximum of 20% of the variance in accident rate, or a factor found to be relevant in one case was found to be irrelevant in others. The concept of accident proneness is discussed in detail in Shaw and Sichel (1971) who conclude that there is little statistical evidence for the trait.

Simpson (1988) reviewed studies which considered individual differences in risk perception and the effects of these differences on behavior. A study by Verhaegen et al. (1985) looked at three groups of workers in wire mills. The first group comprised those who had been directly involved in events which led to the accident (the "active" group). The second group ("passive") were those who had only been involved indirectly ("innocent bystanders") and the third group were a control group who had not been involved in accidents at all.

A series of interviews and questionnaires was given to a sample from each group to address the following issues:

1. Extent of risk-taking behavior
2. Perceived danger of work (risk)
3. Use of personal protective equipment
4. Discomfort of personal protective equipment
5. Positive attitude toward safety department
6. Perception that accidents were random in nature

The results indicated significant differences among the groups for issues 1, 2, and 5. The "active" group had a significantly higher score on risk taking behavior and a lower score for perceived danger of the work (risk) compared with the other two groups. Both active and passive accident groups had a more positive view of the safety department (presumably because of their involvement following accidents). These results suggest a definite relationship among risk perception, risk taking, and an increased likelihood of accidents.

From the perspective of the CPI, this result suggests that it would be valuable to carry out a survey of the perceptions of the workforce with regard to the risks associated with different aspects of plant operations (both field and control room tasks). These perceptions could then be compared with objectively based measures (from risk assessments and accident reports). Where discrepancies exist, appropriate training and information could be provided to ensure that the subjective risk perceptions of personnel were in line with the actual levels of risk associated with the plant operations.

3.6.2.3. Risk Homeostasis Theory (RHT)

The somewhat controversial theory of risk homeostasis is relevant to a discussion of risk taking. RHT was developed initially in the area of driving behavior (Wilde, 1984). The theory states that accident rates are not determined by actual levels of intrinsic risk but by the levels of risk acceptable to individuals in the situation. The theory implies that people adjust their risk-taking behavior to maintain a constant level of perceived risk. Thus, if improved safety measures are introduced (e.g., better guarding, improved protection systems), then individuals will behave in a more risky fashion in order to maintain their accustomed levels of risk.

The basis of RHT is set out in Figure 3.4. Individual levels of accepted risk are said to be determined by the costs and benefits of risky and cautious behavior, as set out in box a.

This target level of risk is compared against two sources of information. The first of these is the perceived effect of some risk reducing intervention in the work environment, that is, a change in the system's PIFs such as design changes, as opposed to a change in motivation to behave more safely (see box c). The second source of information against which the individual compares target levels of risk are his or her perceptions of the general levels of risk associated with the job being performed (box d). On the basis of these perceptions of risk, the worker is then said to modify his or her behavior to maintain the level of risk at the same target value as it was prior to the interventions (box f). Taken across a large number of individuals these changes in behavior have an effect on the overall accident rate in the population, for example, within a particular facility (box g). Following a time delay (box h) this in turn will be perceived as a change in the general levels of accident risk, via box d, thus completing the overall control loop.

The implications of RHT, if it proved to be universally true, would be disturbing from the perspective of human factors. The implication is that any interventions to change systems factors, as indicated by the systems induced error view set out in Chapters 1 and 2, would be canceled out by increased risk taking by workers. Needless to say, RHT has provoked considerable controversy among human factors specialists (see, e.g., Wilde, 1984; McKenna, 1985). Most of the debate has centered around differing interpretations of the evidence for reductions in accident levels following the introduction of improved safety systems. Opponents of RHT have pointed to extensive studies showing that people are generally very poor at estimating the magnitude of risk (e.g.,

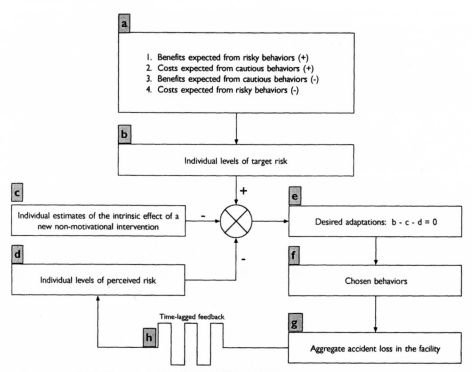

FIGURE 3.4:   **Risk Homeostasis Model (Wilde, 1982).**

Slovic et al., 1981), and hence are unlikely to be able to modify their behavior on the basis of objective changes in risk potential. Because of the difficulty of accurately assigning causes to effects, with the sources of data available, it is probable that the theory cannot be proved or disproved on the basis of data alone.

A major difficulty in assessing the applicability of RHT to tasks in the CPI is that most of the technical work which has been carried out on the theory has been in the driving domain. For example, a major focus has been on whether or not the introduction of seatbelts has actually led to a decrease in fatalities or has been compensated for by riskier driving. There are reasons for believing that RHT is unlikely to apply directly to CPI tasks such as control room operations, maintenance or field operations. First, unlike driving, systems interventions that will increase the effectiveness of human performance (e.g., improved training, better display of process information, clearer procedures) will not necessarily encourage the worker to operate the plant "closer to the limits." Even in areas such as maintenance, where the worker is closer to the hardware and has more discretion with regard to how a job is performed, it is difficult to see how improvements in the factors discussed in this book would lead to greater risk taking. In addition, because of the fact that there are considerable differences in the CPI between different processes and the way in which plants are operated, it would be difficult for a worker to arrive at an acceptable level of risky behavior purely on the basis of feedback from the accident rate in the CPI as a whole.

It could be argued that the presence of enhanced protection systems could lead to a plant being operated to its operational limits in order to obtain better yields in the expectation that, if the process entered a dangerous state, it would be tripped automatically. However, the loss of availability that could arise from such a strategy would discourage this type of behavior

In summary, the application of the RHT model to the CPI may by questionable. Certainly, it provides no compelling arguments against the measures for optimizing human reliability which are proposed in this book.

3.6.2.4. Locus of Control

The term "locus of control" refers to the tendency of individuals to ascribe the causes of things that happen to them either to external or to internal events. Such individuals are referred to as "externals" or "internals" respectively. Some research results point to the relevance of this dimension to an process worker's response under stress. "Internals" are more likely to seek information about a problem and to attempt to control it themselves. "Externals," on the other hand, are more likely to assume that the problem is out of their immediate control and attempt to get assistance from their colleagues. In an emergency situation, "internals" would be expected to respond better than "externals" because they may have a built-in coping mechanism (i.e., they feel their actions can significantly affect what happens to them). "Externals," on the other hand, may feel their actions can do little to control the situation. A study by Gertman et al. (1985) has provided support for the superior performance of "internals" during nuclear power emergencies. This finding may also apply to CPI operations.

3.6.2.5. Emotional Control

This is defined as the tendency to inhibit emotional responses during a crisis (Roger and Nesshoever, 1987). The scale which measures this concept has four factors, namely:

- *Rehearsal*—a preoccupation to ruminate on past events
- *Emotional inhibition*—a tendency to conceal emotions
- *Aggressive control*—a tendency to inhibit aggressive responses
- *Benign control*—a tendency not to say upsetting things

Emotional control is likely to maintain good team communications, particularly at times when the team receives negative feedback about its performance.

3.6.2.6. Type A versus Type B Personality Type
Type B personality is characterized by a relaxed, unhurried, satisfied approach to life and work, in which strivings for achievement tend to flow with the stream of life rather than against it. A type A personality is related to strivings for achievement, and preoccupation with time and success even if against the flow of the environment (Friedman and Rosenman, 1974). A type A personality is considered to be less effective under stress than type B, as the former is characterized by preoccupation with time and success, plus restlessness, and feelings of being pressured (Orpen, 1982)

It is worth pointing out, however, that personality traits which do not provide people with adequate resources to maintain performance under stress, may compensate by supporting other activities during normal operations. For instance, "externals" may be more cautious than "internals" and take no chances to risk plant safety, while type A personalities may have a greater motivation to progress in their jobs and perfect their skills than type B personalities. Depending on the type of task, some personality traits will produce better performance than others. More research would be needed to develop a better understanding of the relationships between types of task and preferable personality styles.

3.6.3. Physical Condition and Age

Conditions of health and age play an important role in human performance. Job demands will determine the general fitness and age of the workers to be employed for a particular job. Recent illness can affect the level of alertness, the required concentration on the job, and the capability to cope with high workload.

A considerable area of research has focused on the way in which age can affect performance. This has been prompted by the increasing age of the general workforce. In general, the effects of age on performance will be determined by two factors, namely, characteristics of the particular task and level of experience with it. Literature reviewed by Murrell (1965) has identified four biological changes which take place with age, namely:

- A decrease in visual acuity and speed of discrimination which may affect the size of detail which can be seen and the ability to read fine scales
- A decrease in the capacity to process information on the control panel
- A loss of working memory which may affect the amount of information that can be retained for long time periods
- A tendency for greater manual variability which affects performance of machine-paced tasks, particularly in the manufacturing industry

Although these impairments in the performance of older personnel can be the result of biological changes due to age, the level of experience with the job may counteract these changes. Continual practice of a particular job role may cause these age differences to disappear. In addition, older personnel may develop more efficient methods of work and thus minimize the demands of the job.

Griew and Tucker (1958) found that in a machine shop, older men appeared to achieve the same results with fewer control movements than younger men working on similar machines. In a study of pillar drilling (Murrell et al., 1962), the performance of older inexperienced workers was substantially worse than young inexperienced workers, but the performance of older professional drillers obtained from industry was slightly better than that of young drillers. This demonstrates the role of experience in compensating for increasing age. However, this compensation only occurs up to a point, and good management should identify those aspects of the task which make the greatest demands upon the older worker and if possible modify the tasks accordingly. An extensive review of the effects of age on performance is available in Small (1987).

3.7. ORGANIZATIONAL AND SOCIAL FACTORS

The various PIFs discussed so far provide a basis for the control of human error at the level of the individual. This section will consider various factors related to the performance of the team and the management practices related to safety.

3.7.1. Teamwork and Communications

Modern process plants grow increasingly complex with highly coupled unit processes. A result of this tendency is that tasks now often require a team rather than individual effort. Team training becomes increasingly important for the safe and efficient operation of plants. The aim of this section is to identify those PIFs which play a critical role in the collective efforts and communications of process workers.

Given the limited resources that a plant can provide for training, a critical question arises concerning emphasis which must be given to individual or team training. Many accident scenarios involve well-trained personnel who failed to work collectively under the particular conditions of the task. We need, therefore, some guidelines to judge the relevant importance of individual versus team performance for different types of tasks.

Blum and Naylor (1968) reviewed the literature on group versus individual training and proposed a useful rule. For tasks which are highly interrelated and which require a great deal of communication and cooperation among members, it is best to employ team training. With tasks which only place low or moderate communication demands on team members, team training is best if the subtasks are fairly simple, but individual training would be best if the subtasks are quite complex. The method of dividing task demands in task organization and task complexity is useful in examining the role of individual versus team training in accident scenarios.

To judge the quality of team performance it is necessary to examine the following PIFs: distribution of workload, clarity of responsibilities, communications, team structure and leadership, and finally, group planning and orientation.

3.7.1.1. *Distribution of Workload*

The distribution of workload among the members of the team will determine the extent of task overload or task underload for each person. It is widely recognized that reliability decreases when people have too much or too little to do. The incident which is described below occurred because of suboptimal allocation of tasks to team members.

Example 3.10. Effects of Overload Due to Poor Organization of Work (Kletz, 1994b)

Plant foremen sometimes suffer from task overload, in that they are expected to handle more jobs than one person can reasonably cope with. For example, two jobs had to be carried out simultaneously in the same pipe trench, 60 feet apart. At 8:00 A.M., the foreman gave permission to the welders to work on the construction of a new pipeline. At 12:00 noon, he signed a work permit for removing a blind from an oil line, making the assumption that the welders would by this time be more than 50 feet from the site of the slip-plate. As he was already very busy on the operating plant, he did not visit the pipe trench, which was about 1500 feet away. Although the pipeline had been emptied, a few gallons of light oil remained and ran out when the slip-plate was broken. The oil spread over the surface of the water in the pipe trench and was ignited by the welders. The man removing the slip-plate was killed. It was unrealistic to expect a foreman to look after normal operations and simultaneously supervise construction work at a distant site.

On the other hand, when workers are seriously under-loaded, they might not be very alert to changing process conditions. Many of the problems of plant automation are common to other situations of task underload. To increase the level of activity in monitoring tasks, additional tasks can be assigned, such as calculating the consumption of fuels, the life of a catalyst, the efficiency of the furnace and so on. Meister (1979) provides a summary of research on team organization.

3.7.1.2. Clarity of Responsibilities

Specifying the amount of workload appropriate for a worker is not enough. The kind of responsibilities assigned must be clearly specified in both everyday duties and emergency situations. In this context, one can distinguish between two situations, namely, "role ambiguity" and "role conflict." Role ambiguity exists (Kahn, 1974a) when an individual has inadequate information about his role at work. This may reflect a lack of clarity about work objectives, about colleagues" expectations, and about the scope and responsibilities of the job. Kahn et al. (1964) and Kahn and French (1970) have defined role conflict as "the simultaneous occurrence of two or more sets of pressures such that compliance with one would make compliance with another more difficult." For instance, conflict may arise when a manager instructs the worker to carry out a particular action which is at variance with instructions given by the worker's foreman.

Responsibility for each item of equipment should be clearly defined at manager, foreman, and worker level and only the men responsible for each item should operate it. If different workers are allowed to operate the same equipment then sooner or later an accident will occur (see Example 1.27).

3.7.1.3. Communications

Even when responsibilities have been assigned in a clear manner, people may fail to tell their colleagues what they need to know, or may misunderstand a message. The following two incidents were due to failures of communication.

Example 3.11. An Accident Due to Misunderstood Communications (Kletz, 1994b)

In one incident, the laboratory staff were asked to analyze the atmosphere in a tanker to see if any hydrocarbon was present. The staff regularly analyzed the atmosphere inside LPG tank trucks to see if any oxygen was present. Owing to a misunderstanding they assumed that an oxygen analysis was required on this occasion and reported over the phone that "none had been detected." The worker assumed that no hydrocarbon had been detected and sent the tank truck for repair. Fortunately the garage carried out their own check analysis.

Example 3.12. Absence of Communications (Kletz, 1994b)

In another incident, a maintenance foreman was asked to look at a faulty cooling water pump. He decided that, to prevent damage to the machine, it was essential to reduce its speed immediately. He did so, but did not tell any of the operating team immediately. The cooling water rate fell, the process was upset and a leak developed in a cooler.

3.7.1.4. Authority and Leadership

The type of power and social relationships in a group will also affect the overall performance. Although a formal status hierarchy is specified for each team by the plant management, it is well documented that groups have their own informal status structure which may be different from the formal one. In everyday duties it might be difficult to detect any contradictions between formal and informal status hierarchies. In an emergency situation, however, where different interpretations of the situation may be reached, such status problems may create difficulties with regard to whose opinion is followed.

The way that a group handles staff disagreement is also very critical. Performance may be hampered by what has often been called "reactance." The notion is that an individual with a high sense of competence will require freedom to express that ability. If this is denied and the competent person is "relabeled" in a subordinate position, performance will be severely impaired by a tendency to prove "how much better things would have been, if they had been done his or her way."

3.7.1.5. Group Planning and Orientation

In an emergency situation, the team will have to spend some time in planning the right strategy to attack the problem and then allocate responsibilities to team members. The extent of group planing and task orientation in the beginning of a process transient will determine the success of the overall performance. This is not an easy task, since the most common human response to stress is to neglect planning and rush into matters with potentially disastrous results.

3.7.2. Management Policies

Management policies have an all pervasive effect on the activities of individuals at every level in the organization. The safety-related factors at the management level which have been considered in the organizational systems perspective in Chapter 2, will be summarized here to complete the general classification scheme of PIFs.

3.7.2.1. Management Commitment

Not surprisingly, management commitment emerges as the dominant factor influencing safety performance. Commitment needs to be present in a tangible form and not merely espoused as part of a company's mission statement. Real commitment is demonstrated by a number cf indicators. For example, line management in each function, operations, engineering, etc. must be responsible for safety performance of the line function. A safety function in an advisory and audit role should be a distinct organizational function and not put under another grouping where its importance is likely to be diluted. Safety matters should be regularly included in plant operating decisions and top management officials should visit the work areas and keep daily contact with supervisors and line workers. This will ensure that policies that are promulgated by senior management with regard to safety are actually being implemented at the operational level. Another demonstration of management commitment is the resources that they are prepared to expend on the safety function as compared with production

The general safety management policy that exists in an organization needs to be assessed proactively and continuously. Several systems are available— the International Safety Rating System (ISRS)—which attempt to provide a comprehensive audit of safety management activities. Further evidence of a commitment to proactive safety methods is the use of extensive "what-if" and simulation exercises in order to determine the weak points in the defenses of an organization. The existence of such exercises indicates that the organization is actively examining its safety capabilities

3.7.2.2. Dangers of a "Rule Book" Culture

Many organizations that have evolved over a long period of time come to believe that the system of safety rules that they have developed is invulnerable to human error. The existence of a "rule book" culture can produce a compla- cent attitude which assumes that if the rules are followed then accidents are impossible. This is based on the belief that a rigid set of rules will cover every contingency and that interpretation by individuals to cover unanticipated situations will never be required. Of course, all rules will at some time require such interpretation, and the need for this should be accepted and built into the system.

Although rules and procedures are a necessary and indeed essential aspect of safety, they need to be regularly reviewed and updated in the light of feedback from operational experience. Unfortunately, such feedback loops become less and less effective with time, and hence need to be reviewed regularly, preferably by an independent third party

3.7.2.3. Overreliance on Technical Safety Methods

In order to achieve the high levels of safety necessary in high risk industries, predictive assessment techniques such as chemical process quantitative risk analysis (CPQRA), hazard and operability studies (HAZOPs), and failure modes effects and criticality analysis (FMECA) are often used. Although these approaches have considerable value, they need to be supplemented with two other perspectives in order to be effective. The first of these is an explicit recognition that human as well as technical failures need to be modeled and assessed, with particular emphasis on "higher level" human functions such as diagnostic and decision making errors. Failures of this type can have substan- tial effects on the safety of hazardous systems because of their capacity to overcome engineering safeguards. It is also necessary to be aware that any predictive technical analysis of a system makes certain (usually implicit) assumptions about the way the plant will be operated, what sort of quality assurance systems will be in operation and so on. These assumptions relate to human aspects of the system such as the way it is managed, and the operating philosophy with regard to safety versus profitability that is applied. If these assumptions are incorrect (e.g., there may have been a change in management policy) then the technical analysis may no longer be valid. It is therefore necessary to explicitly state the assumptions underlying any technical assessments of risk, and to constantly review these assumptions in the light of possible changes in organizational policies and practices. Effective incident reporting systems are also necessary to reveal sources of risk not considered in the safety analyses.

3.7.2.4. *Organizational Learning*
It has been stated that "organizations have no memory" (Kletz, 1993) or, to paraphrase George Santayana (in *Life of Reason*, 1905), that "organizations that cannot learn from the past are condemned to repeat their errors in the future." Learning from the past means not only taking specific actions to deal with a problem that has caused a significant injury or loss of property, but also learning to identify the underlying causes of error and the lessons that can be learned from near misses. Near misses are usually far more frequent than actual accidents, and they provide an early warning of underlying problems that sooner or later will lead to an accident.

Nearly all major disasters provide ample evidence of the failures of organizations to learn from their own or other organizations' experience. In the case of Three Mile Island for example, a similar accident had occurred some months before at the similarly designed Davis Besse plant, but correct worker intervention had averted an accident.

In these and many other cases, there are several reasons why organizations did not learn from experience. Incident reporting systems almost always concentrate on the what rather than the why of what happened. Thus, there is little possibility of identifying recurrent root causes so that countermeasures can be developed. Where effective reporting systems do exist, their findings may not be brought to the attention of policy makers, or it may be that the underlying causes are recognized but incorrect trade-offs are made between the cost of fixing the problems and the risks of maintaining profitability by continuing to operate the system. Example 1.28 illustrates the effects of information on incidents not being widely distributed. Another frequent cause of failing to learn lessons is a "blame culture" which discourages individuals from providing information on long standing system problems which cause frequent near misses

Chapter 6 discusses the ways in which feedback for operational experience can be enhanced by improved data collection and root cause analysis techniques. An effective method of learning from operational experience is the analysis of accidents and near misses to identify the root causes of human errors. However, this cannot be achieved unless a comprehensive communication system exists for transmitting the findings of accident analysis and incident reports to higher levels in the organization. For example, the results of causal analyses of accidents should be provided for the developers of procedures and operating instructions, and should provide inputs to both initial and refresher training. It is important that senior management is provided with feedback from operational experience, even if this is in summary form, so that they are aware of the underlying problems that may potentially compromise safety.

3.8. INTERACTION OF PERFORMANCE-INFLUENCING FACTORS

The various PIFs listed so far have been considered individually from the point of view of their potential to affect human reliability. In a real CPI environment, however, the individual is working under a combination of PIFs of different qualities. The overall influences of a combination of PIFs may be different than the sum of the influences. It should be noted that PIFs are not automatically associated with human error. PIFs such as quality of procedures, level of time stress, and effectiveness of training, will vary on a continuum from the best practicable (e.g., an ideally designed training program based on a proper training needs analysis) to worst possible (corresponding to no training program at all). When all the PIFs relevant to a particular situation are optimal then performance will be also optimal and error likelihood will each individual PIF, since these factors may interact with each other in complex ways. The result of this interaction can amplify or attenuate the individual effects of the factors on performance.

We have seen, for instance, how worker experience can compensate for increasing age. Management factors such as commitment to safety can also affect the way that workers will trade-off productivity and safety and thus make use of safety procedures and work permits. Other examples can be drawn from the interaction of control panel design and procedures or training. Grouping of process information, for instance, is related to the type of strategy that is adopted, which in turn is dependent on the type of procedures and training provided. The indicators of the same pressure valve on two different reactors are, in one sense, highly similar. Yet, in another sense, their similarity is low when compared to the similarity between the valve indicator and the pressure indicator on the input side of a reactor. The latter indicators, belonging to a single system, are more likely to be causally related in a failure and thus belong to the same fault cluster. The optimum way of structuring control panel information will depend on the style and type of strategies adopted by the different individuals.

Although the issue of PIF interactions has long been recognized by human factors researchers, little has been done to develop practical recommendations. This is partially a result of the large number of possible PIF combinations and the complexity of their interactions. One of the most effective ways of studying this interaction is through an in-company human factors study which will use operational feedback to evaluate the results of design and human factors innovations.

3.9. VARIABILITY OF HUMAN PERFORMANCE DURING NORMAL AND EMERGENCY SITUATIONS

This section examines the role of PIFs in human reliability during emergency situations as compared to everyday duties. In general, any deficiencies in the quality of PIFs can maximize the adverse effects on performance, because the workers are operating under pressure to acquire information, interpret the implications for the safety of the plant, and reach the right decision as quickly as possible before any serious consequences ensue. A number of phenomena which occur under stress such as rigidity of problem solving, and polarization of thinking, can change the effects of PIFs because they can make the worker more vulnerable to error. It is necessary, therefore, to understand how people behave under conditions of high stress in order to evaluate the role of each PIF.

An emergency situation may display the following general characteristics:

- High-risk environment
- High time pressure
- High task loading, task complexity
- Unfamiliar process conditions
- High noise level due to alarms
- Long working hours to complete the task

The extent to which a particular combination of such "operating environment" factors will be perceived by the workers as being stressful will depend on the available resources such as the quality of the control panel, procedures, training, organizational and social factors, and, finally, the individual characteristics of the workers. The outcome of this transaction between stress factors and coping resources will influence the onset of worker stress. Situations are not stressful merely because of the presence of a number of external stressors, but because they are perceived as such by workers.

The definition of what constitutes a stressor is also an important issue. So far, we have considered only external stressors stemming from the demands of the operating environment. Deficiencies in the design of the control panel, procedures, training, and problems in the area of teamwork and safety management can also cause stress. Such internal stressors can produce conflicting or ambiguous information, worker overload, production-safety conflicts, ambiguity in the role of team members, and poor communication and team coordination. This in turn can have an adverse effect on human reliability. It is the quality of these PIFs which will determine whether they will have a negative or positive effect. Workers will be placed under high stress when they perceive their resources as insufficient to cope with the emergency situation.

Studies of performance under stress have taken three approaches. The first source of data comes from laboratory-based studies which have investigated the effects of only a single external stressor (e.g., noise or heat), upon relatively simple tasks, that is, choice reaction tasks (see Hartley et al., 1989, for a comprehensive review). The second and possibly richest source of data comes from the analysis of real accidents. Studies by Kletz (1994b), Reason and Mycieska (1982), and Dixon (1976, 1987) belong to this approach. Typically, such analyses depend on the level of detail supplied in the reports or the accuracy of the memory of the participants. The retrospective analyses may also be subject to the effects of the rationalizing "hindsight" bias. The final source of data comes from the use of high fidelity plant simulators (Woods, 1982; Norros and Sammatti, 1986; Reinartz, 1989). The difficulties of this approach include the high costs involved in using the simulator and employing experienced teams as subjects, and the degree of stress induced by artificial simulations.

A study by Kontogiannis and Lucas (1990) has reviewed these approaches and developed a classification of cognitive phenomena which occur under high stress. This is presented in Figure 3.5. The classification was developed by examining a number of incidents from various industrial sectors. The cognitive phenomena illustrate in a practical manner the psychological mechanisms which can precipitate errors under stress.

They can also explain why the role of PIFs can vary in normal versus emergency situations depending upon the set of cognitive phenomena that will be brought into play. Because these phenomena can be unique for each individual, greater differences in human performance during an emergency will be found than in a normal situation. Finally, the classification of cognitive phenomena is useful in narrowing down those aspects of PIFs which play a greater role in human performance under stress. For instance, "grouping of information" and "overview of critical parameters" are two aspects of control panel design which can be optimized to reduce the likelihood of the worker developing "cognitive tunnel vision." With respect to procedures design, the quality of checks and the specification of entry and exit conditions can also prompt the worker to consider alternative hypotheses.

PHENOMENA	FEATURES
Defensive avoidance	Can take a number of forms. For instance, a person could become selectively inattentive to threatening cues and avoid thinking about the dangers through distracting activities. Another form of defensive avoidance is "passing the buck" where someone else is relied upon to make the decision.
Reinforced group conformity	The tendency of a group to protect its own consensus by putting pressure on those members who disagree, and by screening out external information which might break the complacency of the group.
Increased risk taking	Individuals tend to take greater risks when they operate within a group rather than alone. Various explanations have been suggested, namely: the illusion that the system they control is invulnerable, the diffusion of responsibility for any potential problems, the presence of persuasive persons who may take risky positions and the increased familiarization of the problem through discussions.
Dwelling in the past	Groups under stress tend to concentrate on explaining facts which have already been superseded by more recent events.
Tendency to overcontrol the situation	People tend to try to overcontrol the situation rather than delegate responsibility.
Adopt a "wait and see" strategy	As consequences of the crisis become more critical, people appear to be more reluctant to make an immediate decision, and wait to obtain redundant information.
Temporary mental paralysis	The short lived incapacitation of the capability of making use of available information. Postulated as being due to the sudden switch from under- to overstimulation at times of crises
Reduced concentration span	Concentration, that is, the ability to deploy attention on demand decreases with stress.
Cognitive "tunnel vision"	This is also known as "hypothesis anchoring" because the worker tends to seek information which confirms the initially formulated hypothesis about the state of the process, and to disregard information which dis-confirms it.
Rigidity of problem-solving	The tendency to use off-the-shelf solutions which are not necessarily the most efficient.
Polarization of thinking	The tendency to explain the problem by a single global cause rather than a combination of causes.
Encystment and thematic vagabonding	Thematic vagabonding refers to a case where a person's thoughts flit among issues, treating each superficially. Encystment occurs when topics are dwelt upon to excess and small details are attended to while other more important issues are disregarded.
Stereotype takeover	Reversion to an habitual or preprogrammed mode of behaviour derived from past experience with a similar, yet in some respects different, situation.
Hypervigilance	Panic occurs leading to disruption of a person's thoughts. A person may fail to recognize all the alternatives open to him and latch onto a hastily contrived approach that appears to offer and immediate solution.

FIGURE 3.5. **Individual and Cognitive Phenomena under Stress (Kontogiannis and**

151

3.10. SUMMARY

This chapter has reviewed various PIFs which determine the likelihood of human error in the CPI. The list of PIFs in Table 3.1 can be used by engineers and managers to evaluate and audit existing work systems, analyze process incidents and generate error reduction strategies in conjunction with the techniques described in Chapters 4 and 5.

Throughout this chapter it has been argued that the effects of PIFs on human performance will be determined by the characteristics of the task (e.g., process monitoring, procedures-following, diagnosis, planning, manual control). However, many process control tasks involve a combination of such features, and making it difficult to identify their precise effects. To overcome such problems, Chapter 4 presents a number of task analysis methodologies which redescribe complex control tasks into more detailed task elements whose characteristics can be more easily identified and classified in accordance with the previous dimensions. The methodology described in Chapter 4 will assist in applying the knowledge of the effects of PIFs on specific process control tasks. The use of the PIF evaluation approach in the assessment of existing systems can be achieved using the systematic procedures associated with the TRIPOD, HFAM, and HSE approaches described in Chapter 2.

4

Analytical Methods for Predicting and Reducing Human Error

4.1. INTRODUCTION

The previous chapters described various approaches to understanding how human errors arise and provided a comprehensive overview of the wide range of factors that can influence the likelihood of human error. The methods described in this chapter draw upon these insights to provide a comprehensive set of tools that can be used by engineers to evaluate and reduce human error in their plants. These methods can be applied proactively, as part of design and audit procedures to identify and eliminate error-inducing characteristics of a system before an incident occurs. They can also be used "after the event" to understand the underlying causes of an incident and to prescribe suitable measures to prevent a recurrence (see Chapter 6). The use of methods within an overall error-management framework is described in Chapter 8.

The various analytical methods for predicting and reducing human error can be assigned to four groups or sections. In order to make a start on any form of analysis or prediction of human error, it is obviously necessary to gather information. The first section therefore describes a number of techniques that can be applied to acquire data about what the worker does, or what happened in an accident.

The second section describes various task analysis (TA) techniques. Task analysis is a fundamental methodology that is widely used by human factors specialists for a variety of purposes including procedures development, training specification, and equipment design. Task analysis methods organize the information generated by the data acquisition process into a variety of forms and representations, depending on the purpose of the analysis. For example, if the analyst is primarily interested in the design of the human–machine interface, the TA technique will focus on the input and output of information to the worker, the design of the information displays and on the thinking processes involved in operating the plant. In many cases there is a considerable overlap between data acquisition and TA methods.

The third category of methods addressed in this chapter are error analysis and reduction methodologies. Error analysis techniques can either be applied in a proactive or retrospective mode. In the proactive mode they are used to predict possible errors when tasks are being analyzed during chemical process quantitative risk assessment and design evaluations. When applied retrospectively, they are used to identify the underlying causes of errors giving rise to accidents. Very often the distinction between task analysis and error analysis is blurred, since the process of error analysis always has to proceed from a comprehensive description of a task, usually derived from a task analysis.

The last category of techniques are various forms of checklists of factors that can influence human reliability. These are used mainly in a proactive auditing mode. They have the advantage that they are quick and easy to apply. However, considerable training may be necessary to interpret the results and to generate appropriate remedial strategies in the event that problems are identified.

4.2. DATA ACQUISITION TECHNIQUES

The following techniques can be used to collect data about human performance in CPI tasks and provide input to task analysis methods described in Section 4.3. These data can include process information critical for the task, control strategies used by the workers, diagnostic plans etc. A distinction can be made among data collection methods that provide qualitative data (such as interviews, observations, and sources of documentation) and methods that can be used to measure aspects of performance (such as activity analysis, simulations, and information withholding). The latter methods can provide more precise data which can be quantified.

4.2.1. Discussions and Interviews with "Experts"

Analyzing complex tasks is usually best done in collaboration with a task expert. Anybody knowledgeable about a particular job might be described as an "expert." This includes process workers, supervisors, engineers, trainers, safety specialists, and managers. Discussions and structured interviews are likely to emerge at any stage during a task analysis activity. They can either be used during the analysis to collect basic information about the task or at the end of the analysis to check the accuracy of information that has been collected. The interviewer needs to be trained in order to make the task expert feel relaxed and not threatened or embarrassed by the situation. This is not always easy to achieve because people may get the impression that their expertise is being evaluated or compared with that of other experts. For this reason, the objective of the interview and the relevance of each question should be explained in advance to the interviewee. It is useful, therefore to structure the interview beforehand in terms of the aspects of human performance that are of interest to the study. This will also make the whole exercise more economical in terms of the period of time that task experts are taken away from their jobs.

A variant of individual interviews is verbal protocol analysis. In this technique, the person is asked to "think aloud" while carrying out a particular task. These self-commentaries are made while the task is being undertaken in order to avoid the inevitable distortion or forgetting that could occur if the reporting is left until afterwards. The main aim is to gather information on the psychological processes that underlie performance, which are not directly observable. It is essential that the process of providing a verbal commentary should not affect the way in which the task is carried out. To prevent people from elaborating on, or rationalizing their thought processes in any way, it is important to encourage a continuous, flowing commentary.

Useful protocols can only be produced for information that is coded in a verbal form in memory. Tasks that rely on visual imagery for their execution, or have become "automated" due to over-practice will be very difficult to verbalize. Hence the technique may provide little useful information and may even produce misleading reports for these tasks. To encourage task verbalization some coaching should be given to the task experts and the goals of the study should be explained so that they can make greater efforts to report on aspects of the task which are of particular importance.

Some form of audio recording will be essential to collect all the verbal information about the task. To help the analysis of the protocols, the analyst can link the protocol to the state of the chemical process at that time by noting the time and the values of particular indicators. Another technique is to make video recordings of the operations at the same time as the verbal protocols are collected. These can subsequently be played back with the individual who provided the original verbal protocol, in order to gain further insights into the reasons why certain strategies where used. After the tape recordings have been transcribed into a written form, the analyst can structure the available information to examine its content and draw the required inferences.

One way of analyzing the data is to use a columnar format, with columns such as Displays Used, Control Used, Action, Decision, Goal Pursued, etc. which are filled in directly from the protocol information. A useful discussion of the application of the technique to process control tasks is given by Bainbridge (1974), and Ainsworth and Whitfield (1984). Apart from collecting data about the task, discussions and interviews with the workers can get their direct commitment to a project and can make them feel that they "own" any proposed new work system.

4.2.2. Observation

Discussions and interviews with the task experts can be supplemented with observations of their actual performance, for example, taking notes on certain aspects of the task or taking video or audio recordings. Observational techniques can reveal information that may be difficult to acquire in any other way. Detailed physical task performance data can be recorded, and major environmental influences (e.g., noise, light, interruptions) can all be faithfully represented. Observations can also provide an insight into the way that the team members communicate, allocate job responsibilities, and make use of operating procedures and other resources.

Observations are appropriate for recording physical task sequences, or verbal interactions among several people. They are not suitable for collecting precision performance data, or studying cognitive tasks which involve covert mental processing.

It is a good practice to try and predict what level of information is expected to be extracted from the data before conducting sessions relying on observation. For instance, problems posed by movement and interaction among individuals, and the inability of a video system to capture extremely detailed events, must all be considered in advance. If certain aspects of the task are videotaped, the recording process itself should be as unobtrusive as possible. The minimum requirement is that it does not get in the way. Also, some people may react negatively to being observed and recorded. For this reason, the workers should be briefed about the objectives of the observational study in advance.

4.2.3. Critical Incident Technique

This technique sets out to collect data about near-incidents or critical events that have been experienced by the operating team but that are unlikely to be documented. The basic premise of the technique is that events that could have led to serious consequences would tend to be remembered by the workers. Through individual or group interviews, significant events are recalled which are then analyzed in order to generate useful information about the difficulties involved in the performance of a task, the adequacy of the operating procedures, any problems with the equipment or control panel design and so on. The technique can be used in three areas:

- To identify changes to be made in the system to ameliorate operational problems
- To provide data for task analysis methods concerning the difficulties involved in the performance of a task
- To provide data for error analysis methods by pinpointing error-likely situations

The critical incident technique was first described by Flanagan (1954) and was used during World War II to analyze "near-miss incidents." The war time studies of "pilot errors" by Fitts and Jones (1947) are the classic studies using this technique. The technique can be applied in different ways. The most common application is to ask individuals to describe situations involving errors made by themselves or their colleagues. Another, more systematic approach is to get them to fill in reports on critical incidents on a weekly basis. One recent development of the technique has been used in the aviation world, to solicit reports from aircraft crews in an anonymous or confidential way, on incidents in aircraft operations. Such data collection systems will be discussed more thoroughly in Chapter 6.

A degree of rapport must be built between the analyst and the worker in order for them to feel that their commentary will be treated confidentially. This is important in situations where an incident has not been reported in the past and the workers do not wish to open themselves or their colleagues to potential sanctions. Under such conditions, it may be appropriate for the analyst to provide the overall results of the study, rather than the actual content in terms of events etc.

The results should be treated with caution because the technique is subject to loss from memory, of detail, fabrication, and recall of anecdotal events.

4.2.4. Documentation

Documents such as job descriptions, operating manuals, emergency procedures, accident, and "near-accident" records, can be useful sources of information about the task to be studied. Pipework and instrumentation diagrams can also be used to gain an insight into the complexity of the process, the type of control loops installed, and the process parameters to be manually controlled by the workers.

Reference to such documents may be useful at early stages in the task analysis to inform the analyst about the overall nature and breadth of tasks carried out. Later, as the detail of the task is becoming established, such documents serve to provide crucial information. The use of experts in helping with the interpretation of documents is usually necessary, unless the analyst is directly involved with the system on a regular basis.

4.2.5. Activity Analysis

Data about the plans and routines used by workers in controlling a process can be obtained by means of an "activity analysis," a type of input–output analysis. A chart can be made showing how certain process indicators change over time in response to changes of the control settings. From this chart it is possible to determine the type of process information that workers use to carry out their tasks, the size of adjustment of the various control settings, their sequence of adjustment and so forth. The activity analysis usually results in a qualitative description of the workers' control strategies.

There are various types of charts that can be used to record an activity analysis. For tasks requiring continuous and precise adjustments of process variables, a chart displaying the graphs of these variables and the appropriate control settings will fulfill the objectives of the activity analysis. Figure 4.1 shows an activity chart of a subtask for a machine operator in a papermaking plant. This describes how to adjust the weight of a given area of paper to the desired value for each successive customer order and ensure that it remains within the specified limits until the order is completed.

The value of the "basis weight" can be obtained either by removing and weighing a sample, a procedure that can only be carried out during a reel change, or (less precisely) by means of a beta-ray gauge situated at the "dry end" of the machine. In the latter case, the value of the basis weight is controlled by means of a "stuff valve" which controls the flow of pulp into the "wet end" of the machine. Its value also changes with the overall speed of the machine. For a full description of the task see Beishon (1967), and Crossman, Cooke, and Beishon (1974).

For tasks that rely on decision-making rather than on fine manipulations, the activity chart can assume a columnar format, with columns recording process information attended and subsequent changes of discrete control settings.

4.2.6. Simulators and Mock-ups

Under this heading a variety of techniques are available which involve the development and use of some form of simulation of systems ranging from simple mock-ups of a piece of equipment to sophisticated computer-driven plant simulators. The simulation would be typically used to establish appropriate working methods, ergonomics of control layout and design, identification of potential sources of error, or to derive training recommendations. The technique can be used when the real equipment or system is not yet available for study or when the tasks to be examined are critical and operator error could give rise to hazardous conditions. Tabletop simulations, where individuals talk through their responses to emergencies, are used to research the responses of a team in terms of decision making and problem solving.

A range of other data collection techniques are used in conjunction with process simulation such as interviews, the verbal protocols described earlier, walk-throughs and questionnaires. An appropriate analysis of the task is necessary in order to determine the nature of the simulation to be used. An

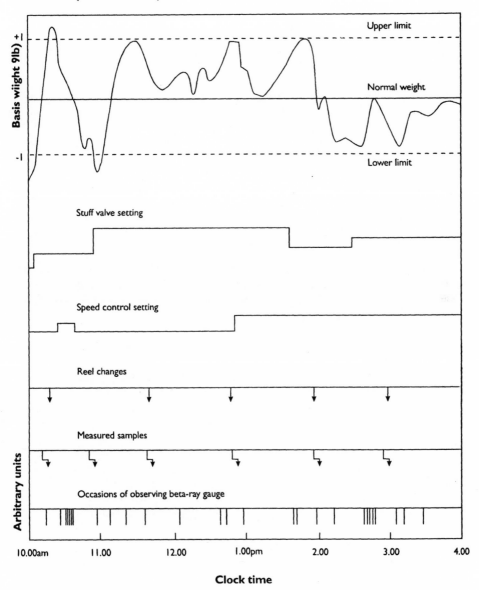

FIGURE 4.1. **Activity Analysis for the Control of "Substance" in Paper Making (Crossman et al., 1974).**

important issue to consider is which aspects of the tasks should be simulated and how faithful the representation will be. Against this has to be weighed the cost of the simulation. This will rise dramatically as more and more fidelity dimensions are built in. Stammers (1981) has offered a useful description of the different dimensions along which the fidelity of a simulator can vary.

Disadvantages may arise because the behavior observed may not be fully realistic. A static simulation, for instance, may not reveal the true nature of operators' dynamic interaction with the system. There is also the possible disadvantage of behavior in a simulator not fully replicating that found in the real situation. This can happen because of the absence of real stressors found in the actual task, for example, risk to life, criticality of the process, and presence of other workers and supervisors.

4.2.7. Withholding Information

The withheld information technique is used to explore the manner in which operators select and use information in process abnormalities. A particular abnormal process event is represented in a control panel "mock-up" or a "low-fidelity" simulator, and information is withheld from the worker until it is requested. This technique has been developed by Marshall et al. (1981) and has been used to elicit the diagnostic plans used by experienced workers during various process transients in a crude distillation unit. There are three main applications of this technique:

- To elicit the knowledge of experienced workers that cannot be verbalized easily
- To design control panels in a way that the search for process information is optimized
- To evaluate training programs and examine how new workers use control panel information to perform a task

To prepare for a withheld information session, the analyst must go through the following stages:

1. Write down the sources of information that the worker might use in the real situation including information provided verbally by his colleagues.
2. Prepare a list of events that need to be studied.
3. Prepare an event-symptom matrix showing the status of each information source for each event.
4. Ask the worker to use this information to diagnose the plant failure.

In this fashion, the way in which the workers reach decisions and deal with problems can be recorded. The problem with this technique is that the representation of the event is artificial and this may distort the data collection. The main objection is that the information offered to the worker is usually limited to easily identified information sources. It is quite feasible that workers can encode several sources of information in a display in a quite novel way which they cannot describe and which the analyst cannot anticipate.

4.3. TASK ANALYSIS

Task analysis is a fundamental methodology in the assessment and reduction of human error. A very wide variety of different task analysis methods exist, and it would be impracticable to describe all these techniques in this chapter. Instead, the intention is to describe representative methodologies applicable to different types of task. Techniques that have actually been applied in the CPI will be emphasized. An extended review of task analysis techniques is available in Kirwan and Ainsworth (1993).

4.3.1. Purpose of Task Analysis

The term *Task Analysis* (TA) can be applied very broadly to encompass a wide variety of human factors techniques. Nearly all task analysis techniques provide, as a minimum, a description of the observable aspects of operator behavior at various levels of detail, together with some indications of the structure of the task. These will be referred to as **action oriented approaches.** Other techniques focus on the mental processes that underlie observable behavior, for example, decision making and problem solving. These will be referred to as **cognitive approaches.**

In addition to their descriptive functions, TA techniques provide a wide variety of information about the task that can be useful for error prediction and prevention. To this extent, there is a considerable overlap between Task Analysis and Human Error Analysis (HEA) techniques described later in this chapter. HEA methods generally take the result of TA as their starting point and examine what aspects of the task can contribute to human error. In the context of human error reduction in the CPI, a combination of TA and HEA methods will be the most suitable form of analysis.

4.3.2. Applications of Task Analysis

Task analysis methods can be used to eliminate the preconditions that give rise to errors before they occur. They can be applied at the design stage when a new system is being developed, or when an existing system is being modified to ensure that the new configuration will not induce errors. They can also be used as part of an audit of an existing plant, in order to identify problem areas.

It is often assumed that TA cannot be applied during design, because until the plant has been fabricated the tasks to be performed by workers cannot be defined in sufficient detail. In fact, many TA techniques can be used to specify the nature of the tasks to achieve the required process plant functions, even before the exact configuration of the system has been finalized. This point will be elaborated later in the context of hierarchical task analysis.

An important aspect of a design process to minimize human error is the correct allocation of functions between human activities and automatic systems such as computer control, trips etc. From a consideration of the strengths of humans (e.g., their adaptability to cope with unpredictable situations) compared with automated systems, decisions can be made with regard to how much control should be allocated to the human in, for example, plant emergencies. A detailed discussion of allocation of function issue is provided in Price (1985) and Kantowitz and Sorkin (1987). The TA also provides information that is essential for a number of other aspects of human–machine system design. The comprehensive task description derived from the TA is a major input to the content of training and operating instructions or procedures. The results of the TA are also essential for the design of information presentation and control at the human–machine interface.

When used in the audit mode, TA can be used to develop the most efficient operating procedure for achieving the goals of a task. In many process plants it is common to find that there are wide discrepancies among the ways in which different workers or shifts carry out the same task. This is often due to inadequate or outdated operating instructions, and the absence of a culture that encourages the sharing of information about working practices. A systematic task analysis method provides the means for gathering and documenting information from different shifts and workers in order to develop the most efficient operating method from the point of view of safety, quality, and cost effectiveness.

Task analysis can also be used in a retrospective mode during the detailed investigation of major incidents. The starting point of such an investigation must be the systematic description of the way in which the task was actually carried out when the incident occurred. This may, of course, differ from the prescribed way of performing the operation, and TA provides a means of explicitly identifying such differences. Such comparisons are valuable in identifying the immediate causes of an accident.

4.3.3. Action Oriented Techniques

4.3.3.1. Hierarchical Task Analysis (HTA)
Hierarchical task analysis is a systematic method of describing how work is organized in order to meet the overall objective of the job. It involves identifying in a top down fashion the overall goal of the task, then the various subtasks and the conditions under which they should be carried out to achieve that goal. In this way, complex planning tasks can be represented as a hierarchy of **operations**—different things that people must do within a system—and **plans**—the conditions which are necessary to undertake these operations. HTA was developed by Annett et al. (1971) and further elaborated by Duncan (1974) and Shepherd (1985) as a general method of representing various

industrial tasks involving a significant planning component. Although the technique was developed in the context of process control training, it has also been used in a number of other applications such as display design, development of procedures and job aids, work organization, and human error analysis. A case study of applying the method to procedures design is given in Chapter 7.

Hierarchical Task Analysis commences by stating the overall objective that the person has to achieve. This is then redescribed into a set of suboperations and the plan specifying when they are carried out. The plan is an essential component of HTA since it describes the information sources that the worker must attend to, in order to signal the need for various activities. Each suboperation can be redescribed further if the analyst requires, again in terms of other operations and plans.

Figure 4.2 shows an example HTA for the task of isolating a level transmitter for maintenance. Redescribing operations into more detailed plans and suboperations should only be undertaken where necessary, otherwise a great deal of time and effort is wasted. Since the description is hierarchical the analyst can either leave the description in general terms or take it to greater levels of detail, as required by the analysis.

The question of whether it is necessary to break down a particular operation to a finer level of detail depends on whether the analyst believes that a significant error mode is likely to be revealed by a more fine grained analysis. For example, the operation "charge the reactor" may be an adequate level of description if the analyst believes that the likelihood of error is low, and/or the consequences of error are not severe. However, if this operation was critical, it could be further redescribed as shown below:

1. Charge reactor
   Plan: Do 1, if pressure >20 psig, wait 5 minutes then do 2–6 in order.
   1.1 Ensure pressure in reactor is less than 20 psig
   1.2 Open charging port
   1.3 Charge with reactant X
   1.4 Charge with reactant Y
   1.5 Ensure seal is properly seated
   1.6 Close and lock charging port

If the consequences of not waiting until the pressure had dropped were serious and/or omitting to check the pressure was likely, then it would be necessary to break down the operation "charge reactor" to its component steps. This approach to deciding on the level of decomposition is called the $P \times C$ rule (where P is the probability of failing to carry out an operation and C the cost of the consequences). The size of the product $P \times C$ determines whether or not to describe the operation in more detail (Shepherd, 1985).

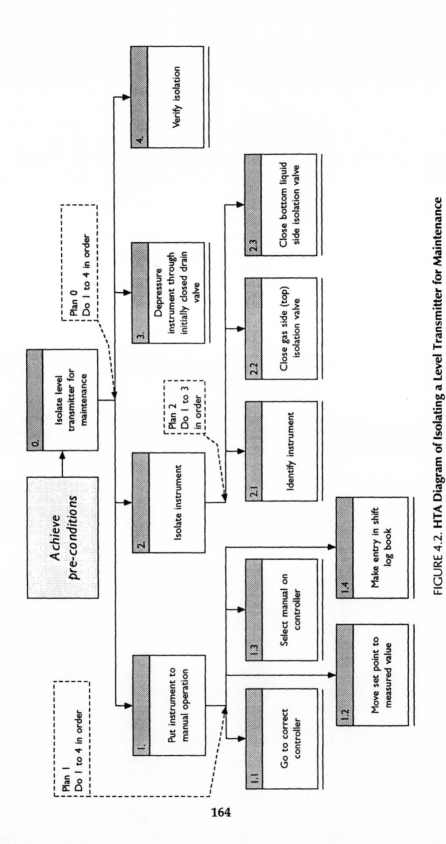

FIGURE 4.2. HTA Diagram of Isolating a Level Transmitter for Maintenance

164

This approach suffers from two major disadvantages.

- Both P and C are difficult to determine, as will be seen in Chapter 5 which reviews techniques for quantifying the likelihood of errors.
- Until the analyst has broken down the operation further, it is difficult to envision how a suboperation at the next lower level of breakdown might fail, and what the consequences of this failure might be.

In practice, a consideration of the general quality of the PIFs (performance-influencing factors) (e.g., training, supervision, procedures) in the situation being evaluated will give a good indication of the overall likelihood of error in the specific operation being evaluated. Similarly, the consequences of errors can be evaluated in terms of the overall vulnerability to human error of the subsystem under consideration. By considering these factors together, it is usually obvious where the analysis should be terminated. Differing levels of detail may be necessary for different purposes, for example, risk analysis, training specification or procedures design.

There are two main ways for representing a HTA description: the diagrammatic and tabular format. Diagrams are more easily assimilated but tables often are more thorough because detailed notes can be added. It is possible to start with a diagrammatic format and finally record the analysis in step by step format. This allows other aspects of the task to be considered such as information about the human–machine interface, communications with other team members, time characteristics, side-effects caused by failure to follow the correct plan, and the knowledge required to carry out a plan. An example of this format is provided in Figure 4.3 for the task step of optimizing a high pressure in a distillation column. Including this information in the task analysis will be very useful for gaining an insight into the workload imposed by various task components, the various points where performance may degrade, and finally into the methods that are likely to optimize human performance.

Analyzing complex tasks that entail considerable skill is usually done in collaboration with people who are knowledgeable about the job such as the workers, the supervisors, or the engineers. Information can be collected from a variety of sources including verbal protocols, activity analysis, operating procedures, emergency procedures, and records of critical incidents. It is rarely a good idea to rely on observing performance as a prime source of task information, especially in tasks involving substantial decision making, since the individual's intentions and information seeking strategies are seldom apparent. Because of the necessity to rely on cooperation of operating personnel, who have other demands on their time, it is useful to agree at the outset with the client how much time is likely to be required to ensure that such cooperation will be forthcoming.

TASK STEPS	INPUTS (CUES)	OUTPUTS (ACTIONS)	FEEDBACK	COMMUNICATIONS	TIME CHARACTERISTICS	TASK DEPENDENCIES	SECONDARY DUTIES, DISTRACTIONS	COMMENTS
4 Optimize pressure value (PR5) in column within 1–1.5 atm.	Pressure recorder indicates PR5 > 1.5 atm (No alarm indications)	If PR5 > 1.8 atm increase cooling rate in condenser (4.2) and decrease heating rate in reboiler (4.3) If PR5 < 1.8 atm do 4.2 only (CRO monitors recorders and OSO adjusts valves on site)	Pressure recorder Temperature recorder Condensate level in reflex drum	Radio communications between control room operator (CRO) and outside operator (OSO)	Optimization should start **not later than** 2 min from the initiation of the abnormal event	Fluctuations of temperature will degrade quality of products	CRO busy with other tasks in the control room	CRO can fail to detect PR5 increase OSO can omit to adjust rates of cooling to the required levels Hazards: Danger of explosion due to accumulation of vapors inside column OSO should wear protective clothing

FIGURE 4.3. **Tabular HTA Showing How to Optimize a High Pressure in a Distillation Column**

The advantages and disadvantages of the technique can be summarized as follows:

Advantages of Hierarchical Task Analysis
- HTA is an economical method of gathering and organizing information since the hierarchical description needs only to be developed up to the point where it is needed for the purposes of the analysis.
- The hierarchical structure of HTA enables the analyst to focus on crucial aspects of the task that can have an impact on plant safety.
- When used as an input to design, HTA allows functional objectives to be specified at the higher levels of the analysis prior to final decisions being made about the hardware. This is important when allocating functions between personnel and automatic systems.
- HTA is best developed as a collaboration between the task analyst and people involved in operations. Thus, the analyst develops the description of the task in accordance with the perceptions of line personnel who are responsible for effective operation of the system.
- HTA can be used as a starting point for using various error analysis methods to examine the error potential in the performance of the required operations.
- For application in chemical process quantitative risk analysis (CPQRA), the hierarchical format of HTA enables the analyst to choose the level of event breakdown for which data are likely to be available. This is useful for human reliability quantification (see the discussion in Chapter 5).

Disadvantages
- The analyst needs to develop a measure of skill in order to analyze the task effectively since the technique is not a simple procedure that can be applied immediately. However, the necessary skills can be acquired reasonably quickly through practice.
- In order to analyze complex decision making tasks, HTA must be used in combination with various cognitive models of performance. Also HTA presents some limitations in describing tasks with a significant diagnostic component.
- Because HTA has to be carried out in collaboration with workers, supervisors, and engineers, it entails commitment of time and effort from busy people.

4.3.3.2. *Operator Action Event Trees (OAET)*
Operator action event trees are treelike diagrams that represent the sequence of various decisions and actions that the operating team is expected to perform when confronted with a particular process event. Any omissions of such

decisions and actions can also be modeled together with their consequences for plant safety. OAETs are described in Hall et al. (1982) and Kirwan and Ainsworth (1993), and have many similarities with the event trees used for the analysis of hardware reliability.

Figure 4.4 gives an example of an OAET for events that might follow release of gas from a furnace. In this example a gas leak is the initiating event and an explosion is the final hazard. Each task in the sequence is represented by a node in the tree structure. The possible outcomes of the task are depicted as "success" or "failure" paths leading out of the node. This method of task representation does not consider how alternative actions (errors of commission) could give rise to other critical situations. To overcome such problems, separate OAETs must be constructed to model each particular error of commission.

By visual inspection of an OAET it is possible to identify the elements of a process control task which are critical in responding to an initiating event. An important issue in the construction of OAETs is the level of task breakdown. If the overall task is redescribed to very small subtasks it might be difficult to gain insights from the OAET because it can become relatively unwieldy. Hierarchical Task Analysis provides a useful framework for the

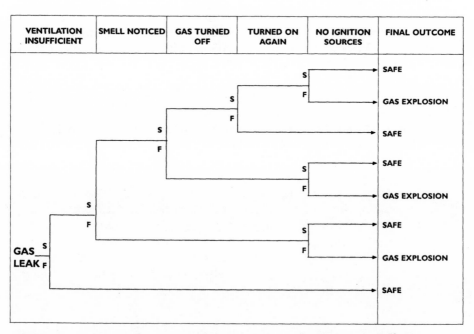

FIGURE 4.4. **Event Tree for a Gas Leak from a Furnace (S=Success; F=Failure).**

identification of required tasks, and also help the analyst clarify the appropriate level of task decomposition.

Care should also be taken in the use of recovery factors, because these can exert a significant effect. In general, recovery paths are appropriate where there is a specific mechanism to aid error recovery, that is an alarm, a supervising check, or a routine walk round inspection.

While OAETs are best used for the qualitative insights that are gained, they can also be used as a basis for the quantitative assessment of human reliability. By assigning error probabilities to each node of the event tree and then multiplying these probabilities, the probability of each event state can be evaluated (see Chapter 5).

The advantages and disadvantages of OAETs are as follows:

Advantages
- The OAET is a logical method of structuring information concerning operator actions resulting from a particular initiating event.
- OAETs help to identify those tasks which are important in responding to particular initiating events.
- OAETs readily interface with system analysis techniques that are commonly used by engineers in CPQRA applications.

Disadvantages
- The approach is not a satisfactory method of identifying mistaken intentions or diagnostic errors.
- OAETs are best suited to represent errors of omission. The important errors of commission (i.e., alternative actions that may be performed) are difficult to include satisfactorily.
- No assistance is provided to guarantee that the data used in the modeling process is complete and accurate. Therefore, the comprehensiveness of the final OAET will be a function of experience of the analyst. (This criticism applies to all HRA techniques.)
- The OAET approach does not address error reduction or make any attempt to discover the root causes of the human errors represented.

4.3.3.3. Decision/Action Flow Diagrams
These are flow charts that show the sequence of action steps and questions to be considered in complex tasks that involve decision-making. Decision/action flow diagrams are similar to the flow charts used in computer program development. Both charts are based on binary choice decisions and intervening operations. In general, the binary decision logic in decision/action charts expedites communications through the use of simple conventions and provides for easy translation of decision/action charts into logic flow charts for computerized sections of the system.

Decision/action charts can be learned easily and workers usually find them useful in formulating for the analyst their mental plans which may involve decision-making, time-sharing, or complex conditions and contingencies. Figure 4.5 shows a decision/action chart for a furnace start-up operation. Decision/Action charts have only a single level of task description, and when complex tasks are analyzed the diagrams become unwieldy and difficult to follow. Also, it is possible to lose sight of the main objectives of the task. To this extent, HTA is more appropriate because the task can be represented in varying degrees of detail and the analyst can get a useful overview of the main objectives to be achieved during the performance of the task.

A general problem in task analysis is how to describe tasks that involve diagnosis of system failures. Duncan and Gray (1975) have described diagnostic tasks in terms of decision trees that guide personnel through a number of checks to various system failures. Decision trees are very much like decision/action charts. Figure 4.6 shows a decision/action chart for diagnosing faults in a crude distillation unit.

Although little training is required to learn the technique, decision/action charts should be verified by different operators to ensure that a representative view of the decision task is obtained. The advantages and disadvantages of the technique are summarized as follows:

Advantages
- Decision/action charts can be used to represent tasks that involve decision-making, time-sharing, or complex conditions and contingencies.
- Workers find it easy to express their work methods in terms of flow diagrams. This representation can then provide input to other task analysis methods.
- They can be used to identify critical checks that the workers have to carry out to complete a process control task.
- For fault-diagnostic tasks, they can help the analyst to identify whether new staff members make effective use of plant information.

Disadvantages
- Decision/action charts are linear descriptions of the task and provide no information on the hierarchy of goals and objectives that the worker is trying to achieve.
- For complex tasks, the diagrams can become unwieldy.
- They offer no guidance concerning whether or not a particular operation or decision should be redescribed in more detail.

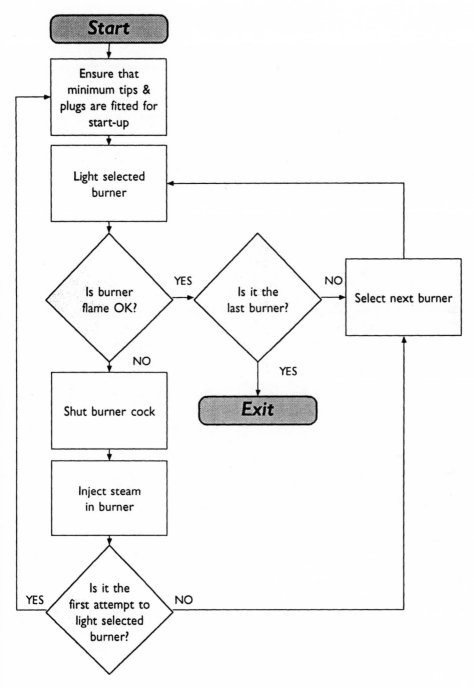

FIGURE 4.5. **Decision/Action Flow Diagram of a Furnace Start-Up Operation.**

4.3.3.4. *Operational Sequence Diagrams (OSDs)*

Operational sequence diagrams are flcw-charting techniques that represent any sequence of control movements and information collection activities that are executed in order to perform a task. Various activities in the diagram are represented with a symbolic notation, supported where necessary by a text description. For the majority of simple applications, OSDs assume a linear flow drawn from top to bottom with a limited degree of branching and looping. The symbols used are usually tailored to fit the type of task being studied and its level of analysis.

The three significant OSD attributes are its sequential flow, its classification of activity type, and its ability to describe interactions between people and machines. In these respects, OSDs are similar to the Decision/Action charts, but more complex. The OSD can be seen as a static simulation of the system operations. This is also the reason why OSDs can become tedious to develop in the analysis of complex systems.

Operational sequence diagrams provide a versatile method representing the timing relationships among operations, functional requirements of human–machine interfaces, and spatial relationships among items of equipment on which operations are performed. Depending on the characteristics of the task being studied, the analyst can use one of the three OSD derivatives, namely temporal OSDs, partitioned OSDs, spatial OSDs, or a combination of these. Tasks with a high cognitive component produce particular problems of classification and identification of discrete operations. Such cognitive tasks will generally not allow the production of OSDs. Also complex tasks can cause problems, as is the case with most graphical methods of representation, because operational sequences very soon become incomprehensible, particularly if they are not highly linear.

The type of OSDs to be used depends on the data to be represented. The three main forms of OSDs will be considered in more detail below.

Temporal OSDs

These diagrams focus on the temporal or time relationships of operations and they can be used to solve resource allocation problems, to determine whether there is any potential for time stress, and to consider alternative work methods in the execution of a procedure. An example drawn from traditional industrial engineering methods is shown in Figure 4.7. The chart is used to analyze the interaction between people and equipment. As indicated in the summary portion of this chart, there is a high proportion of idle time which would probably indicate the use of alternative procedures in the execution of this task. The chart enables the analyst to see the relationships among the activities of the different components in planning such alternatives.

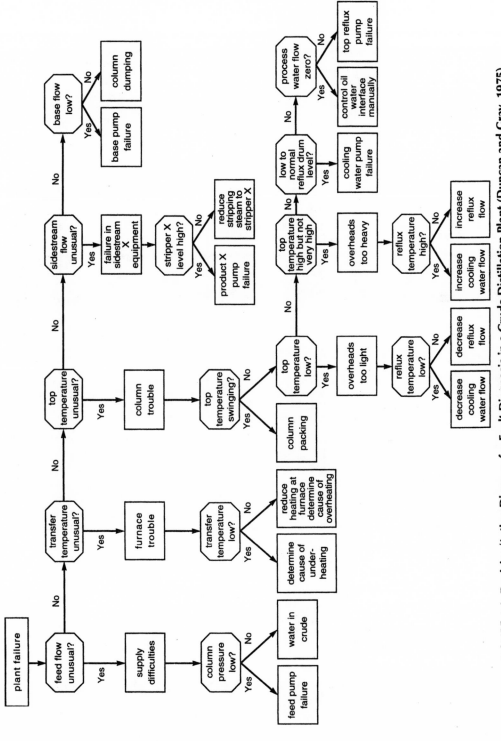

FIGURE 4.6. Decision/Action Diagram for Fault Diagnosis in a Crude Distillation Plant (Duncan and Gray, 1975).

173

OPERATION: Slitting Coated Fabric				OP. NO. S46	
PART NAME: Coated Fabric				PART NO. F261	
MACHINE NAME: Slitting Machine (Special)				MACH.NO. S431	
OPERATOR NAME: J. S. Wilson S. K. Smith (Helper)				DATE:	
OLD METHOD: ☒ IMPROVED METHOD: ☐				CHART BY: J. S. K.	

Operator	Time*	Helper	Time*	Help	Time*
Run Machine	2.2	Prepare wrappers and labels	.9	Slit stock	2.2
		Wait for machine	1.3		
Wait for helper	.7	Wrap rolls	.9	Idle	3.0
Label rolls	.6	Wait for operator	.7		
Open winder	.3				
Wait for helper	.8	Remove rolls	.8		
Start machine	.6	Place on skid	.6		

*Time in minutes

Summary			
	Operator	Helper	Machine
Idle Time	1.5 min	2.0 min	3.0 min
Working time	3.7	3.2	2.2
Total cycle time	5.2	5.2	5.2
Utilzation in per-cent	Operator utilization = $\frac{3.7}{5.2} = 71\%$	Helper utilization = $\frac{3.2}{5.2} = 62\%$	Machine utilization = $\frac{2.2}{5.2} = 42\%$

FIGURE 4.7. **Temporal Operational Sequence Diagram in a Slitting Coated Fabric Operation (from Barnes, 1980).**

174

Partitioned OSDs

In this case, the operations within a sequence are further defined according to various criteria such as whether they involve reception or transition of information, storage of information, manual control responses, inspections, and decisions. However, some other dimensions of an operation may require particular emphasis such as whether information is transmitted electronically, by external communication etc. A type of vertical axis can still be used to represent sequential order and if required this can incorporate the same timing information as temporal OSDs.

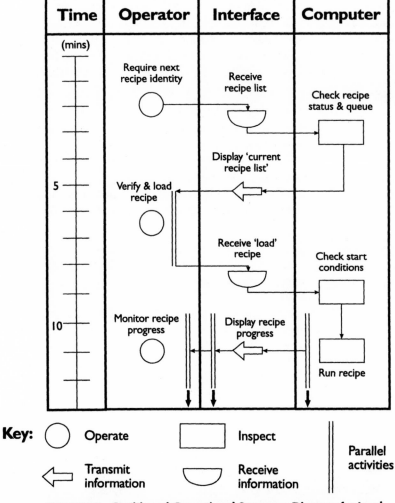

FIGURE 4.8. **Partitioned Operational Sequence Diagram for Loading a Recipe to a Computer Controlled Reactor (adapted from Kirwan and Ainsworth, 1993).**

Figure 4.8 shows a specific example of this type of diagram which includes some symbols. The diagram shows the tasks that the operator and the computer must perform in a computer controlled reactor. The central column is used to show any functional requirements of the human–computer interface.

Spatial OSDs

In spatial OSDs the flow of events and symbols is overlaid on a map of all items of equipment with which the operator interacts during the task. The map itself does not have to be very accurate, provided that the general geographical relationships among items of equipment are shown. The spatial OSD thus provides a graphical description of the perceptual–motor load a particular task imposes on the performance of the worker. For multiperson tasks, the operational sequences for several workers can be coded in different colors and superimposed onto the same equipment map. This can generate useful information for the distribution of tasks to different members of the operating team.

In summary, OSDs have the following advantages and disadvantages:

Advantages
- Operational sequence diagrams are very useful to show temporal, spatial, and even conceptual relationships among operations that are difficult to show by textual or tabular representations.
- To some extent, more than one type of relationships can be shown but this can give rise to excessive complexity.
- They can be used for solving resource allocation problems, looking at aspects of time-stress, and designing the human computer interface.

Disadvantages
- Operational sequence diagrams can become cluttered and confusing when used for complex or highly conditional tasks. It is particularly important that the analyst is working to the level of detail that is most appropriate.
- Spatial OSDs can also become very difficult to read if individual pieces of equipment are used many times.
- Operational sequence diagrams cannot represent highly cognitive tasks because it is difficult to assign cognitive components to discrete symbols.
- Although OSDs can be used to optimize general operator performance, they are limited to the extent that they can identify human errors.

4.3.3.5. Signal-Flow Graph Analysis

This technique is derived from a method developed by electrical engineers to facilitate the analysis of electrical networks. It has been applied to process operator studies by Beishon (1967). The method describes the process to be controlled in terms of "manually controlled" variables, "displayed" variables and "hidden" variables which can be deduced from those displayed or from calculations. By tracing the signal-flow graph (SFG) from the "controlled" to the "displayed" variables, it is possible to identify the control loops available to the worker and the types of deductions required to understand and control the system. SFG analysis is a method that represents "how the system works" rather than "how the worker should perform the task."

Signal-flow graphs are particularly useful in two respects. First, they make the process designer examine in considerable detail the dynamic structure and functioning of the process. Second, the nature of the interface between person and machine can be seen more clearly. The variables that are displayed in a system are, of course, available for study, but workers frequently respond to derivative functions of variables or "hidden" variables that must be deduced. Given that the process variables to be displayed will influence the worker's control strategy and that the number of deductions to be made will affect the mental workload involved, a process designer can select the type and amount of process information which will enhance performance of the task.

A study of paper making carried out by Beishon (1969) illustrates the part an SFG can play in the design of control panel information and specification of control strategies. The top part of Figure 4.9 shows a continuous paper making machine controlled by a worker. The paper is formed from a liquid containing fibers, the stock, which is spread out onto an endless belt of wire mesh. The water drains or is sucked through the mesh, leaving a sheet of paper that can be lifted on to endless belts of felt for pressing and drying. Part of the worker's job is to produce paper of different weights, or "substance values." In order to understand the complex factors that determine the important time relations in the process, a fairly complete SFG was drawn (see bottom part of Figure 4.9). The SFG was used to select appropriate process variables to be displayed to the worker to assist in improving his performance.

Signal-flow graphs are useful in another sense; they provide an objective representation of "how the system works" which can be used to evaluate the worker's subjective mental representation of the system. The influence modeling and assessment (IMAS) technique, which is described in subsequent sections, can also be used to elicit the worker's representation of the system. Both techniques, IMAS and SFG, can therefore be used for training personnel.

Advantages
- The SFG is a useful technique to represent the process variables that affect system performance.
- They can be used for designing the human–machine interface.
- They provide useful data for evaluating the worker's understanding of how the system functions.

Disadvantages
- Signal-flow graphs cannot explicitly identify the error potential for particular action steps.

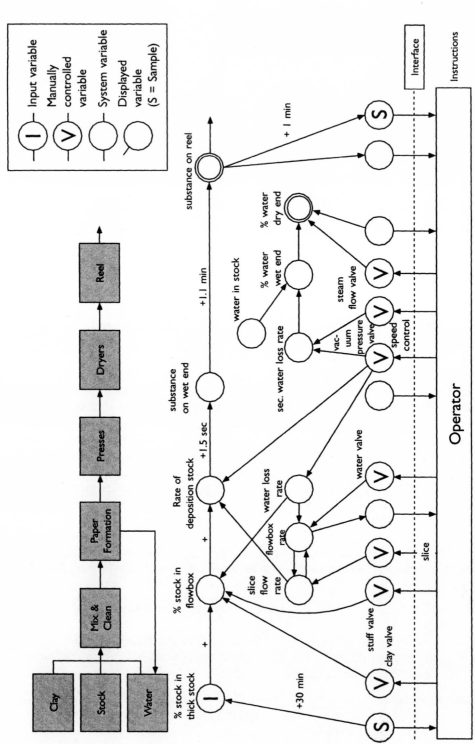

FIGURE 4.9. Block Diagram and Signal-Flow Graph for "Substance" Control System in Paper-Making (from Beishon, 1969).

178

- Signal-flow graphs do not provide a complete description of the process in a control task. The process may go through a sequence of stages, in which different variables are being altered, so that different control loops are important. The control task in such a sequence of stages can best be described by a sequence of SFGs, each of which shows the subset of process variables that are important at a particular stage.
- For a complete description of the task it is also necessary to specify the targets and tolerances to which the process should be controlled.

4.3.4. Cognitive Task Analysis Techniques

The task analysis techniques described in the previous section are mainly oriented toward observable actions, although hierarchical task analysis (HTA) allows it to address functional requirements as well as the specific actions that are required to satisfy these requirements.

Cognitive task analysis techniques attempt to address the underlying mental processes that give rise to errors rather than the purely surface forms of the errors. This is particularly important where the analysis is concerned with those aspects of process plant operation that require higher level mental functions such as diagnosis and problem solving. As plants become more automated, the job of the process plant worker is increasingly concerned with these functions and it is therefore necessary to develop analytical methods that can address these aspects of plant control. For example, the worker is often required to deal with abnormal plant states that have not been anticipated by the designer. In the worst case, the worker may be required to diagnose the nature of a problem under considerable time stress and develop a strategy to handle the situation. It is clearly desirable in these situations to provide appropriate decision support systems and training to improve the likelihood of successful intervention. It is also necessary to be able to predict the types of decision errors that are likely to occur, in order to assess the consequences of these failures for the safety of the plant. In all of these areas, task analysis techniques that address the covert thinking processes, as opposed to observable actions, are necessary.

The problems associated with the analysis of cognitive processes are much greater than with action oriented task analysis methods. The causes of "cognitive errors" are less well understood than action errors, and there is obviously very little observable activity involved in decision making or problem solving. These difficulties have meant that very few formal methods of cognitive task analysis are available, although several researchers have developed specialized methods when studying process control skills (see, e.g., Bainbridge, 1974).

Despite these difficulties, the issue of cognitive errors is sufficiently important that we will describe some of the approaches that have been applied to process industry systems. These techniques can be used in both proactive and retrospective modes, to predict possible cognitive errors (i.e., "mistakes" as opposed to "slips" using the terminology of Chapter 2) during predictive risk assessments, or as part of an incident investigation.

4.3.4.1. Critical Action and Decision Evaluation Technique (CADET)

This method is based on the Rasmussen stepladder model described in Chapter 2. It was first described in Embrey (1986). The basic units of CADET are the critical actions or decisions (CADs) that need to be made by the operator usually in response to some developing abnormal state of the plant. A CAD is defined in terms of its consequences. If a CAD fails, it will have a significant effect on safety, production or availability.

The following approach is then used to analyze each CAD. The first stage consists of identifying the CADs in the context of significant changes of state in the system being analyzed. The approach differs from the OAET (Section 4.3.3.2) in that it does not confine itself to the required **actions** in response to critical system states, but is also concerned with the decision making that precedes these actions. Having identified the CADs that are likely to be associated with the situation being analyzed, each CAD is then considered from the point of view of its constituent decision/action elements. These are derived from the Rasmussen stepladder model discussed in Chapter 2 and reproduced in linear form in Figure 4.10. The potential failures that can occur at each of these elements are then identified.

To illustrate how CADET can be applied to decision analysis Figure 4.11 describes a hypothetical example an experienced worker who has to diagnose a plant failure (e.g., top reflux pump failure in a distillation column). A column is created for each decision/action element of the Rasmussen decision ladder to allow an extensive description of how the worker processes diagnostic information and eliminates an initial set of possible equipment failures to arrive at the actual problem. CADET presents the analyst with a structured list of questions about potential diagnostic errors. The protocol in Figure 4.11 shows a good diagnostic strategy in which the worker is looking initially for spurious indications before drawing any conclusions about the state of process equipment. CADET can be used both to evaluate and to support human performance in terms of training exercises.

Lucas and Embrey (1987) further extended the CADET concept as a practical analysis tool by developing a structured questionnaire for use by an analyst when interacting with plant personnel. For each CAD the analyst is prompted to consider a possible failure at each of the stages in the Rasmussen model described in Figure 4.10.

The CADET technique can be applied both proactively and retrospectively. In its proactive mode, it can be used to identify potential cognitive errors, which can then be factored into CPQRA analyzes to help generate failure scenarios arising from mistakes as well as slips. As discussed in Chapter

DECISION/ACTION ELEMENT	OBJECTIVE	TYPICAL ERROR PATTERNS
Initial Alert	Alerting/Signal Detection of initial stages of problem	Distraction/Absent-Mindedness/Low Alertness
Observation	Observation/Data Collection from instruments	Unjustified Assumptions/Familiar Associations
Identification	Identify System State	Information Overload Time Delay
Interpretation	Interpret what has happened and its implications	Failure to Consider Alternative Causes/Fixation on the Wrong Cause
Evaluation	Evaluation and Selection of Alternative Goals	Failure to Consider Side Effects/Focusing on Main Event
Planning	Plan success path	Wrong Task May be Selected due to Shortcuts in Reasoning and Stereotyped Response to Familiar State
Procedure Selection/Formulation	Choosing or formulating a procedure to achieve required objective	Procedural Steps Omitted/Reversed (Particularly if "Isolated")
Execution	Executing chosen procedure	Reversals of Direction or Sign (Up/Down Left/Right) when carrying out action. Habit Intrusion
Feedback	Observe change of state of system to indicate correct outcome of actions	Feedback ignored or misinterpreted

FIGURE 4.10. **Decision/Action Elements of the Rasmussen Model (Embrey, 1986).**

2, errors arising from misdiagnosis can be particularly serious, in that they are unlikely to be recovered. They also have the potential to give rise to unplanned operator interventions based on a misunderstanding of the situation. These error modes need to be explicitly identified by CPQRA analysts. Another proactive use of CADET is in the development of error reduction strategies based on the possible error root causes identified by the questionnaire. The technique can also be applied retrospectively to identify any cognitive errors implicated in accidents.

Pew et al. (1981), developed a series of "Murphy diagrams" (named after the well-known Murphy's Law: *If something can go wrong, it will*). Each decision element in the Rasmussen model has an associated Murphy diagram, which specifies possible direct "proximal") causes of the internal malfunction. Each of these causes are then considered in terms of indirect "distal") causes which could influence or give rise to the primary cause. A Murphy diagram for the

TIME	SIGNAL DETECTION	DATA COLLECTION	IDENTIFICATION	INTERPRETATION	GOAL SELECTION	CADET ANALYSIS
t_1	Column temperature alarm		Not a complete indication at this stage. It may be a spurious alarm	Cross-examine related indicators		**Data collection:** Can operator acquire irrelevant or insufficient data? Can operator fail to cross-check for spurious indications?
t_2		TR14 = High (new) TR15 = Very High (check)	Inadequate cooling of column or thermal conditions of input are disturbed	Distinguish between the two. Examine flow rate and temperature of input		**Identification/Interpretation:** Can operator fail to consider all possible system states and causes of problem? Can operator fail to perform a correct evaluation? Can operator fixate on the wrong cause?
t_3		FI1 = Normal (new) FR15 = Normal (check) TRC8 = Normal (new)	Conditions are as specified. It must be inadequate cooling of column	Possible causes: • Cooling water pump failure • Top reflux pump failure		**Goal Selection:** Can operator fail to consider possible side-effects? Can operator fail to consider alternative goals? Can operator fixate on the wrong goal?
t_4		LIC3 = High (new) Drum sight glass = High (check)	Conditions are as specified. It must be inadequate cooling of column	Level in drum is high, thus condensation is OK. It must be failure of the top reflux pump		
t_5		FIC8 = No Flow (new)	Conditions are as specified. It must be inadequate cooling of column	Top reflux pump failure (confirmed)	**Alternative goals:** • Reduce heating in reboiler • Reduce flow rate of input • Increase cooling in condenser	

TR14, TR15 = Column Temperature; LIC3 = Level in Reflux Drum; FIC8 = Reflux Flow; FI1, FR15 = Crude Flow at Entry Point; TRC8 = Crude Temperature at Entry Point.

FIGURE 4.11. **CADET analysis of a fault-diagnostic task in an oil refinery.**

182

decision element "Plan Success Path" is given in Figure 4.12. The Murphy diagram can be of considerable value to the analyst because it suggests specific causes of errors which will be amenable to design solutions. Only a relatively small number of decision elements will be associated with each CAD in most cases, which means that the process of analysis is reasonably manageable.

4.3.4.2. The Influence Modeling and Assessment Systems (IMAS)

Reference has already been made to the difficulty of accessing the mental processes involved in diagnosis and decision making. Success in these activities is likely to be dependent on the worker having a correct understanding of the dynamics of what is likely to happen as an abnormal situation develops. This is sometimes referred to as the worker's "mental model" of the situation (see Chapter 2 for a further discussion of this topic). Diagnosis in the event of a plant emergency does not depend only on the absolute values of variables (e.g., flow rates) but also relies upon the changes in these indicators over time. Knowledge of the mental model possessed by the operator can be extremely useful in predicting possible diagnostic failures.

The IMAS technique was originally developed as an on-line decision support system to assist personnel in making diagnoses during plant emergencies (see Embrey and Humphreys, 1985; Embrey, 1985). The technique is used to elicit the mental models of process abnormalities from personnel. These are in the form of graphical representations of the perceptions of the operating team regarding:

- The various alternative causes that could have given rise to the disturbance
- The various consequences that could arise from the situation
- Indications such as VDU displays, meters, and chart recorders available in the control room or on the plant that are associated with the various causes and consequences

A specific example of the representation of the mental model derived by this approach is given in Figure 4.13. This was developed for a process plant in which powders are transferred by a rotary valve to a slurry mix vessel. Because of the flammable nature of the powders, they are covered with a blanket of nitrogen. Any ingress of air into the system can give rise to a potential fire hazard, and hence an oxygen analyzer is connected to the alarm system. Because the system can only be entered wearing breathing apparatus, it is monitored via closed circuit television (CCTV) cameras. The situation under consideration occurs when there is a failure to transfer powder and the model represents the various causes of this situation and some of the possible consequences. Any node in the network can be either a cause or a consequence, depending on where it occurs in the causal chain. It can be seen that the various indicators (given in square boxes) are associated with some of the events that could occur in the situation.

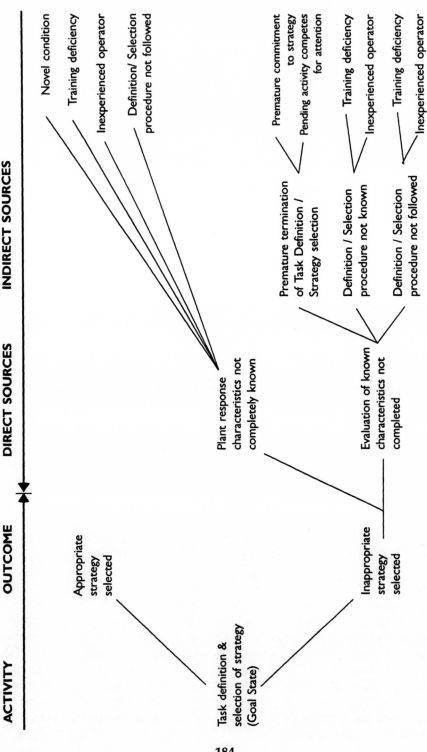

FIGURE 4.12. Murphy Diagram for "Planning" Element of Rasmussen Model (Pew et al., 1981; see Figure 4.10).

184

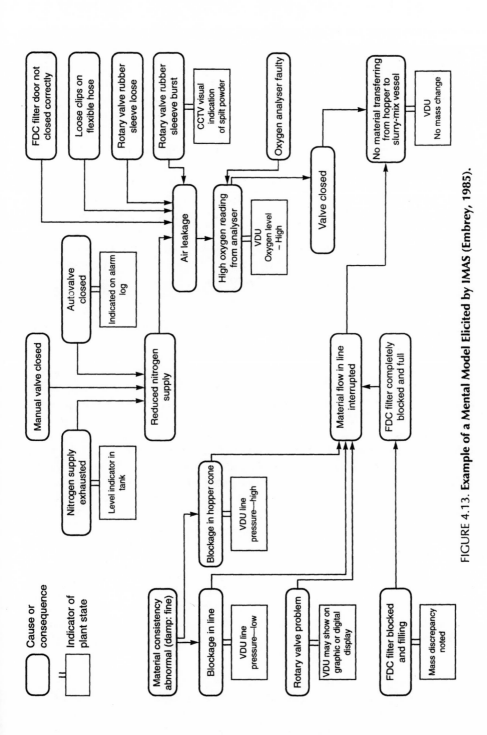

FIGURE 4.13. **Example of a Mental Model Elicited by IMAS (Embrey, 1985).**

185

The model may be developed using the expertise of an individual or several workers in a team. The process of eliciting the model can be performed by hand or with the aid of a computer program called LINKCC (Embrey and Humphreys 1985). In developing the mental model, the analyst begins at a specific point in a process disturbance (e.g., an increase of pressure in a line), and asks the worker what the event **stems from, leads to,** or is **indicated by.** Repeated applications of these questions produce a network representation of the "group model" of the operating team or the individual process worker. As can be seen from Figure 4.13, an event can stem from more than one alternative cause, and lead to more than one outcome. The task of the worker is to identify which of the alternative causes gave rise to the pattern of observed indicators.

It is important to note that the mental model representation elicited by this technique is not a process engineering model, but instead represents the process workers' understanding of the various causes and consequences of the disturbance. This may or may not be in accordance with the actual chemistry or dynamics of the physical process.

Application of IMAS

The mental model representation elicited by LINKCC can be used for a variety of purposes:

- *Evaluation of the Accuracy of the Mental Model of an Operator during Training*

One of the major problems in training personnel to acquire diagnostic skills is the difficulty of knowing whether or not their understanding of process disturbances is sufficiently comprehensive in terms of alternative causes and possible consequences. Elicitation of the mental model at various stages of training enables the trainer to evaluate the development and accuracy of the workers' understanding of a range of process disturbances. A set of representations of the mental models developed using experienced operational teams can be used as standards to define the knowledge requirements to handle critical plant disturbances. Comparison of the trainees' mental models with these representations will indicate where further training is required.

- *Information Requirements for Diagnosis*

Since the mental model elicited by IMAS explicitly identifies the information needed to identify the causes of disturbances (and to distinguish among alternative causes), it can be used to specify the critical variables that need to be readily available to the process controller at the interface. This information can be used as an input to the design and upgrading of interfaces, particularly when new technology is being installed.

- *Modeling of Cognitive Errors for CPQRA*

The traditional approach to CPQRA only considers human failures to perform required functions (usually errors of omission). However, many critical errors arise from misdiagnoses (mistakes) leading to erroneous, inappropriate actions which can have serious consequences for the plant. IMAS can be used to predict possible diagnostic errors by examining the model elicited from the worker and identifying the Performance Influence Factors (e.g., inadequate display of critical process information) that could give rise to misdiagnoses (e.g., where different plant abnormalities could exhibit similar symptoms).

- *Simulation of the Thinking Processes of the Operator during Plant Emergencies*
  IMAS has a facility called EXPLORE allows the analyst to specify which indicators (e.g., temperatures, pressures, valve settings) are present, and which are absent in a particular scenario. EXPLORE then traverses the various links in the mental model representation network and generates a report that simulates the worker's thinking processes. This form of simulation provides useful information to the analyst with regard to the worker's capability to achieve correct diagnoses. Embrey (1985) gives an example of these simulations for the mental model in Figure 4.13.

The IMAS technique described above is useful, in that it addresses aspects of operational skills, that is, diagnostic and problem solving abilities, that are not covered by other techniques. To that extent it can be regarded as a method of cognitive task analysis. It is not essential to use a computer program to obtain useful results. The mental models produced by IMAS can be elicited by pencil and paper methods. Nevertheless interpretation and application of the results require some expertise.

4.3.5. Evaluation of Task Analysis Methods

The TA methods described so far can be evaluated in terms of their focus on different aspects of the human–machine interaction. To facilitate the process of selection of appropriate TA methods for particular applications. Figure 4.14 describes ten criteria for evaluation. These criteria are in terms of the usability of the methods for the following applications :

1. Analyzing actions
2. Analyzing cognitive behavior
3. Identification of critical decisions
4. Description of critical control panel information
5. Description of time related aspects of tasks
6. Identification of side-effects of errors
7. Identification of human–computer interactions
8. Description of team communications
9. Classification of task types
10. Description of the technical system

METHOD EVALUATION	HTA	OAET	DA CHARTS	OSD	SFGS	CADET	IMAS
1 Does the method focus on the observable aspects of operator behavior?	Y	Y	Y	Y	Y	N	N
2 Does the method focus on the mental processes that underlay behavior?	N	N	N	N	N	Y	Y
3 Can the method identify points where critical decisions have to be made?	Y	Y	Y	N	N	Y	Y
4 Can the method identify important information on the control panel?	Y	N	Y	P	Y	Y	Y
5 Does the method describe the temporal characteristics of the task?	P	N	P	Y	N	N	N
6 Can the method identify interactions between task-steps, and possible side effects?	Y	N	Y	N	N	Y	Y
7 Does the method describe the interactions between people and control systems?	Y	P	N	Y	Y	P	P
8 Does the method describe the communication requirements among team members?	Y	N	N	Y	N	N	N
9 Does the method classify tasks into different categories?	N	N	N	P	N	Y	N
10 Does the method provide a qualitative description of the technical system?	N	N	N	N	Y	N	N

P = Criterion is only partially fulfilled

FIGURE 4.14.   **Criteria for Evaluating the Suitability ofVarious T.A. Methods**

In general, HTA, IMAS, and CADET fulfill most of the above criteria, hence they can be used together as a framework for carrying out both action and cognitive task analysis. When particular aspects of the human–machine interaction must be examined in greater detail; for example, the temporal characteristics of the task or the team communications, certain methods can be selected to provide this information—OSDs in this case. Most TA methods are concerned with descriptions of the tasks performed by personnel. However, there may be a need to provide qualitative descriptions of the technical system itself. The last criterion (10) was introduced for this purpose.

Another way of classifying the various TA methods is in terms of the application areas in which they might be seen as most useful. Figure 4.15 provides such a classification in terms of seven human factors applications, namely:

1. Design of operating procedures
2. Training needs analysis
3. Team organization
4. Human–machine allocation of tasks
5. Control panel design
6. Workload analysis
7. Input to human error analysis

It is worth pointing out that Figures 4.14 and 4.15 present only a broad qualitative classification along a number of criteria. It is conceivable that some methods may fulfill a criterion to a greater extent than others.

4.4. HUMAN ERROR ANALYSIS TECHNIQUES

The application of human error analysis (HEA) techniques is to predict possible errors that may occur in a task. The next stage of error analysis is to identify error recovery possibilities implicit within the task, and to specify possible

APPLICATIONS	HTA	OAET	DA CHARTS	OSDS	SFGS	CADET	IMAS
1 Design of operating procedures	Y	N	Y	N	N	P	P
2 Training needs analysis	Y	N	Y	N	N	Y	Y
3 Team organization	Y	N	N	Y	N	P	N
4 Human–machine task allocations	Y	P	P	Y	Y	Y	P
5 Control panel design	Y	N	Y	P	Y	Y	Y
6 Workload analysis	P	N	N	Y	N	Y	N
7 Input to human error analysis	Y	Y	Y	N	N	Y	Y

P = Criterion is only partially fulfilled

FIGURE 4.15. **How to Use Various TA Methods in Human Factors Application**

remedial strategies to eliminate the causes of errors or to enhance their likelihood of recovery before the consequences occur. The consequences of possible unrecovered errors are also often considered error analysis. The requirements for error analysis techniques are therefore as follows:

1. Provide assistance to the analyst in exhaustively identifying possible errors.
2. Identify error recovery opportunities.
3. Develop error reduction strategies (ERS).
4. Consider the consequences of possible errors for risk assessment or for cost–benefit analysis when considering alternative ERS.

There are a wide range of potential applications of HEA techniques (see Kirwan, 1992, for an overview). In a process plant, the various operating modes include normal operating conditions, maintenance, plant disturbances and emergencies. After carrying out a task analysis to define the worker's role in these areas, error analysis can be used to identify possible human failures with significant consequences and to specify appropriate hardware procedures, training, and other aspects of design to prevent their occurrence.

The other main application area for predictive error analysis is in chemical process quantitative risk assessment (CPQRA) as a means of identifying human errors with significant risk consequences. In most cases, the generation of error modes in CPQRA is a somewhat unsystematic process, since it only considers errors that involve the failure to perform some pre-specified function, usually in an emergency (e.g., responding to an alarm within a time interval). The fact that errors of commission can arise as a result of diagnostic failures, or that poor interface design or procedures can also induce errors is rarely considered as part of CPQRA. However, this may be due to the fact that HEA techniques are not widely known in the chemical industry. The application of error analysis in CPQRA will be discussed further in Chapter 5.

Error analysis techniques can be used in accident analysis to identify the events and contributory factors that led to an accident, to represent this information in a clear and simple manner and to suggest suitable error reduction strategies. This is achieved in practice by identification of the causal event sequence that led to the accident and the analysis of this sequence to identify the root causes of the system malfunction. A discussion of accident analysis techniques is included in Chapter 6.

4.4.1. Predictive Human Error Analysis (PHEA)

Predictive human error analysis can be performed manually or by means of a computer software package. Three types of analysis are possible within PHEA.

- **Preconditioned plan analysis:** This addresses errors in the planning of the task or ensuring that the correct preconditions apply.
- **Embedded plan analysis:** This considers errors arising from the plan specified in the HTA (e.g., ignoring the condition in the plan which specifies how the steps should be executed).
- **Task element analysis:** This aspect of the procedure systematically identifies a range of errors (e.g., failing to close a valve, closing the wrong valve) that could arise at each step of the task.

For the purposes of this description the focus will be on the task element analysis. The analysis procedure proceeds through a number of stages:

Task Element Selection
If the whole task being analyzed has already been identified as being highly critical, then it may be necessary to subject every step to a PHEA. However, in most cases only those steps which have a high risk potential if errors occur will be examined in detail. Procedures for identifying critical tasks are described in Chapter 5.

Detailed Analysis
The whole range of error types that could occur at each task step are described in Figure 4.16. The terms *action errors* and *checking errors* are self-explanatory. Retrieval errors refer to the retrieval of information either from an external source (e.g., a chart recorder or a procedure) or from memory. Transmission/ communication errors refer to communications among individuals either directly or via written communications. Selection/choice errors refer to making incorrect choices among alternative operations, for example, manual instead of automatic.

For each subset of task steps that have been defined, the analyst first asks if any steps in the group involve any of the activities implied by the error categories, for example, action, checking, communication etc. If an activity does not occur within the task steps being considered, then this is not considered further at this stage. This enables groups of task steps to be eliminated at an early stage of the analysis, to reduce the number of questions that need to be asked later.

At this stage of the technique, it is necessary for the analyst to make a general assessment of any error-inducing conditions due to poor PIFs in the situation under consideration, to determine if these are likely to give rise to any of the errors that will be considered at the next stage of the analysis. Typical error-inducing conditions such as poor procedures, time stress, inadequate interface design, have already been considered in Chapter 3.

The analyst then decides, for each step if any of the error modes from the complete error classification given in Figure 4.16 are possible. For example:

For task step 12.1:   Open valve V17
       Is it possible that the action could be omitted?
       Is it possible that it may not be opened fully?

Action Errors
A1   Action too long/short
A2   Action mistimed
A3   Action in wrong direction
A4   Action too little/too much
A5   Misalign
A6   Right action on wrong object
A7   Wrong action on right objec
A8   Action omitte
A9   Action incomplete
A10   Wrong action on wrong object

Checking Errors
C1   Checking omitted
C2   Check incomplete
C3   Right check on wrong object
C4   Wrong check on right object
C5   Check mistimed
C6   Wrong check on wrong object

Retrieval Errors
R1   Information not obtained
R2   Wrong information obtained
R3   Information retrieval incomplete

Transmission Errors
T1   Information not transmitted
T2   Wrong information transmitted
T3   Information transmission incomplete

Selection Errors
S1   Selection omitted
S2   Wrong selection made

Plan Errors
P1   Plan preconditions ignored
P2   Incorrect plan executed

FIGURE 4.16.   **Error Classification used in Predictive Error Analysis**

The answers to these questions are clearly dependent on the quality of the PIFs in the situation under consideration, for example, labeling or procedures. The consequences of the error, the factors that will support recovery of the error before the consequences occur, and the error prevention strategies will all be considered during the analysis.

Documentation
Figure 4.17 shows a useful format for documenting the results of error analysis. This is based on the HTA in Figure 4.2. For every critical error (e.g., action omitted) the implications or consequences for the system and the possibilities

of error recovery are described in the same format. This facilitates the development of design or other solutions to prevent the error.

Applications of the Technique

The exhaustive nature of the technique means that it is well suited to the analysis of critical systems where it is essential that all credible error modes are identified. For this reason it is useful as a means of generating error modes for inclusion in CPQRA analyses.

For the purpose of procedures design, the technique can be used to identify errors with significant consequences at particular task steps. Warnings can be included at these steps to alert the worker to the consequences of errors. If the predicted errors have severe consequences and high likelihood of occurrence, then equipment redesign might be indicated. Error analysis also provides an input to training, in that it indicates the aspects of the job which require particular attention during training. The advantages and disadvantages of the PHEA can be summed up as follows:

Advantages

1. The technique is rigorous and exhaustive and hence is likely to ensure that most errors are identified.
2. A validation study of the technique showed that it was capable of predicting a high proportion (98%) of errors with serious consequences that actually occurred in an equipment calibration task over a 5-year period (Murgatroyd and Tait, 1987).

TASK STEP	TASK TYPE	ERROR TYPE	DESCRIPTION	CONSEQUENCES	RECOVERY	ERROR REDUCTION STRATEGY
1.1 Move set point to measured value	Action	Action Omitted	Set point left at original value	System may operate at wrong set point. Process hazard may occur (Moderate)	Noticeable change of value of variable may occur at step 1.2	Introduce check in checklist
	Action	Right action on wrong object	Set point changed on wrong controller	Same as above	Same as above	Clearly label controllers to distinguish among set point controls
	Action	Wrong action on right object	Controller set to wrong value	Same as above	Same as above	Introduce check in checklist

FIGURE 4.17. **Documentation of the Results of Human Error Analysis**

3. It provides a standardized procedure to ensure consistency among analysts. This was tested by carrying out two independent evaluations of the same task. Of the 60 errors identified in the above validation study, 70% were common to both analysts. Of the remainder, 11 differences were due to differences in knowledge of the equipment by the two analysts and 5 were due to different interpretations of the procedures.
4. The method provides an explicit link with the results of task analysis.
5. Some aspects of cognitive errors, that is, planning errors, can be addressed.

Disadvantages

1. The method requires a substantial investment of time and effort if there are a large number of task steps to be analyzed.
2. The success of the method requires a detailed knowledge of the task being evaluated. Time has to be invested to acquire this knowledge.
3. The user of the technique needs to be trained to correctly interpret the questions.
4. A separate evaluation of PIFs needs to be performed in order to predict which error types are likely.

4.4.2. Work Analysis

This is a technique developed by Petersen and Rasmussen. The full documentation of the technique is extensive and only an outline can be provided here. Full details are available in Petersen (1985). The major steps in performing work analysis are as follows:

Analyze the Task Element Sequence:

(a) Define task elements that cannot be omitted or changed without affecting the probability that the goal will be achieved.
(b) Define alternative routes (i.e., alternative plans/task elements) that could also achieve the goal.
(c) Subject each of these routes separately to the following analyses.

Analyze the Task Steps:

(a) Define the criteria for the overall success of the task or subtask under consideration.
(b) Define error recovery points, that is points in the sequence where previously committed errors have a high probability of recovery. This could be because there is considerable observable feedback, or because it would be physically difficult to proceed beyond that point given the occurrence of the earlier error(s).
(c) Define erroneous actions or action sequences for which detection is unlikely, reducing the likelihood of immediate error recovery.

(d) For these actions, identify error mechanisms (see flow charts in Appendix 2B) and resulting errors that could lead to an unacceptable (i.e., irrecoverable) effect on the task.

(e) Evaluate conditions for error detection and recovery at the points identified in (b). Identify errors that will render recovery mechanisms unsuccessful.

(f) Apply quantitative human reliability assessment techniques to evaluate the total task reliability, given the error modes and recovery paths identified in (d) and (e).

(g) If the error recovery probabilities at the point identified in (b) are assessed to be sufficiently high, ignore errors in the actions preceding these points.

(h) If not, repeat step (c) for these sequences [see (f) above].

Analyze Potential Coupled Failures

(a) Note the errors that could have an effect on systems other than those being worked upon (e.g., because they are in close physical proximity or are functionally coupled).

Analyze Effects of Task Disturbances

(a) Evaluate sources of disturbances. These could include unavailability of tools, instruments or personnel, equipment faults, or changes in work scheduling due to anticipated delays. The analysis should attempt to formally categorize the different problems that could occur.

(b) Assess the effects of unavailability of tools, equipment, personnel etc., for each of the task steps not covered by recovery and for the error recovery path assessed.

(c) Assess the likely improvisations that could occur if the disturbances considered under (b) occurred.

(d) For the improvised task sequence identified under (c), repeat the analyses described in the first three sections.

Advantages

- The technique provides a very exhaustive analysis of errors in both normal and disturbed conditions.
- Error recovery is explicitly analyzed.
- The effects of task disturbances are explicitly covered.

Disadvantages

- Because of the depth of analysis involved the technique is very resource intensive.

4.5. ERGONOMICS CHECKLISTS

4.5.1. Application of the Technique

Another method of predicting and reducing human error in the CPI is through the use of ergonomics checklists. These can be used by an engineer to ascertain whether various factors which influence performance of a task meet particular ergonomic criteria and codes of good practice. Items within the checklist can include the design and layout of the control panel, the labeling and location of equipment, the usability of the operating procedures, aspects of training and team communications as well as other PIFs which have been examined in Chapter 3. By applying the checklist several times on different aspects of a CPI task, the engineer can identify work conditions that can induce human error and subsequently specify error reduction strategies. Checklists can be used either retrospectively to audit an existing system or proactively to design a new system.

Although checklists are a useful way of transferring information about human–machine interaction to designers and engineers, they are not a stand-alone tool and they cannot provide a substitute for a systematic design process. The main concern with checklists is that they do not offer any guidance about the relative importance of various items that do not comply with the recommendations, and the likely consequences of a failure due to a noncompliance. To overcome such problems, checklists should be used in combination with other methods of task analysis or error analysis that can identify the complexities of a task, the relationships among various job components, and the required skills to perform the task.

4.5.2. Examples of Checklists

There are several checklists in existence that focus on different aspects of human–machine interaction. Some are intended to assess the overall design of the plant while others focus on more specific issues such as the design of the control panel, the dialogue between operator and VDU interfaces, and the usability of procedures and other job-aids. Depending on the scope of application, the items within a checklist can vary from overall subjective opinions, for example, "have operators been given adequate training in fault-diagnostic skills?" to very specific objective checks, for example, "is the information presented on the screen clear and is contrast in the range of 1 to 5–10?" On many occasions it is necessary to expand or modify an existing checklist to ensure that other standards or codes of practice are being met.

There are many checklists that can be used to identify error-inducing conditions and ensure conformance with particular ergonomic standards, and the following examples illustrate the range of areas covered.

Short Guide to Reducing Human Error in Process Operation
(United Kingdom Atomic Energy Authority, 1987)
This guide is arranged as a checklist of questions on the following five areas of system design that impact upon plant safety: worker–process interface, procedures, workplace and working environment, training, and task design and job organization. The guide could be used in developing new plant designs or making changes to existing plant, auditing existing arrangements or investigating causes of incidents. The list of questions is intended to assess either the overall plant design or the reliability of performing a particular task. Table 4.1 provides an extract from this guide for the evaluation of operating procedures.

The guide is described as a "short guide" because it draws attention to general problems only. A more detailed guide (*The Long Guide*) which provides full explanations for each checklist item is also available (United Kingdom Atomic Energy Authority, 1991).

CRT Display Checklist (Blackman et al., 1983)
This checklist presents criteria for comparing different ways of presenting information on CRT displays.

VDU Checklist (Cakir et al., 1980)
This checklist presents detailed information for assessing VDU terminals and their workplaces. The items concern technical information about VDU characteristics but they do not directly consider the nature of the task performed using the VDU system.

Principles of Interface Design for Computer Controlled Processes
(Bellamy and Geyer, 1988)
This is a list of ergonomic considerations that should be taken into account in the interface design of computer controlled processes. The principles refer essentially to monitoring and control tasks, and they have been derived from a literature review supplemented by the analysis of a number of incidents.
Advantages
- Checklists are quick and easy to apply. The answers to the questions in the checklist provide insights into remedial strategies.

Disadvantages
- Checklists do not provide any assistance to assess the relative importance of different items or to indicate the degree to which items may fail to meet the criteria. Thus, there is a need to undertake some prioritization of checklist failures, in order to avoid misinterpretation of the information.

TABLE 4.1

A Checklist on Procedures Extracted from the "Short Guide to Reducing Human Error" (UK Atomic Energy Authority, 1987)

Concise procedures
There should be no ambiguity about when procedures are to be used.
- Are the procedures available when required?
- Are the conditions in which the procedures must be used clear and unambiguous?
- Is there a simple unambiguous indexing method for choosing the required procedure?

Mandatory procedures
When procedures are mandatory, there should be no incentive to use other methods.
- Are procedures and manually operated safety interlocks sufficiently simple to use?
- Are there no easier, but more dangerous alternatives?
- Is there a convenient area of the workplace for using the procedural documentation?
- Are the documentary procedures routinely checked, compared with operator action and revised as appropriate?

Supporting procedures
Procedures should where possible support the worker's skills and discretion rather than replace them.
- Are the procedures and worker's skills complementary?
- Where the workers are skilled and experienced, and an absolutely standard sequence is not necessary, the procedures should be in the form of reminder checklists with guidance on priorities, rather than detailed instructions.

Correct operational procedures
Procedures should be easy to understand and follow.
- Can the instructions be easily understood and followed, particularly by a person who is unfamiliar with them?
- Is there a mechanism for keeping place in a sequence of instructions, so that it can be returned to after an interruption or distraction?
- Where two or more procedures share a common sequence of operations, or working environment, do they contain checks that the worker is continuing to use the correct procedure?
- Does a different person subsequently make an independent check that mandatory procedures have been carried out?
- Can emergency procedures be implemented whether or not the worker knows what is wrong?

- Checklists generally take no account of the context in which the tasks are carried out. Some form of task analysis or error analysis may also be required to gain an insight into the overall task context.
- Checklist are one-dimensional, and do not provide any guidance with regard to the reasons for the questions.

It is also important that the analyst should take some time to become familiar with the task prior to undertaking the checklist survey, otherwise a considerable amount of time will be devoted to discovering the background of the task rather than assessing the checklist items.

4.6. SUMMARY

The intention of this chapter has been to provide an overview of analytical methods for predicting and reducing human error in CPI tasks. The data collection methods and ergonomics checklists are useful in generating operational data about the characteristics of the task, the skills and experience required, and the interaction between the worker and the task. Task analysis methods organize these data into a coherent description or representation of the objectives and work methods required to carry out the task. This task description is subsequently utilized in human error analysis methods to examine the possible errors that can occur during a task.

The focus of this chapter has been on proactive application of these analytical methods such as safety audits, development of procedures, training needs analysis, and equipment design. However, many of these methods can also be used in a retrospective mode, and this issue deserves further attention in its own right. Chapter 6 describes analytical methods for accident investigations and data collection.

5

Qualitative and Quantitative Prediction of Human Error in Risk Assessment

5.1. INTRODUCTION

There is an increasing requirement by regulatory authorities for companies to conduct formal safety assessments of hydrocarbon and chemical process plants. As part of these assessments, risk and reliability analysts are required to perform evaluations of human reliability in addition to the analyses of hardware systems, which are the primary focus of a typical safety assessment (see Bridges et al., 1994, for techniques for including human error considerations in hazard analyses). Emphasis is being placed by regulators on a comprehensive assessment of the human role in system safety following the occurrence of major disasters in the petrochemical industry (Piper Alpha, Feyzin, Bhopal, Texas City) where human errors were implicated as direct or indirect causes (see CCPS, 1989b, 1992d for further examples).

The usual emphasis in human reliability has been on techniques for the derivation of numerical error probabilities for use in fault trees (see Kirwan et al., 1988, for a comprehensive review of these techniques). However, in many ways, this emphasis on absolute quantification is misplaced. Many practitioners emphasize the fact that the major benefits of applying a formal and systematic technique to risk assessment are the qualitative insights that emerge with regard to the sources of risk, and where resources should be expended in minimizing these risks. Although the quantitative results of the assessment are important in arriving at decisions in specific areas, for example the siting of on-shore plants with potentially hazardous processes, it is widely recognized that there are considerable uncertainties in the data available for inclusion in these analyses.

Given these uncertainties, it becomes even more important that a systematic and comprehensive qualitative method is adopted for identifying the sources of risk and the consequences of failures. Such a procedure must ensure that no significant failures are omitted from the analysis. A comprehensive evaluation of the plant from the perspective of its management, procedures, training, communication, and other systemic factors also provides insights into how generic failure data should be modified for use in the particular risk assessment of interest. The main focus of this chapter is the description of a defensible procedure for qualitative human error prediction that will achieve these objectives.

In addition, the chapter will provide an overview of human reliability quantification techniques, and the relationship between these techniques and qualitative modeling. The chapter will also describe how human reliability is integrated into chemical process quantitative risk assessment (CPQRA). Both qualitative and quantitative techniques will be integrated within a framework called SPEAR (System for Predictive Error Analysis and Reduction).

5.2. THE ROLE OF HUMAN RELIABILITY IN RISK ASSESSMENT

5.2.1. An Illustrative Case Study

Although the main emphasis of this chapter will be on qualitative human reliability methods in risk assessment, this section will illustrate the importance of both qualitative and quantitative methods in CPQRA. An example of a typical assessment, described by Ozog (1985) will be considered. The stages of the risk assessment are as follows:

System Description
The system is a storage tank designed to hold a flammable liquid under a low positive nitrogen pressure (see Figure 5.1). This pressure is controlled by PICA-1. A relief valve is fitted which operates if overpressurization occurs. Liquid is fed to the tank from a tank truck, and is subsequently supplied to the process by the pump P-1.

Hazard Identification
A hazard and operability study (HAZOP) was used to identify potential hazards, the most serious of which is an unrecoverable release from the storage tank.

Construction of the Fault Tree
The fault tree is constructed based on the system description and initiating events identified in the HAZOP. Figure 5.2 shows a portion of an extended version of Ozog's fault tree, taken from CCPS (1989b). The following terminology is used:

B	is a Basic or Undeveloped event
M	is an Intermediate event
T	is the Top event

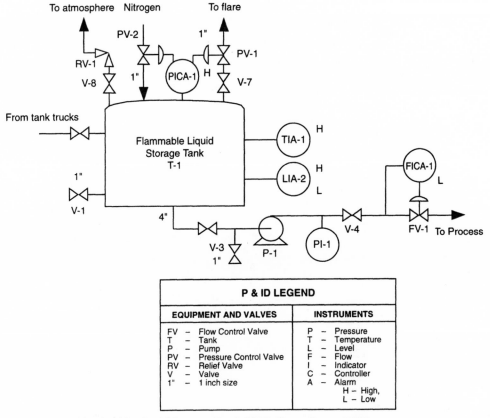

FIGURE 5.1 **Flammable Liquid Storage Tank P&ID (from Ozog, 1985).**

The events that could give rise to the major flammable release are as follows:

M1:	Spill during tank unloading
M2:	Tank rupture due to external event
B1:	Tank drain breaks
M3:	Tank rupture due to implosion (not shown)
M4:	Tank rupture due to overpressure (not shown)

Quantification
The overall frequency of the top event is calculated by combining together the constituent probabilities and frequencies of the various events in the fault tree using the appropriate logical relationships described by the AND and OR gates (the detailed calculation is given in CCPS, 1989b).

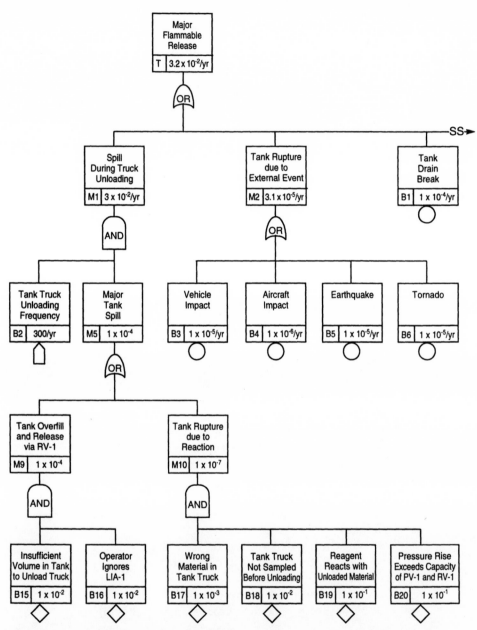

FIGURE 5.2 **Fault tree Analysis of Flammable Liquid Storage Tank (from Ozog, 1985).**

5.2.2. Implications of Human Error for the Analysis

From a human reliability perspective, a number of interesting points arise from this example. A simple calculation shows that the frequency of a major release (3.2×10^{-2} per year) is dominated by human errors. The major contribution to this frequency is the frequency of a spill during truck unloading (3×10^{-2} per year). An examination of the fault tree for this event shows that this frequency is dominated by event B15: Insufficient volume in tank to unload truck, and B16: Failure of, or ignoring LIA-1. Of these events, B15 could be due to a prior human error, and B16 would be a combination of instrument failure and human error. (Note however, that we are not necessarily assigning the causes of the errors solely to the operator. The role of management influences on error will be discussed later.) Apart from the dominant sequence discussed above, human-caused failures are likely to occur throughout the fault tree. It is usually the case that human error dominates a risk assessment, if it is properly considered in the analysis. This is illustrated in Bellamy et al. (1986) with an example from the analysis of an offshore lifeboat system.

These examples suggest that it is critical for the potential human causes of major incidents to be exhaustively identified. Unfortunately, the tools currently used by risk analysts for hazard identification do not adequately address this issue. A commonly used method is the HAZOP approach (Kletz, 1992, CCPS, 1992b) as shown in Figure 5.3. Some of the causes of process deviations generated by a HAZOP analysis may actually be ascribed to human error. However, the team doing the analysis is given no explicit guidance within the HAZOP (or any other hazard identification technique) that would enable them to identify human causes of these process deviations. Although it can be argued that the knowledge and experience of the analyst concerning the system should be sufficient to identify human errors, it is obviously preferable to have a systematic procedure that will ensure a comprehensive identification of possible causes, even if the analyst does not know the system well.

Another danger of an inadequate appreciation of human causes of hazards is that the HAZOP analyst may consider a particular high risk event (identified by a guide word and deviation) to be noncredible, because he or she only takes into account the hardware failures (with an extremely low probability) that could give rise to the event. When human causes are taken into account, the likelihood of the event may actually be quite high.

The framework to be described later in this chapter can be seen as a complementary procedure to hardware orientated hazard identification procedures. Ideally, the two approaches should be applied in parallel to a plant evaluation, in order to benefit from the synergy of considering both perspectives.

PROCESS UNIT: DAP PRODUCTION Node: <u>1</u> Process Parameter: <u>Flow</u>				
GUIDE WORD	**DEVIATION**	**CONSEQUENCES**	**CAUSES**	**SUGGESTED ACTION**
No	No Flow	Excess ammonia in reactor. Release to work area.	1. Valve A fails closed. 2. Phosphoric acid supply exhausted. 3. Plug in pipe; pipe ruptures.	Automatic closure of valve B on loss of flow from phosphoric acid supply
Less	Less Flow	Excess ammonia in reactor. Release to work area, with amount released related to quantitative reduction in supply. Team member to calculate toxicity vs. flow reduction.	1. Valve A partially closed. 2. Partial plug or leak in pipe.	Automatic closure of valve B on reduced flow from phosphoric acid supply. Set point determined by toxicity vs. flow calculation.
More	More Flow	Excess phosphoric acid degrades product. No hazard in work area.	—	—
Part of	Normal flow of decreased concentration of phosphoric acid	Excess ammonia in reactor. Release to work area, with amount released related to quantitative reduction in supply.	1. Vendor delivers wrong material or concentration. 2. Error in charging phosphoric acid supply tank.	Check phosphoric acid supply tank concentration after charging.

FIGURE 5.3. **Sample of HAZOP Worksheet (CCPS, 1985).**

5.2.3. Quantification Aspects

In the preceding section, the importance of a comprehensive human reliability modeling approach has been emphasized from the qualitative perspective. However, such an approach is also critical in order to ensure accurate quantification of risk. If significant human contributors to the likelihood of major accidents occurring are omitted, then the probability of the event occurring may be seriously underestimated. Conversely, the role of the human in enhancing the reliability of a system needs to be taken into account. One reason for including humans in engineered systems is that they have the capability to respond to situations that have not been anticipated by the designers of the system. For example, they can prevent an undesirable outcome (e.g., the major flammable release in the situation described earlier) by taking appropriate action at an early stage in the event.

These two points can be illustrated in the fault tree in Figure 5.2. Taking the branch dealing with the frequency of the spill during truck unloading (event M1 and below), a comprehensive analysis might have revealed that other human errors could give rise to a major tank spill (event M5) in addition to events M9 and M10. For example, an evaluation of the procedures during unloading might indicate that V1 could be accidentally opened instead of the valve from the tank truck (because of similar appearance of the valves, poor labeling and unclear procedures). If this probability was deemed to be high (e.g., 1×10^{-3}) on the basis of the evaluation of the operational conditions, then this event would dominate the analysis. M5 would become about 1.1×10^{-3} and the frequency of the flammable release T would become about 3.2×10^{-1} per year (approximately one release every 3 years) which would be totally unacceptable.

Although risk assessment usually concentrates on the negative effects of the human in the system, the operator also has the capability to reduce risk by recovering from hardware failures or earlier errors. This can be taken into account in the assessment. Consider the scenario where the operator will detect the escape of liquid through the relief valve as soon as overfilling has occurred, and immediately close the valve to the tank truck. (It is assumed that the alternative error of accidentally opening V1, as discussed above, will not occur.) Although it is still likely that some spillage would occur, this would probably not constitute a major tank spill. If the recovery action is given a conservative failure probability of 1×10^{-2} and joined by an AND gate to events B15 and B16, then the probability of M9 and M5 becomes 1×10^{-6}. This considerably reduces the overall frequency of a major flammable release (T) to 3.2×10^{-4}.

The analysis set out above demonstrates the importance of a comprehensive evaluation of the human aspects of a hazardous operation, from the point of view of identifying all contributory events and recovery possibilities. It also indicates the need for a complete evaluation of the operational conditions (procedures, training, manning levels, labeling, etc.) which could impact on these probabilities.

5.3. SYSTEM FOR PREDICTIVE ERROR ANALYSIS AND REDUCTION (SPEAR)

The SPEAR framework to be described in subsequent sections is designed to be used either as a stand-alone methodology, to provide an evaluation of the human sources of risk in a plant, or in conjunction with hardware orientated analyses to provide an overall system safety assessment. The overall structure of the framework is set out in Figure 5.4.

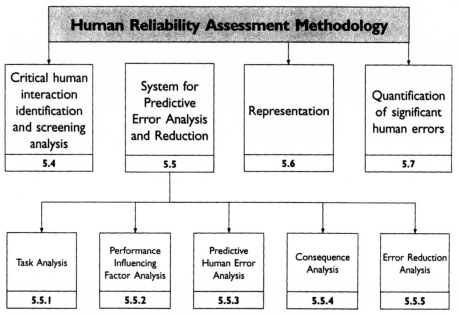

FIGURE 5.4. **System for Predictive Error Analysis and Reduction.**

Critical Human Interaction Identification and Screening (Stage 1)

The process involves identifying and describing human interactions with the system which will have major impact on risk if errors occur. A human interaction can in some cases comprise a single operation, for example, closing a valve or detecting a temperature increase. Usually, however, a human interaction will consist of a task directed at achieving a particular system objective, for example starting up a reactor or responding correctly in an emergency. Human interactions are obviously not confined to operational situations. They may also be involved in maintenance and plant changes. Errors, in these operations, can give rise to latent failures.

Qualitative Analysis of Human Errors (Stage 2)

This stage involves the prediction of errors that could arise on the basis of performance-influencing factors (PIFs) which exist in the situation, the nature of the human interaction with the system (e.g., actions, checking, communication), and the models of error discussed in Chapter 2. Only if human errors are identified that may have significant consequences (loss of life, plant damage, major production loss) will the subsequent stages of the process be performed. This stage therefore includes a consequence analysis, together with an error reduction analysis.

Representation (Stage 3)

This stage involves representing the structure of the tasks in which errors with severe consequences could occur, in a manner that allows the probabilities of these consequences to be generated. The usual forms of representation are event trees and fault trees.

Quantification (Stage 4)

The quantification process involves assigning numerical probabilities or frequencies to the errors (or error recovery opportunities) that have been identified during the preceding stages. Following the quantification process, the error probabilities will be combined with the hardware analyses to allow an overall measure of risk to be calculated. If this expected level of risk is unacceptable, then changes will be made in the human or hardware systems to reduce it (see Figure 5.5). In the case of human errors this may involve consideration of alternative strategies on the basis of cost-effectiveness considerations.

5.4. CRITICAL TASK IDENTIFICATION AND SCREENING ANALYSIS

The purpose of the Critical Task Identification and Screening analysis is to reduce the amount of analysis required by focusing on tasks that have a significant error potential. The screening process essentially asks the following questions:

Is there a hazard present in the area of the plant (e.g., a reactor, or a complete process unit) being considered?

In this context the term *hazard* is taken to mean "potential to cause harm," and would include any substance or plant item with characteristics such as toxicity, flammability, high voltage, mechanical energy, or asphyxiation potential.

Given that there is a hazard present, are there any human interactions with the plant that could cause the harm potential to be released?

Interactions refers to any jobs, tasks, or operations carried out by people who could directly or indirectly cause the hazard to be released. Direct interactions with the plant might involve breaking open pipework, opening reactors, etc. Indirect interactions would include remote activation of valves from a control room, or the performance of maintenance on critical plant items. Errors that might occur during these interactions could allow the harm potential to be released. This could occur directly (for example, a worker could be overcome by a chlorine release if an incorrect valve line-up was made) or indirectly (for example, if a pump bearing in a critical cooling circuit was not lubricated, as in the example in Chapter 1). The procedure as described above

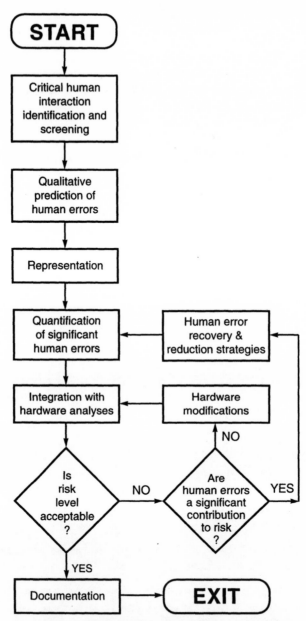

FIGURE 5.5. **Relationship of SPEAR to Human Reliability Assessment Methodology**

210

is analogous to the process performed for hardware failures in a typical HAZOP (see CCPS, 1992b).

Information on the types of human interactions with hazardous systems that occur would be obtained from sources such as plant operating instructions, job safety analyses and similar sources. These interactions are referred to as **critical tasks (CT)**.

Given that workers interact with hazardous systems, how frequently are they likely to make errors in these critical tasks?

The answer to this question will depend on two factors: the frequency with which the CT occur, and the likelihood of errors arising when performing these tasks. The frequency of the interactions can usually be specified relatively easily by reference to plant procedures, production plans, and maintenance schedules. The probability of error will be a function of the PIFs discussed extensively in Chapter 3 and other chapters in this book. In order to obtain a measure of error potential, it is necessary to make an assessment of the most important PIFs for each of the CT.

In summary, at the screening stage of the SPEAR process, the ranking of tasks in order of potential risk is made on the basis of three criteria:

- The known or hazard severity potential (HSP) that is present in the systems with which the worker is interacting
- The extent to which the nature of the task could allow the hazard to cause harm to workers, the public or the environment (hazard release potential., HRP)
- The frequency (F) with which the task is performed

If these functions are each rated from 1 to 5, a scale of task criticallity can be generated ranging from 0 to 1 as follows:

$$\text{Task Criticality Index (TCI)} = [(\text{HP} \times \text{HSP} \times \text{F}) - 1]/124$$

Each task can then be assessed on this basis to produce a ranking of risk potential. Only those tasks above a predetermined level of the TCI will be subjected to a detailed analysis.

5.5. QUALITATIVE HUMAN ERROR ANALYSIS

Qualitative human error prediction is the most important aspect of assessing and reducing the human contribution to risk. For this reason, it will be described in some detail in this section. The qualitative analysis performed in SPEAR involves the following techniques:

- Task analysis
- Performance-influencing factor analysis

- Predictive human error analysis
- Consequence analysis
- Error reduction analysis

Many of these techniques have been described in Chapter 4. They will be illustrated in this chapter with reference to a simple example, the loading of a chlorine tanker.

5.5.1. Task Analysis

As discussed in Chapter 4, task analysis is a very general term that encompasses a wide variety of techniques. In this context, the objective of task analysis is to provide a systematic and comprehensive description of the task structure and to give insights into how errors can arise. The structure produced by task analysis is combined with the results of the PIF analysis as part of the error prediction process.

The particular type of task analysis used in this example is hierarchical task analysis (HTA) (see Chapter 4). This has the advantage that it has been applied extensively in the chemical and other industries. As described in Chapter 4, HTA breaks down the overall objective of a task by successively describing it in increasing detail, to whatever level of description is required by the analysis. At each of the levels, a "plan" is produced that describes how the steps or functions at that level are to be executed.

Figure 5.6 shows an extract from the HTA of the chlorine tanker filling operation which will be used as an example. The first level (numbered 1, 2, 3, etc.) indicates the tasks that have to be carried out to achieve the overall objective. These tasks are then broken down to a further level of detail as required. As well as illustrating the hierarchical nature of the analysis, Figure 5.6 shows that plans, such as those associated with operation 3.2, can be quite complex. The term *operation* is used to indicate a task, subtask, or task step, depending on the level of detail of the analysis.

A practical advantage of HTA compared with other techniques is that it allows the analysis to proceed to whatever level of detail is appropriate. At each level, the questions can be asked "could an error with serious consequences occur during this operation?" If the answer to this question is definitely no, then it is not necessary to proceed with a more detailed analysis.

5.5.2. Performance Influencing Factor Analysis

During this stage of the qualitative analysis, a PIF analysis is performed that considers those factors which will determine the probability of error for the type of task under consideration. A structured form of PIF analysis such as the HFA tool described in Section 2.7.2 will facilitate this process.

---

0. Fill tanker with chlorine
Plan: Do tasks 1 to 5 in order.

1. Park tanker and check documents (not analyzed)

2. Prepare tanker for filling
Plan: Do 2.1 or 2.2 in any order then do 2.3 to 2.5 in order.
2.1 Verify tanker is empty
Plan: Do in order.
    2.1.1 Open test valve
    2.1.2 Test for Cl_2
    2.1.3 Close test valve
2.2 Check weight of tanker
2.3 Enter tanker target weight
2.4 Prepare fill line
Plan: Do in order.
    2.4.1 Vent and purge line
    2.4.2 Ensure main Cl_2 valve closed
2.5 Connect main Cl_2 fill line

3. Initiate and monitor tanker filling operation
Plan: Do in order.
3.1 Initiate filling operation
Plan: Do in order.
    3.1.1 Open supply line valves
    3.1.2 Ensure tanker is filling with chlorine
3.2 Monitor tanker filling operation
Plan: Do 3.2.1, do 3.2.2 every 20 minutes. On initial weight alarm, do 3.2.3 and 3.2.4. On final weight alarm, do 3.2.5 and 3.2.6.

3.2.1 Remain within earshot while tanker is filling
3.2.2 Check road tanker
3.2.3 Attend tanker during last 2–3 ton filling
3.2.4 Cancel initial weight alarm and remain at controls
3.2.5 Cancel final weight alarm
3.2.6 Close supply valve A when target weight reached

4. Terminate filling and release tanker
4.1 Stop filling operation
Plan: Do in order.
    4.1.1 Close supply valve B
    4.1.2 Clear lines
    4.1.3 Close tanker valve
4.2 Disconnect tanker
Plan: Repeat 4.2.1 five times then do 4.2.2 to 4.2.4 in order.
    4.2.1 Vent and purge lines
    4.2.2 Remove instrument air from valves
    4.2.3 Secure blocking device on valves
    4.2.4 Break tanker connections
4.3 Store hoses
4.4 Secure tanker
Plan: Do in order.
    4.4.1 Check valves for leakage
    4.4.2 Secure locking nuts
    4.4.3 Close and secure dome
4.5 Secure panel (not analyzed)

5. Document and report (not analyzed)

FIGURE 5.6. **Chlorine Tanker Task Analysis.**

5.5.3. Predictive Human Error Analysis

Predictive human error analysis (PHEA) is the process via which specific errors associated with tasks or task steps are predicted. The process also considers how these predicted errors might be recovered before they have negative consequences. The inputs to the process are the task structure and plans, as defined by the task analysis, and the results of the PIF analysis. The basic procedure of the PHEA is as follows:

5.5.3.1. Decide on the Level of Detail to Conduct Analysis

The hierarchical structure of the HTA allows errors to be predicted at a variety of different levels. For example, consider Section 2 of the HTA in Figure 5.6. The subtask: **Prepare tanker for filling** requires subtasks 2.1 to 2.5 to be performed. There are a number of ways in which these subtasks could fail to be performed correctly **at this level**. For example subtasks 2.3 to 2.5 could be carried out in the wrong order. If there were multiple tankers, 2.1: **verify tanker is empty** could be carried out on the wrong tanker. It should be noted that this analysis may be quite independent of an analysis at the next lower level, where individual task steps would be analyzed.

5.5.3.2. Perform Planning Error Analysis

The failure to perform the operations required at the particular level of the HTA being analyzed could occur because of deficiencies in the plan. The categories of plan failure are shown in Figure 5.7.

If the procedures were not regularly updated or were otherwise incorrect, or if training was inadequate, P1 errors could occur. P2 errors would often arise as a result of misdiagnosing a situation, or if the entry conditions for executing a sequence of operations were ambiguous or difficult to assess and therefore the wrong procedure was selected. It is important to note that if a planning error occurs, then this implies that a detailed analysis needs to be conducted of the alternative course of action that could arise.

5.5.3.3. Perform Operation Error Analysis

This analysis is applied to each operation at the particular level of the HTA being evaluated. In most cases the analysis is performed at the level of a step, for example, **Open valve 27B**. For each operation, the analyst considers the likelihood that one or more of the error types set out in classification in Figure 5.7 could occur. This decision is made on the basis of the information supplied by the PIF analysis, and the analyst's knowledge concerning the types of error likely to arise given the nature of the mental and physical demands of the task and the particular configuration of PIFs that exist in the situation. The different error categories are described in more detail below:

Operation Errors

Operation errors are errors associated with one or more actions that change the state of the system, for example, steps such as open valve A, secure blocking device. These errors can also apply at the level of whole tasks, for example, disconnect or secure tanker (tasks 4.2 and 4.4 in Figure 5.6).

Action		Retrieval	
A1	Action too long / short	R1	Information not obtained
A2	Action mistimed	R2	Wrong information obtained
A3	Action in wrong direction	R3	Information retrieval incomplete
A4	Action too little / too much		
A5	Misalign	**Transmission**	
A6	Right action on wrong object	T1	Information not transmitted
A7	Wrong action on right object	T2	Wrong information transmitted
A8	Action omitted	T3	Information transmission incomplete
A9	Action incomplete		
A10	Wrong action on wrong object	**Selection**	
		S1	Selection omitted
Checking		S2	Wrong selection made
C1	Checking omitted		
C2	Check incomplete	**Plan**	
C3	Right check on wrong object	P1	Plan preconditions ignored
C4	Wrong check on right object	P2	Incorrect plan executed
C5	Check mistimed		
C6	Wrong check on wrong object		

FIGURE 5.7 **Error Classification.**

Checking Errors

These are errors such as failing to perform a required check, which will usually involve a data acquisition process such as verifying a level or state by visual inspection, rather than an action.

Retrieval Errors

These are concerned with retrieving information from memory (e.g., the time required for a reactor to fill), or from a visual display or a procedure.

Communication or Transmission Errors

These errors are concerned with the transfer of information among people, either directly or via written documents such as permit systems. These errors are particularly pertinent in situations where a number of people in a team have to coordinate their activities.

Selection Errors

These are errors that occur in situations where the operator has to make an explicit choice among alternatives. These may be physical objects (e.g., valves, information displays) or courses of action. It should be emphasized that the categorization of errors in Figure 5.7 is generic, and may need to be modified for specific industries.

The first stage of the operation error analysis is to determine if any of the error categories in Figure 5.7 apply to the task, subtask, or task step being analyzed. For example, at the level of individual task steps, operations would be actions performed at each step. If a particular step (e.g., checking a level in a sight glass), did not actually involve actions, then it would not be necessary to consider this category of errors further. The appropriate category in this case would be checking errors. Other applicable categories are retrieval, communication, or selection errors.

Once certain categories of error have been ruled out, the analyst decides whether or not any of the errors in the remaining applicable categories could occur within the task, subtask, or task step being evaluated.

5.5.3.4. *Perform Recovery Analysis*

Once errors have been identified, the analyst then decides if they are likely to be recovered before a significant consequence occurs. Consideration of the structure of the task (e.g., whether or not there is immediate feedback if an error occurs) together with the results of the PIF analysis, will usually indicate if recovery is likely.

5.5.4. Consequence Analysis

The objective of consequence analysis is to evaluate the safety (or quality) consequences to the system of any human errors that may occur. Consequence Analysis obviously impacts on the overall risk assessment within which the human reliability analysis is embedded. In order to address this issue, it is necessary to consider the nature of the consequences of human error in more detail.

At least three types of consequences are possible if a human error occurs in a task sequence:

- The overall objective of the task is not achieved.
- In addition to the task not achieving its intended objective, some other negative consequence occurs.
- The task achieves its intended objective but some other negative consequence occurs (either immediate or latent), which may be associated with some other system unrelated to the primary task.

Generally, risk assessment has focused on the first type of error, since the main interest in human reliability was in the context of human actions that were **required** as part of an emergency response. However, a comprehensive Consequence Analysis has to also consider other types, since both of these outcomes could constitute sources of risk to the individual or the plant.

One example of a particularly hazardous type of consequence in the second category is where, because of misdiagnosis, the operator performs some alternative task other than that required by the system. For example, a rise of pressure in a reactor may be interpreted as being the result of a blockage in an output line, which would lead to attempts to clear the line. If, instead, it was due to impurities causing an exothermic reaction, then failure to attend to the real cause could lead to an overpressurization accident. With regard to the third category, the operator may achieve the final required objective by a route that has an impact on another part of the process. For example, pipework may be connected in such a way that although the main task succeeds, an accident may occur when another process is started that uses the same pipework.

5.5.5. Error Reduction Analysis

For those errors with significant consequences where recovery is unlikely, the qualitative analysis concludes with a consideration of error reduction strategies that will reduce the likelihood of these errors to an acceptable level. These strategies can be inferred directly from the results of the PIF analysis, since this indicates the deficiencies in the situation which need to be remedied to reduce the error potential.

5.5.6. Case Study Illustrating Qualitative Analysis Methods in SPEAR

This example illustrates the qualitative aspects of SPEAR, using the chlorine tanker loading case study as a basis.

5.5.6.1. Select Task Steps on the Basis of Screening Analysis
The task analysis is performed on tasks 2, 3, and 4. Tasks 1 and 5 were eliminated from the analysis because they did not involve any direct exposure to hazardous substances (from the initial screening analysis described in Section 2.1). The analysis considers operations 2.1 to 2.5, 3.1 to 3.2 and 4.1 to 4.5 in Figure 5.6.

5.5.6.2. Perform Task Analysis
The task analysis is shown in Figure 5.6.

5.5.6.3. Perform PIF analysis
For the purpose of this example, it will be assumed that the PIFs which influence performance in all tasks are identical, that is,

- Time stress score (score 7, ideal value 1)
- Experience /training of operators score (score 8, ideal value 9)
- Level of distractions score (score 7, ideal value 1)
- Quality of procedures /checklists (score 5, ideal value 9)

These PIFs represent the major factors deemed by the analyst to influence error probability for the operations (coupling hoses, opening and closing valves) and planning activities being carried out within the tasks analyzed at this level. In practice, the analyst would need to consider if different types of PIFs applied to the different tasks 2, 3, and 4.

The numbers appended to the PIFs represent numerical assessments of the quality of the PIFs (on a scale of 1 to 9) across all task steps being evaluated. The ratings indicate that there are negative influences of high time stress and high levels of distractions. These are compensated for by good training and moderate (industry average) procedures. Again, in some cases, these ratings could differ for the different tasks. For example, the operator may be highly trained for the types of operations in some tasks but not for others. It should be noted that as some factors increase from 1 to 9, they have a negative effect on performance (time stress and level of distractions), whereas for the other factors, an increase would imply improved performance (quality of procedures and experience/training).

5.5.6.4. Perform Detailed Predictive Human Error Analysis (PHEA)
A selection of the results of the PHEA is shown in Figure 5.8 for task elements 2.3, 3.2.2, 3.2.3, and 3.2.5. The possible errors are predicted by considering all the possible error types in Figure 5.7 for each element. Planning errors are not included in Figure 5.8, but would be predicted using the appropriate planning error category. Possible error recovery routes are also shown in Figure 5.8.

5.5.6.5. Evaluate Consequences
Consequence analyses are set out in Figure 5.8.

5.5.6.6. Error Reduction Analysis
Figure 5.9 illustrates some of the possible error reduction strategies available. Apart from the specific strategies set out in Figure 5.9, the PIF analysis also indicates which PIFs should be modified to reduce the likelihood of error. In the case of the chlorine loading example, the major scope for improvements are the reduction of time stress and distractions and the development of better quality procedures.

The error reduction analysis concludes one complete cycle of the qualitative human error analysis component of the methodology set out in Figure 5.4. The analyst then decides if it is appropriate to perform a more detailed analysis on any of the operations considered at the current level. As a result of this process, operations 3.2: **Monitor tanker following operation**, 4.1: **Stop filling operation**, 4.2: **Disconnect tanker**, and 4.4: **Secure tanker** are analyzed in more detail (see Figure 5.6).

The qualitative human error analysis stages described above are applied to the task steps in subtask 3.2. Examples of the results of this analysis are shown in Figure 5.8. The corresponding error-reduction strategies are shown in Figure 5.9.

STEP	ERROR TYPE	ERROR DESCRIPTION	RECOVERY	CONSEQUENCES AND COMMENTS
2.3 Enter tanker target weight	Wrong information obtained (R2)	Wrong weight entered	On check	Alarm does not sound before tanker overfills
3.2.2 Check tanker while filling	Check omitted (C1)	Tanker not monitored while filling	On initial weight alarm	Alarm will alert operator if correctly set. Equipment fault, e.g.,leaks not detected early and remedial action delayed
3.2.3 Attend tanker during last 2–3 ton filling	Operation omitted (O8)	Operator fails to attend	On step 3.2.5	If alarm not detected within 10 minutes tanker will overfill
3.2.5 Cancel final weight alarm	Operation omitted (O8)	Final weight alarm taken as initial weight alarm	No recovery	Tanker overfills
4.1.3 Close tanker valve	Operation omitted (O8)	Tanker valve not closed	4.2.1	Failure to close tanker valve would result in pressure not being detected during the pressure check in 4.2.1
4.2.1 Vent and purge lines	Operation omitted (O8) Operation incomplete (O9)	Lines not fully purged	4.2.4	Failure of operator to detect pressure in lines could lead to leak when tanker connections broken
4.4.2 Secure locking nuts	Operation omitted (O8)	Locking nuts left unsecured	None	Failure to secure locking nuts could result in leakage during transportation

FIGURE 5.8 **Results of Predictive Human Error Analysis.**

5.6. REPRESENTATION

If the results of the qualitative analysis are to be used as a starting-point for quantification, they need to be represented in an appropriate form. The form of representation can be a fault tree, as shown in Figure 5.2, or an event tree (see Bellamy et al., 1986). The event tree has traditionally been used to model simple tasks at the level of individual task steps, for example in the THERP (Technique for Human Error Rate Prediction) method for human reliability

| | ERROR REDUCTION RECOMMENDATIONS | | |
STEP	PROCEDURES	TRAINING	EQUIPMENT
2.3 **Enter tanker** target weight	Independent validation of target weight.	Ensure operator double checks entered date. Recording of values in checklist	Automatic setting of weight alarms from unladen weight. Computerize logging system and build in checks on tanker reg. no. and unladen weight linked to warning system. Display differences between unladen and current weights
3.2.2 Check Road Tanker while filling	Provide secondary task involving other personnel. Supervisor periodically checks operation	Stress importance of regular checks for safety	Provide automatic log-in procedure
3.2.3 Attend tanker during filling of last 2–3 tons (on weight alarm)	Ensure work schedule allows operator to do this without pressure	Illustrate consequences of not attending	Repeat alarm in secondary area. Automatic interlock to terminate loading if alarm not acknowledged. Visual indication of alarm.
3.2.5 Cancel final weight alarm	Note differences between the sound of the two alarms in checklist	Alert operators during training about differences in sounds of alarms	Use completely different tones for initial and final weight alarms
4.1.3 Close tanker valve	Independent check on action. Use checklist	Ensure operator is aware of consequences of failure	Valve position indicator would reduce probability of error
4.2.1 Vent and purge lines	Procedure to indicate how to check if fully purged	Ensure training covers symptoms of pressure in line	Line pressure indicator at controls. Interlock device on line pressure.
4.4.2 Secure locking nuts	Use checklist	Stress safety implication of training	Locking nuts to give tactile feedback when secure

FIGURE 5.9. **Error Reduction Recommendations Based on PHEA**

assessment, Swain and Guttmann (1983) (see Section 5.7.2.1). It is most appropriate for sequences of task steps where few side effects are likely to occur as a result of errors, or when the likelihood of error at each step of the sequence is dependent on previous steps.

Figure 5.10 shows a detailed fault tree for an offshore drilling operation. The top event of the fault tree is **Failure to use shear rams to prevent blowout.** As with the fault tree in Figure 5.2, the representation combines both hardware

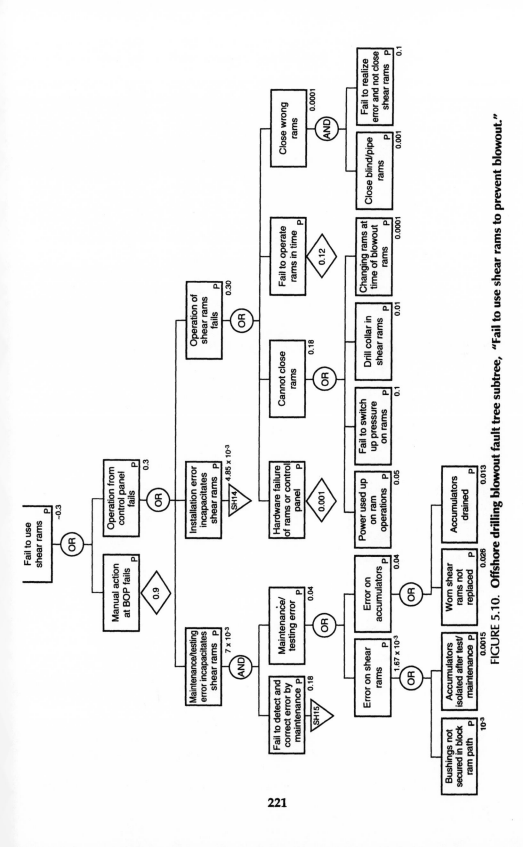

FIGURE 5.10. Offshore drilling blowout fault tree subtree, "Fail to use shear rams to prevent blowout."

221

and human failures. Figure 5.11 is an event tree representation of operator actions involved in an offshore emergency shutdown scenario (Kirwan, 1990). This type of event tree is called an operator action event tree (OAET) because it specifically addresses the sequence of actions required by some initiating event. Each branch in the tree represents success (the upper branch) or failure (the lower branch) to achieve the required human actions described along the top of the diagram. The probability of each failure state to the right of the diagram is the product of the error and/or success probabilities at each node of branch that leads to the state. The overall probability of failure is given by summing the probabilities of all the failure states. The dotted lines indicate recovery paths from earlier failures.

In numerical terms, the probability of each failure state is given by the following expressions (where SP is the success probability and HEP the human error probability at each node):

F1 = [SP 1.1 + HEP 1.1 × SP 1.2] × SP 1.3 × SP 1.5 × SP 1.6 × SP 1.7 × HEP 1.8
F2 = [SP 1.1 + HEP 1.1 × SP 1.2] × SP 1.3 × SP 1.5 × SP 1.6 × HEP 1.7
F3 = [SP 1.1 + HEP 1.1 × SP 1.2] × SP 1.3 × SP 1.5 × HEP 1.6
F4 = [SP 1.1 + HEP 1.1 × SP 1.2] × SP 1.3 × HEP 1.5
F5 = [SP 1.1 + HEP 1.1 × SP 1.2] × HEP 1.3 × HEP 1.4
F6 = HEP 1.1 × HEP 1.2

Total failure probability T is given by

$$T = F1 + F2 + F3 + F4 + F5 + F6$$

Further details about fault tree and event tree applications in quantitative risk assessment (QRA) are given in CCPS (1989b).

5.7. QUANTIFICATION

Because most research effort in the human reliability domain has focused on the quantification of error probabilities, a large number of techniques exist. However, a relatively small number of these techniques have actually been applied in practical risk assessments, and even fewer have been used in the CPI. For this reason, in this section only three techniques will be described in detail. More extensive reviews are available from other sources (e.g., Kirwan et al., 1988; Kirwan, 1990; Meister, 1984). Following a brief description of each technique, a case study will be provided to illustrate the application of the technique in practice. As emphasized in the early part of this chapter, quantification has to be preceded by a rigorous qualitative analysis in order to ensure that all errors with significant consequences are identified. If the qualitative analysis is incomplete, then quantification will be inaccurate. It is also important to be aware of the limitations of the accuracy of the data generally available

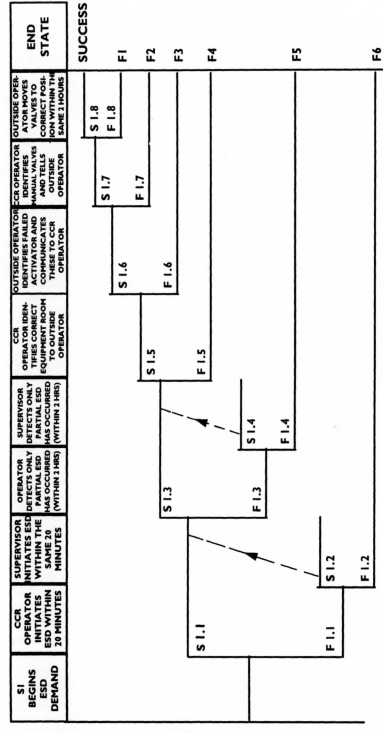

FIGURE 5.11. Operator Action Tree for ESD Failure Scenario (Kirwan, 1990).

223

for human reliability quantification. This issue is discussed in more detail in Chapter 6.

5.7.1. The Quantification Process

All quantification techniques follow the same four basic stages:

5.7.1.1. Modeling the Task

This involves analyzing the task of interest and identifying which aspects should be quantified. In some cases, the analyst will be interested in a probability for a discrete human action, for example, "what is the likelihood that the control room operator will close the feed supply valve within 30 seconds of an alarm?"

In other cases, the interest will be in quantifying a complete task, for example, "What is the probability that a lifeboat will be successfully launched?" In this case, quantification can be carried out at the global level of the whole task, or the task can be broken down to task elements, each of which is quantified (the decomposition approach). The overall probability of success or failure for the whole task is then derived by combining the individual task elements in some way.

Quantification at a global task level is essentially the same process as with a single discrete operation. A single probability is assigned without explicit reference to the internal structure of the task. There are arguments for and against both the global and the decomposition approach. The advantages of the decomposition approach are as follows:

- It can utilize any databases of task element probabilities that may be available.
- Recovery from errors in individual task steps can be modeled.
- Consequences to other systems arising from failures in individual task steps (e.g., the results of **alternative** actions as opposed to simply omitted actions) can be modeled and included in the assessment.
- Effects of dependencies among task steps can be modeled.

Advocates of the global approach would argue that human activities are essentially goal-directed (the cognitive view expressed in Chapter 2), and that this cannot be captured by a simple decomposition of a task into its elements. They also state that if an intention is correct (on the basis of an appropriate diagnosis of a situation), then errors of omission in skill-based actions are unlikely, because feedback will constantly provide a comparison between the expected and actual results of the task. From this perspective, the focus would be on the reliability of the cognitive rather than the action elements of the task.

On the whole, most quantification exercises have employed the decomposition approach, partly because most engineers are more comfortable with the analysis and synthesis approach, and partly because of the rather mechanistic model of human performance that has been the basis for most work in human reliability assessment.

5.7.1.2. Representing the Failure Model

The decomposition approach is used, it is necessary to represent the way in which the various task elements and other possible failures are combined to give the failure probability of the task as a whole. Generally, the most common form of representation is the event tree (see Section 5.7). This is the basis for THERP, which will be described in the next section. Fault trees are only used when discrete human error probabilities are combined with hardware failure probabilities in applications such as CPQRA (see Figure 5.2).

5.7.1.3. Deriving Error Probabilities for Task Steps

Error probabilities that are used in decomposition approaches are all derived in basically the same manner. Some explicit or implicit form of task classification is used to derive categories of tasks in the domain addressed by the technique. For example, typical THERP categories are selections of switches from control panels, walk-around inspections, responding to alarms and operating valves.

A basic error probability is then assigned to tasks in each category or subcategory. This probability may be derived from expert judgment or empirical data. It usually represents the error likelihood under "average" conditions. This probability is then modified by specifying a set of factors which tailor the baseline probability to the specific characteristics of the situation being assessed. Thus, a baseline probability of, say, 10^{-3} for the probability of correctly operating a valve under normal conditions may be degraded to 10^{-1} under the effects of high stress.

5.7.1.4. Combining Task Element Probabilities to Give Overall Task Failure Probabilities

During the final stage of the decomposition approach, the task element probabilities in the event tree are combined together using the rules described in Section 5.3.3 to give the overall task failure probability. At this stage, various corrections for dependencies among task elements may be applied.

5.7.2. Quantitative Techniques

To illustrate contrasting approaches to quantification, the following techniques will be described in detail in subsequent sections:

THERP	Techniques for human error rate prediction
SLIM	Success likelihood index method
IDA	Influence diagram approach

These techniques were chosen because they illustrate contrasting approaches to quantification.

5.7.2.1. Technique for Human Error Rate Prediction (THERP)

History and Technical Basis
This technique is the longest established of all the human reliability quantification methods. It was developed by Dr. A. D. Swain in the late 1960s, originally in the context of military applications. It was subsequently developed further in the nuclear power industry. A comprehensive description of the method and the database used in its application, is contained in Swain and Guttmann (1983). Further developments are described in Swain (1987). The THERP approach is probably the most widely applied quantification technique. This is due to the fact that it provides its own database and uses methods such as event trees which are readily familiar to the engineering risk analyst. The most extensive application of THERP has been in nuclear power, but it has also been used in the military, chemical processing, transport, and other industries.

The technical basis of the THERP technique is identical to the event tree methodology employed in CPQRA. The basic level of analysis in THERP is the task, which is made up of elementary steps such as closing valves, operating switches and checking. THERP predominantly addresses action errors in well structured tasks that can be broken down to the level of the data contained in the THERP Handbook (Swain and Guttmann, 1983). Cognitive errors such as misdiagnosis are evaluated by means of a time–reliability curve, which relates the time allowed for a diagnosis to the probability of misdiagnosis.

Stages in Applying the Technique
PROBLEM DEFINITION. This is achieved through plant visits and discussions with risk analysts. In the usual application of THERP, the scenarios of interest are defined by the hardware orientated risk analyst, who would specify critical tasks (such as performing emergency actions) in scenarios such as major fires or gas releases. Thus, the analysis is usually driven by the needs of the hardware assessment to consider specific human errors in predefined, potentially high-risk scenarios. This is in contrast to the qualitative error prediction methodology described in Section 5.5, where all interactions by the operator with critical systems are considered from the point of view of their risk potential.

QUALITATIVE ERROR PREDICTION. The first stage of quantitative prediction is a task analysis. THERP is usually applied at the level of specific tasks and the steps within these tasks. The form of task analysis used therefore focuses on the operations which would be the lowest level of a hierarchical task analysis such as that shown in Figure 5.6. The qualitative analysis is much less formalized than that described in Section 5.5. The main types of error considered are as follows:

- Errors of omission (omit step or entire task)
- Errors of commission
- Selection error
  —selects wrong control
  —mispositions control
  —issues wrong command
- Sequence error (action carried out in wrong order)
- Time error (too early / too late)
- Quantitative error (too little / too much)

The analyst also records opportunities to recover errors, and various performance shaping factors (called performance-influencing factors in this book) which will subsequently be needed as part of the quantification process.

REPRESENTATION. Having identified the errors that could occur in the execution of the task, these are then represented in the form of an event tree (Figure 5.12). This event tree is taken from Swain and Guttmann (1983). The branches of the tree to the left represent success, and to the right, failures. Although the event tree in Figure 5.12 is quite simple, complex tasks can generate very elaborate event trees. Error recovery is represented by a dotted line as in the event tree shown in Figure 5.11.

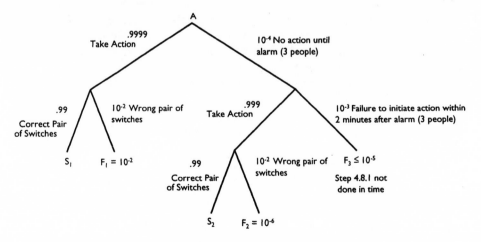

$$F_T = F_1 + F_2 + F_3 \approx 10^{-2}$$

FIGURE 5.12. **THERP Event Tree (Swain and Guttman, 1983).**

QUANTIFICATION. Quantification is carried out in the THERP event tree as follows:

- Define the errors in the event tree for which data are required. In Figure 5.12, these errors are:
  —No action taken until alarm (action omitted)
  —Failure to initiate action within 2 minutes of alarm
  —Wrong pair of switches chosen
- Select appropriate data tables in Swain and Guttmann (1983). This handbook contains a large number of tables giving error probabilities for operations commonly found in control rooms or plants, for example, selecting a switch from a number of similar switches. Because the handbook was originally written for the nuclear industry, the data reflect the types of operations frequently found in that industry. The source of these data is not defined in detail by the authors, although it appears to be partly based on the American Institute for Research human error database (Munger et al., 1962) together with plant data extrapolated and modified by the authors' experience.
- Modify the basic data according to guidelines provided in the handbook, to reflect differences in the assumed "nominal" conditions and the specific conditions for the task being evaluated. The major factor that is taken in to account is the level of stress perceived by the operator when performing the task.
- Modify the value obtained from the previous stage to reflect possible dependencies among error probabilities assigned to individual steps in the task being evaluated. A dependence model is provided which allows for levels of dependence from complete dependence to independence to be modeled. Dependence could occur if one error affected the probability of subsequent errors, for example if the total time available to perform the task was reduced.
- Combine the modified probabilities to give the overall error probabilities for the task. The combination rules for obtaining the overall error probabilities follow the same addition and multiplication processes as for standard event trees (see last section).

INTEGRATION WITH HARDWARE ANALYSIS. The error probabilities obtained from the quantification procedure are incorporated in the overall system fault trees and event trees.

ERROR REDUCTION STRATEGIES. If the error probability calculated by the above procedures leads to an unacceptable overall system failure probability, then the analyst will reexamine the event trees to determine if any PIFs can be modified or task structures changed to reduce the error probabilities to an acceptable level.

5.7.2.2. THERP Case Study

The case study that follows is reproduced with permission from the Chemical Manufacturers Association publication *Improving Human Performance in the Chemical Industry: A Manager's Guide*, Lorenzo (1990). Another CPI case study that uses THERP is documented in Banks and Wells (1992).

Assume that the system described below exists in a process unit recently purchased by your company. As the manager, the safety of this unit is now your responsibility. You are concerned because your process hazard analysis team identified the potential for an operator error to result in a rupture of the propane condenser. You have commissioned a human reliability analysis (HRA) to estimate the likelihood of the condenser rupturing as the result of such an error and to identify ways to reduce the expected frequency of such ruptures

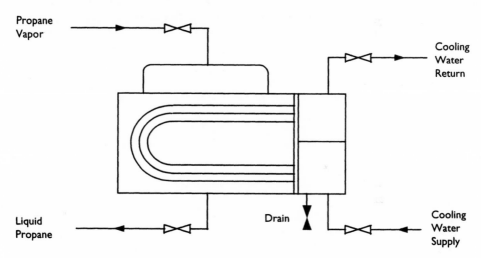

FIGURE 5.13. **Propane Condenser Schematic (Lorenzo, 1990).**

System Description

Four parallel propane condensers, one of which is illustrated in Figure 5.13, are designed with a 450-psig shell pressure rating and a 125-psig tube pressure rating. The propane vapor pressure is controlled at 400 psig; the cooling water flowing through the condenser tubes is normally maintained at 75 psig. Liquid propane flows out of the condenser as soon as it condenses; there is no significant inventory of liquid propane in the condenser. The two propane isolation valves for each condenser are rising-stem gate valves with no labels. The two water isolation valves for each condenser are butterfly valves with no labels. Their handwheel actuators have position indicators.

A tube has failed in one of the four condensers about once every three years. If a condenser tube fails, the affected condenser can be removed from service by closing four isolation valves (propane vapor inlet valve), liquid propane outlet valve, cooling water supply valve, and cooling water return valve). However, if a tube fails, it is essential that the operator close the two propane isolation valves before closing the two water isolation valves. Closing the two water valves first would allow pressure to build on the tube side of the condenser and rupture the tube head.

Analyzed System Conditions
- A tube has failed in the condenser.
- The low depropanizer pressure alarm has sounded in the control room.
- The experienced field operator has observed water and gas being emitted from the hydrocarbon vent at the cooling tower. The field operator shouts over the radio that a propane vapor cloud appears to be forming and moving towards the control room.
- The control room operator has directed the field operator to isolate the failed condenser as quickly as possible so that a unit shutdown will not be necessary.
- The operator must close the valves by hand. If a valve sticks, there is no time to go get tools to help close the valve—the process must be shut down.
- The field operator has correctly identified the condenser with the failed tube by the sound of the expanding propane and the visible condensation/frost on the shell.

Qualitative HRA Results
The first step of the analysis is to identify the human actions and equipment failures that can lead to the failure of interest. An HRA event tree (Figure 5.14) is then constructed to depict the potential human errors (represented by capital English letters) and the potential equipment failures (represented by capital Greek letters). The series of events that will lead to the failure of interest is identified by an F_1 at the end of the last branch of the event tree. All other outcomes are considered successes even though the propane release is not isolated in outcomes S_2 and S_3, so the process must be shut down.

Inspection of the HRA event tree reveals that the dominant human error is Error A: the operator failing to isolate the propane valves first. The other potential human errors are factors only if a propane isolation valve sticks open. Based on these qualitative results alone, a manager might decide to periodically train operators on the proper procedure for isolating a failed condenser and to ensure that operators are aware of the potential hazards. The manager might

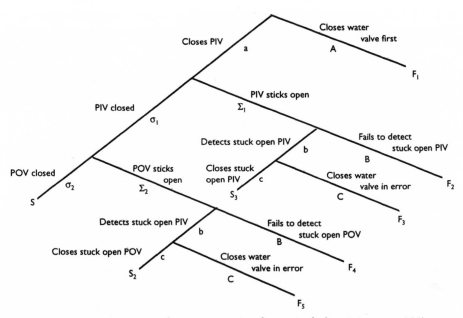

FIGURE 5.14. **HRA Event Tree for Improper Condenser Isolation (Lorenzo, 1990).**

also decide to require regular preventive maintenance on the propane isolation valves to help ensure that they will properly close when required.

Quantitative HRA Results
This manager requested quantitative results, so the analyst must estimate the
probability of each failure or error included in the event tree. Data for all the
failures and errors in this particular problem are available in tables in the
Handbook, Swain and Guttman (1983). The analyst must modify these data as
necessary to account for specific characteristics of the work situation, such as
stress levels, equipment design features, and interoperator dependencies.
Table 5.1 summarizes the data used in this problem.

There is a written procedure for condenser isolation, but it is normally a
simple step-by-step task that is second nature to the operator and is performed
from memory. However, under the threat of a potential vapor cloud explosion,
the operator may forget to close the propane valves first (Error A). The HEP
in Handbook *Table 20-7 #5 footnote (.01) is increased by a factor of 5 per* Handbook
Table 20-16 #6a to account for stress.

The probability of a valve sticking open is unaffected by the operator's
stress level, but the probability of the operator failing to detect the stuck valve
(Error B) is increased. The HEP in *Handbook* Table 20-14 #3 is increased by a
factor of 5 per *Handbook* Table 20-16 #6a.

TABLE 5.1
Events Included in the HRA Event Tree (Lorenzo, 1990)

FAILURE SYMBOL	FAILURE DESCRIPTION	ESTIMATED RROBABILITY	DATA SOURCE
A	Operator fails to close the propane valves first	.05	T20-7 #5 footnote × 5, per T20-16 #6a
Σ_1	Propane inlet valve sticks open	.001	T20-14 footnote
Σ_1	Propane outlet valve sticks open	.001	T20-14 footnote
B	Operator fails to detect a stuck valve	.025	T20-14 #3 × 5, per T20-16 #6a
C	Operator chooses to close the cooling water valves to stop the propane release	.25	T20-16 #7a

The third potential human error (Error C) is that the operator will decide to close the cooling water valves even though he/she diagnoses that a propane valve is not closed. The likelihood of such an error (a dynamic decision in a threatening situation) is listed as 0.25 in *Handbook* Table 20-16 #7a.

The analyst can then calculate the total probability of failure (F_T) by summing the probability of all failure paths (F_{1-5}). The probability of a specific path is calculated by multiplying the probabilities of each success and failure limb in that path. (*Note:* The probabilities of success and failure sum to 1.0 for each branch point. For example, the probability of Error B is 0.025 and the probability of Success b is 0.975.) Table 5.2 summarizes the calculations of the HRA results, which are normally rounded to one significant digit after the intermediate calculations are completed.

TABLE 5.2
Human Reliability Analysis Results (Lorenzo, 1990)

$F_1 = A$	$= 5.0 \times 10^{-2}$
$F_2 = a\Sigma1B$	$= 2.4 \times 10^{-5}$
$F_3 = a\Sigma1bC$	$= 2.3 \times 10^{-4}$
$F_4 = a\sigma1\Sigma2B$	$= 2.4 \times 10^{-5}$
$F_5 = a\sigma1\Sigma2bC$	$= 2.3 \times 10^{-4}$
$F_T = F_1 + \cdots + F_5$	$= .05$

Finally, the HRA analyst would calculate the expected frequency of condenser ruptures as a result of improper isolation. The frequency of condenser tube failures is 0.33 per year (1 every 3 years), and the calculated probability of improper isolation is 0.05. Multiplying these two numbers shows the expected frequency of improper isolation of a failed condenser is 0.017 per year, or about once every 60 years. The manager can use this number to help compare the costs and benefits of improvements proposed as a result of the HRA or other studies.

For example, the same process hazards review team that spurred the manager's original concern might have suggested (1) installing a pressure relief device on the tube side of the exchanger, or (2) removing the propane isolation valves (which would require that the unit be shut down in the event of a condenser tube failure). In addition, the HRA team may have suggested (3) increasing operator training and (4) more frequent maintenance of the propane isolation valves. Based on the quantitative HRA results and estimates of the consequences of a condenser rupture, the manager can decide whether the benefits of the proposed changes outweigh their costs. The manager can then choose the best way to apply loss prevention resources.

5.7.2.3. The Success Likelihood Index Method (SLIM)
History and Technical Basis
The SLIM technique is described in detail in Embrey et al. (1984) and Kirwan (1990). The technique was originally developed with the support of the U.S. Nuclear Regulatory Commission but, as with THERP, it has subsequently been used in the chemical, transport, and other industries. The technique is intended to be applied to tasks at any level of detail. Thus, in terms of the HTA in Figure 5.6, errors could be quantified at the level of whole tasks, subtasks, task steps of even individual errors associated with task steps. This flexibility makes it particularly useful in the context of task analysis methods such as HTA.

The basic premise of the SLIM technique is that the probability of error associated with a task, subtask, task step, or individual error is a function of the PIFs in the situation. As indicated in Chapter 3, an extremely large number of PIFs could potentially impact on the likelihood of error. Normally the PIFs that are considered in SLIM analyses are the direct influences on error such as levels of training, quality of procedures, distraction level, degree of feedback from the task, level of motivation, etc. However, in principle, there is no reason why higher level influences such as management policies should not also be incorporated in SLIM analyses.

In the SLIM procedure, tasks are numerically rated on the PIFs which influence the probability of error, and these ratings are combined for each task to give an index called the success likelihood index (SLI). This index is then converted to a probability by means of a general relationship between the SLI and error probability which is developed using tasks with known probabilities and SLIs. These are known as calibration tasks.

Stages in Applying the Technique
PROBLEM DEFINITION, QUALITATIVE ERROR PREDICTION AND REPRESENTATION. The recommended problem definition and qualitative error prediction approach for use with SLIM has been described in Section 5.3.1 and 5.3.2. The fact that PIFs are explicitly assessed as part of this approach to qualitative error prediction means that a large proportion of the data requirements for SLIM are already available prior to quantification. SLIM usually quantifies tasks at whatever level calibration data are available, that is, it does not need to perform quantification by combining together task element probabilities from a data base. SLIM can therefore be used for the global quantification of tasks. Task elements quantified by SLIM may also be combined together using event trees similar to those used in THERP.

QUANTIFICATION PROCEDURE. In order to illustrate the SLIM quantification method, the case study developed in the earlier part of the chapter based on the chlorine tanker filling example will be used. The following operations from Figure 5.6 will be used to illustrate the method.

2.1.3   Close test valve
4.1.3   Close tanker valve
4.4.2   Secure locking nuts
4.2.3   Secure blocking device on valves

- **Form groups of homogenous operations.**
The first stage is to group together operations that are likely to be influenced by the same PIFs. The four operations in the above set all involve physical actions for which there is no immediate feedback when incorrectly performed. Two of the operations, 4.1.3 and 4.4.2 are noted in Figure 5.8 as having significant consequences if they occur. It is legitimate to assume therefore, that the error probability will be determined by the same set of PIFs for all the operations in this set.

- **Decide on the relevant PIFs.**
Ideally, data bases will have been developed within a company such that predetermined PIFs are associated with particular categories of task. If this is not the case, the analyst decides on a suitable set of PIFs. In this example, it is assumed that the main PIFs which determine the likelihood of error are time stress, level of experience, level of distractions, and quality of procedures. (See Section 5.3.2.6.)

- **Rate each operation on each PIF.**
A numerical rating on a scale of 1 to 9 is made for each operation on each PIF. Normally the ends of the scale represent the best or worst PIF conditions.

For example, a high level of time stress would be represented by a rating of 9, which would imply an increased level of errors. However, in the case of level of experience, 9 would represent the optimal rating corresponding to a highly experienced operator. The fact that the same rating value can have a different significance with different PIFs needs to be taken into account by the analyst. With the computer program that is available for the SLIM technique, Embrey (1994), these adjustments are made automatically. The ratings shown in Table 5.3 are made for the operations.

These ratings can be interpreted as follows. In the case of the Time Stress PIF, all the operations have a high level of time stress, apart from **close test valve**, where stress is low. The operators are very experienced in carrying out all the tasks. Distractions are moderately high for **close test valve**, but otherwise low. Procedures are poor for **secure locking nuts** and **secure blocking device**, but above average for the other two tasks.

- **Assign weights if appropriate**

Based on the analyst's experience, or upon error theory, it is possible to assign weights to the various PIFs to represent the relative influence that each PIF has on all the tasks in the set being evaluated. In this example it is assumed that in general the level of experience has the least influence on these types of errors, and time stress the most influence. The relative effects of the different PIFs can be expressed by the following weights:

Time Stress	0.4
Distractions	0.3
Procedures	0.2
Experience	0.1

It should be noted that the analyst should only assign weights if he or she has real knowledge or evidence that the weights are appropriate. The assignment of weights is not mandatory in SLIM. If weights are not used, the technique assumes that all PIFs are of equal importance in contributing to the overall likelihood of success or failure.

TABLE 5.3
PIF Ratings

OPERATION	TIME STRESS	EXPERIENCE	DISTRACTIONS	PROCEDURES
Close test valve	4	8	7	6
Close tanker valve	8	8	5	6
Secure locking nuts	8	7	4	2
Secure blocking device	8	8	4	2

TABLE 5.4
Rescaled Ratings and SLIs

OPERATIONS	PIFs				SLIs
	TIME STRESS	EXPERIENCE	DISTRACTIONS	PROCEDURES	
Close test valve	0.63	0.88	0.25	0.63	0.54
Close tanker valve	0.13	0.88	0.50	0.63	0.41
Secure locking nuts	0.13	0.75	0.63	0.13	0.34
Secure blocking device	0.13	0.88	0.63	0.13	0.35
Weights	0.4	0.1	0.3	0.2	

- **Calculate the Success Likelihood Indices**

The SLI is given by the following expression:

$$SLI_j = \sum R_{ij} W_i$$

where SLI_j is the SLI for task j; W_i is the normalized importance weight for the ith PIF (weights sum to 1); and R_{ij} is the rating of task on the ith PIF. The SLI for each task is the weighted sum of the ratings for each task on each PIF.

In order to calculate the SLIs, the data in Table 5.3 have to be rescaled to take into account the fact that the some of the ideal points are at different ends of the rating scales. Rescaling also converts the range of the ratings from 1 to 9 to 0 to 1. The following formula converts the original ratings to rescaled ratings:

$$RR = [1 - ABS\ (R - IP)]/[4 + ABS\ (5 - IP)]$$

where RR is the rescaled rating; R is the original rating, and IP is the ideal value for scale on which the rating is made.

The accuracy of this formula can be verified by substituting the values 1 and 9 for scales where the ideal point is either 1 or 9. The formula converts the original ratings to 0.0 or 1.0 as appropriate. Values of ratings between 1 and 9 are converted in the same way.

Using this formula on the ratings in Table 5.3 produces Table 5.4, which contains the rescaled ratings, the assigned weights for the PIFs and the calculated Success Likelihood Indices for each task.

- **Convert the Success Likelihood Indices to Probabilities**

The SLIs represent a measure of the likelihood that the operations will succeed or fail, relative to one another. In order to convert the SLI scale to a probability scale, it is necessary to calibrate it. If a reasonably large number of operations in the set being evaluated have known probabilities (for example, as a result of incident data having been collected over a long period of time), then it is possible to perform a regression analysis that will find the line of best fit between the SLI values and their corresponding error probabilities. The resulting regression equation can then be used to calculate the error probabilities for the other operations in the group by substituting the SLIs into the regression equation.

If, as is usually the case, there are insufficient data to allow the calculation of an empirical relationship between the SLIs and error probabilities, then a mathematical relationship has to be assumed. The usual form of the assumed relationship is log-linear, as shown below:

$$\log(\text{HEP}) = A \text{ SLI} + B \tag{1}$$

where HEP is the human error probability and A and B are constants

This assumption is based partly on experimental evidence that shows a log-linear relationship between the evaluation of the factors affecting performance on maintenance tasks, and actual performance on the tasks, Pontecorvo (1965). In order to calculate the constants A and B in the equation, at least two tasks with known SLIs and error probabilities must be available in the set of tasks being evaluated.

In the example under discussion, it is found that there were few recorded instances of the test valve being left open. On the other hand, locking nuts are often found to be loose when the tanker returns to the depot. On the basis of this evidence and the frequency that these operations are performed, the following probabilities were assigned to these errors:

Probability of test valve left open = 1×10^{-4}
Probability of locking nuts not secured = 1×10^{-2}

These values, and the corresponding SLIs for these tasks (from Table 5.4), are substituted in the general equation (1). The resulting simultaneous equations can be used to calculate the constants A and B. These are substituted in the general equation (1) to produce the following calibration equation:

$$\log(\text{HEP}) = -2.303 \text{ SLI} + 3.166 \tag{2}$$

If the SLI values from Table 5.4 for the other two tasks in the set are substituted in this equation, the resulting error probabilities are as follows:

Task A: Probability of not opening tanker valve = 1.8×10^{-3}
Task B: Probability of not securing blocking device = 7.5×10^{-3}

- **Perform Sensitivity Analysis**
  The nature of the SLIM technique renders it very suitable for "what if" analyses to investigate the effects of changing some of the PIF values on the

resulting error probabilities. For example, there are high levels of time stress for both of the above tasks (rating of time stress = 8, best value = 1). The effects of reducing time stress to more moderate levels can be investigated by assigning a rating of 5 for each task. This changes the SLI, and if the new SLI value is substituted in equation (2) the probabilities change as follows:

Task A: Probability of not opening tanker valve = 5.6×10^{-5}
Task B: Probability of not securing blocking device = 2.4×10^{-4}

An alternative intervention would be to make the procedures ideal (rating = 9). Changing the ratings for procedures to this value for each task (instead of reducing time stress) produces the following results.

Task A: Probability of not closing tanker valve = 3.2×10^{-4}
Task B: Probability of not securing blocking device = 1.3×10^{-4}

Thus the effect of making the procedures ideal is an order of magnitude greater for Task B compared with Task A (see Table 5.5). This is because the procedures for Task A were already highly rated at 6, whereas there was room for improvement with Task B which was rated 2 (see Table 5.3).

TABLE 5.5
Effects of Improvements in Procedures on Error Probabilities Calculated Using SLIM

	ORIGINAL ERROR PROBABILITY	AFTER IMPROVEMENTS IN PROCEDURES	RATIO BEFORE/ AFTER IMPROVEMENTS
Task A	1.8×10^{-3}	3.2×10^{-4}	5.6
Task B	7.5×10^{-3}	1.3×10^{-4}	57

Conclusions
The SLIM technique is a highly flexible method that allows considerable freedom in performing what-if analyses. In common with most human reliability quantification techniques, it requires defensible data, preferably from a plant environment, to be effective. In the absence of such data, the calibration values have to be generated by expert judgments made by experienced plant personnel.

5.7.2.4. The Influence Diagram Approach
History and Technical Basis. The influence diagram approach (IDA) (also known as the sociotechnical approach to human reliability (STAHR) (see Phillips et al., 1990) is a technique that is used to evaluate human error probabilities as a

function of the complex network of organizational and other influences that impact upon these probabilities. Unlike most other techniques, IDA is able to represent the effects of not only the direct influences of factors such as procedures, training, and equipment design on error likelihood but also the organizational influences and policy variables which affect these direct factors. As described in Phillips et al. (1990), it is possible to construct a generic Influence Diagram to represent these relationships. In the case study that will be used to illustrate the application of the influence diagram to human error probability evaluation, a more specific diagram (Figure 5.15) will be used, based on a study by Embrey (1992).

The basic steps in carrying out an IDA session are described in Phillips et al. (1990). A group of subject matter experts are assembled who have a detailed knowledge of the interactions between indirect and direct PIFs which determine error probability. The influence diagram is then constructed using insights from this expert group. Once the diagram has been developed, the experts are asked to assess the current state of the lowest level factors (i.e., project management and assignment of job roles in Figure 5.15). The assessment made is the probability (or "balance of evidence") that the factor being considered is positive or negative in its effects on error. This evaluation is performed on all the bottom level influences in the diagram, using scales similar to those used to evaluate PIFs described in Figure 3.1. Once these individual factors have been evaluated, based on an objective evaluation of the situation being assessed, the next stage is to evaluate the combined effects of the lowest level influences on higher level influences, as specified by the structure of the influence diagram.

This process is repeated for combinations of these variables on the factors that **directly** impact on the probability of success or failure for the scenario

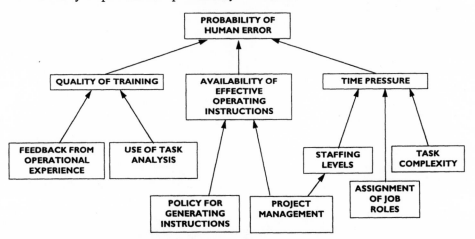

FIGURE 5.15 **Influence Diagram (Embrey, 1992).**

being evaluated. These numerical assessments are combined to give weights which are then used to modify unconditional probability estimates that the error will occur given various positive or negative combinations of the influences assessed earlier. The unconditional probability estimates have to be derived by another technique such as absolute probability judgment, SLIM, or from any field data that may be available.

Case Study

In the influence diagram for operator errors given in Figure 5.15, the main level 1 factors influencing the probability of error are quality of training, availability of effective operating instructions and time pressure on the operator. Two factors are specified as influencing the quality of training. These are the extent to which task analysis was employed to generate the training specification, and the use of feedback to modify the existing training regime in the light of operational experience. The availability of effective operating instructions is modeled as being dependent upon two policy factors. The first of these is the policy for developing instructions, which ensures that procedures are kept up to date, and are designed according to accepted standards. The other policy factor is project management, since this influences the early definition of work required, so that appropriate instructions will be available at the workplace when required.

Project management also influences the likelihood that staffing levels will be adequate for the tasks required. This latter factor, together with the extent to which appropriate jobs are assigned to individuals, and the complexity of the jobs, all influence the level of time pressure likely to be felt by the operator. The detailed calculations, which show how the probability of human error is influenced by changes in the sociotechnical factors in the situation, are given in Appendix 5A.

5.8. SUMMARY

This chapter has provided an overview of a recommended framework for the assessment of human error in chemical process risk assessments. The main emphasis has been on the importance of a systematic approach to the qualitative modeling of human error. This leads to the identification and possible reduction of the human sources of risk. This process is of considerable value in its own right, and does not necessarily have to be accompanied by the quantification of error probabilities.

Some examples of major quantification techniques have been provided, together with case studies illustrating their application. It must be recognized that quantification remains a difficult area, mainly because of the limitations of data, which will be discussed in Chapter 6. However, the availability of a systematic framework within which to perform the human reliability assessment means that despite data limitations, a comprehensive treatment of human reliability within CPQRA can still yield considerable benefits in identifying, assessing, and ultimately minimizing human sources of risk.

5.9. APPENDIX 5A: INFLUENCE DIAGRAM CALCULATIONS

Commentary on the Calculations

This commentary is provided to clarify the calculations in the following tables. In Table 1, the assessment team is asked to evaluate the evidence that feedback from operational experience is used to develop training. In order to make this evaluation, they will be provided with an "indicator" in the form of a scale specifying the nature of the evidence that should be taken into account. For example, the end of the scale defining the ideal situation would include conditions such as: "Results from operational experience fed directly to the training department," and "evidence that training regime is modified as a result of feedback." The other end of the scale would describe the worst case situation, for example, "No feedback from operational experience into training." In the example cited, the evidence strongly indicates that feedback is not used effectively in developing training.

1 What is the weight of evidence for feedback from operational experience in developing training?	
Good	Poor
.20	.80

2 What is the weight of evidence for use of task analysis in developing training?	
Used	Not Used
.20	.80

3			For Quality of Training			
If	**and**	**then**	*weight of evidence that Quality of training is*		Joint Weight (feedback x Task Analysis)
feedback is:	*Task Analysis is:*		*high is:*	*low is:*	
Good	Used		.95	.05	.04 (.20 x .20)
Good	Not Used		.80	.20	.16 (.20 x .80)
Poor	Used		.15	.85	.16 (.80 x .20)
Poor	Not Used		.10	.90	.64 (.80 x .80)
Unconditional Probability (weighted sum) that Quality of Training is high vs. low is:			.254	.746	

4	What is the weight of evidence that Policy for generating instructions is:
Effective	Ineffective
.30	.70

5	What is the weight of evidence that Project Management is:
Effective	Ineffective
.10	.90

6	For Availability of Effective Operating Instructions				
If *Policy for generating instructions is:*	**and** *Project Management is:*	**then**	*weight of evidence that operating instructions are*		Joint Weight (Policy x Project Management)
			available is:	*not available is:*	
Effective	Effective		.90	.10	.03 (.30 x .10)
Effective	Ineffective		.60	.40	.27 (.30 x .90)
Ineffective	Effective		.50	.50	.07 (.70 x .10)
Ineffective	Ineffective		.05	.95	.63 (.70 x .90)
Unconditional Probability (weighted sum) that Effective Operating Instructions are available vs. not available is:			.255	.744	

Table 2 contains a similar assessment to Table 1 but for the use of task analysis. As illustrated in Table 3, the assessment team is then asked to evaluate the weight of evidence that the quality of training will be high (or low) given various combinations of the influencing factors feedback and use of task analysis. Of course, such evaluations are difficult to make. However, they utilize whatever expert knowledge is possessed by the evaluation team, and factor this into the analysis. They also allow the assessors to factor into their evaluations any interactions among factors. For example, the combined effects of poor feedback and nonuse of task analysis may degrade the quality of training more strongly than either influence in isolation. Each of the conditional assessments is then weighted by the results of stages 1 and 2 and the products added together to give an estimate of the unconditional probability that the training is adequate.

Similar assessments are performed to evaluate the probability that effective operating instructions are available (Table 6) that staffing levels are adequate (Table 9) and that time pressure will be high or low (Table 10). In this latter case, since three influences impact upon time pressure, eight joint assessments need to be made.

7	What is the weight of evidence for Assignment of Job Roles?
Good	Poor
.50	.50

8	What is the weight of evidence for Task Complexity?
High	Low
.60	.40

9	For Staffing Levels			
If *Project Management is:*	**then**	*weight of evidence that Staffing Levels are*		**Weight (Project Management) (from 5)**
		adequate is:	*inadequate is:*	
Effective		.60	.40	.10
Ineffective		.20	.80	.90
Unconditional Probability (weighted sum) that Staffing Levels are adequate vs. inadequate is:		.24	.76	

10	For Time Pressure					
If *Staffing levels are:*	**and** *Assignment of Job Roles is:*	**and** *Project Management is:*	**then**	*weight of evidence for time pressure being*		**Joint Weight (staffing levels x job roles x task complex.)**
				high is:	*low is:*	
Adequate	Good	High		.95	.05	.072 (.24 x .50 x .60)
Adequate	Good	Low		.30	.70	.048 (.24 x .50 x .40)
Adequate	Poor	High		.90	.10	.072 (.24 x .50 x .60)
Adequate	Poor	Low		.25	.75	.048 (.24 x .50 x .40)
Inadequate	Good	High		.50	.50	.023 (.76 x .50 x .60)
Inadequate	Good	Low		.20	.80	.015 (.76 x .50 x .40)
Inadequate	Poor	High		.40	.60	.023 (.76 x .50 x .60)
Inadequate	Poor	Low		.01	.99	.015 (.76 x .50 x .40)
Unconditional Probability (weighted sum) that Time Pressure is high vs. low is:				.3981	.6019	

Although these combined assessments are arduous, it should be noted that the evaluations of the effects of combinations of influences may be regarded as applicable across a range of systems, and hence would only need to be performed once for a generic model. The system specific evaluations would then be the simpler level 2 assessments set out in Tables 1, 2, 4, 5, 7, and 8. As discussed earlier, guidance for performing these assessments could be provided by the use of PIF scales delineating the conditions for the least and most favorable ends of the scales. Similar scales can be used to make direct evaluations of the level 1 influences, if the assessments described earlier are judged to be too difficult. Even if the full assessments are made, it is useful to compare these with the indirect assessments to check convergence.

The final stage of the procedure is to generate an overall unconditional probability of human error (Table 11). This is achieved by assigning probabilities of error to combinations of the three first level influences quality of training, availability of operating instructions and time pressure. These conditional probabilities are generic, in that they could apply to any system. They are made specific to the situation under consideration by multiplying them by the assessed probabilities of the level 1 influences, as derived from the earlier analyses. These products are then summed to give the overall unconditional probability of error occurrence in the situation being evaluated.

11			For the task modeled			
If	**and**	**and**	**then**	**the**		
	Effective			*probability of*		**Joint Probabilities**
Quality of	*Operating*	*Time*				**(training quality**
Training	*Instructions*	*Pressure*		*success*	*failure*	**x instructions x time**
is:	*are:*	*is:*		*is:*	*is:*	**pressure.)**
High	Available	Low		.99	.01	.0390 (.25 x .26 x .60)
High	Available	High		.978	.022	.0258 (.25 x .26 x .40)
High	N. available	Low		.954	.046	.1137 (.25 x .74 x .60)
High	N. available	High		.90	.10	.0752 (.25 x .74 x .40)
Low	Available	Low		.90	.10	.1145 (.75 x .26 x .40)
Low	Available	High		.78	.22	.076 (.75 x .26 x .40)
Low	N. available	Low		.54	.46	.3341 (.75 x .74 x .60)
Low	N. available	High		.00	1.00	.2209 (.75 x .74 x .40)
Assessed Unconditional Probability of success vs. failure is:				.58	.42	

The SLIM method described earlier is particularly suitable for the derivation of the conditional probabilities in Table 11, since it evaluates probabilities as a function of variations in PIFs that correspond to the level 1 factors used in this example. Each of the eight conditions in Table 11 can be treated as a separate task for evaluation by SLIM, using common weights for each factor across all conditions, but differing ratings to reflect the differing conditions in each case. SLIM requires calibration data to be supplied for the two end-point conditions, but this is considerably less onerous than evaluating probabilities for all conditions. Another source of probabilities to include in Table 11 would be laboratory experiments where the first level influencing factors were varied systematically.

6

Data Collection and
Incident Analysis Methods

6.1. INTRODUCTION

The preceding chapters of this book have focused on how accidents due to human error can be prevented at source. These preventive measures include systematic design strategies, techniques to identify potential errors with serious consequences, and audits of performance-influencing factors in existing systems to specify opportunities for improvements. To complement these proactive strategies, it is important to have feedback systems in place so that lessons can be learnt effectively from minor incidents, near-misses and from major accident investigations. This chapter describes a range of techniques and systems to achieve these objectives.

To most plant managers, the term *data collection,* at least in the context of safety, refers to the collection of statistical data in areas such as lost-time accidents and other reportable injuries. Because such data are required by law, and because they are perceived to have a major impact on accident prevention via their motivational effects, considerable resources are expended every year to produce these data. They constitute the "bottom line" that will be used to justify the safety performance of the organization to the public, the regulators, and to its shareholders. Although the central importance of this aspect of data collection is acknowledged, this chapter will describe a much wider range of data collection activities that need to be carried out in order to maximize the effectiveness of error reduction programs in organizations.

Another publication produced by the Center for Chemical Process Safety, *Guidelines for Investigating Chemical Process Incidents* (CCPS, 1992d), is directed at achieving similar objectives but from a differing perspective and with differing emphasis. Both sources of information can be used in a complementary manner to improve the quality of data collection and incident analysis in the CPI.

This chapter is divided into the following sections:

Overview of Data Collection Systems (6.2)
This section provides an overall structure within which the different aspects of data collection and incident analysis methods can be integrated. The importance of effective data collection systems as part of the continuous improvement process in Total Quality Management.

Types of Data Collection System (6.3)
The major categories of data collection systems are described. These include:

- *Incident reporting systems,* designed to identify underlying and direct causes for larger numbers of incidents with relatively minor causes
- *Near-miss reporting systems*
- *Root cause analysis systems,* intended to provide in-depth evaluations of major incidents
- *Quantitative human reliability data collection systems* for generating human error probabilities for use in quantitative risk assessment.

Organizational and Cultural Aspects of Data Collection (6.4)
This section discusses the company culture that is necessary to support effective data collection and root cause analysis.

Types of Data Collected (6.5)
The types of data required for incident reporting and root cause analysis systems are specified. Data Collection practices in the CPI are described, and a detailed specification of the types of information needed for causal analyses is provided.

Methods of Data Collection, Storage, and Retrieval (6.6)
This section provides information on the personnel who should be involved in data collection and the design of reporting forms. The specific data needs for major incident analyses are discussed, together with the storage and retrieval of data for the purpose of analysis.

Data Interpretation (6.7)
The need for a causal model to guide data collection is emphasized. This makes the connection between the nature of the error and the PIFs in the situation.

Root Cause Analysis Techniques (6.8)
A range of techniques is described for analyzing the structure of incidents and the causal factors involved.

Implementing and Monitoring the Effectiveness of Error Reduction Measures (6.9)
The specification and implementation of error reduction measures arises directly from the identification of causes. The data collection system needs to be able to evaluate the effectiveness of such measures.

Setting up a Data Collection System in a Chemical Plant (6.10)
This section sets out a step-by-step procedure for setting up a data collection system, including the important issues of gaining workforce acceptance and management support.

6.2. AN OVERVIEW OF DATA COLLECTION SYSTEMS

The function of this section is to provide an overall framework within which to describe the important aspects of data collection systems in the CPI. As mentioned in the introduction, the emphasis in this chapter will be on methods for identifying the causes of errors that have led to accidents or significant near misses. This information is used to prevent reoccurrence of similar accidents, and to identify the underlying causes that may give rise to new types of accidents in the future. Data collection thus has a proactive accident prevention function, even though it is retrospective in the sense that it is usually carried out after an accident or near miss has already occurred.

In an overall proactive error management system, data collection provides feedback information on the effectiveness of specific interventions that have been made to reduce error potential. However, in most plants in the CPI such proactive error management strategies will not be in existence. Therefore, the setting up of a data collection system which addresses human error causes will often be the first stage of an error management program. The advantages of this are twofold. First, both company and regulatory requirements mean that some form of data collection system, even if it only fulfills the most basic of statutory requirements, will probably already be in existence. It is therefore possible to build upon this to develop a more comprehensive system designed to address the underlying causes of incidents. Setting up a data collection system as the first stage of an error management program provides insights into where the major problems lie, and hence allows subsequent proactive interventions to be targeted at the areas where the most rapid benefits will be obtained.

Figure 6.1 provides an overview of the structure of a data collection system. As with all aspects of human error management, the attitudes and beliefs held by the company and plant management to safety in general, and human factors in particular, will be critical in developing a successful data collection system. Management will influence the effectiveness of data collection systems in three ways. First, they control the resources required to set up and maintain the system. Second, management will be responsible for determining the culture that exists in the plant. As will be discussed in more detail in Section 6.5, if management encourages a culture which emphasizes blame and punishment for errors, then it is unlikely that a data collection system which is intended to address the underlying causes of incidents will ever be successful. Third, the attitudes of management will determine the "model" of

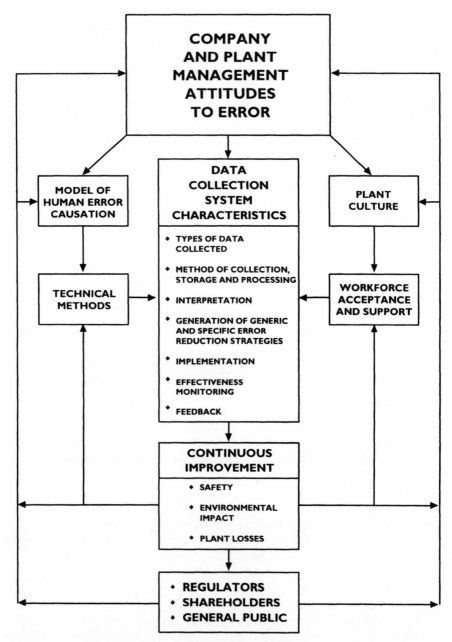

FIGURE 6.1. **Overall Structure of Data Collection System**

250

error causation that drives the data collection effort. Thus, the traditional view of human error which emphasizes individual rather than system causes of error (see Chapter 2) will lead to the use of data collection and analysis methods which focus on these factors. The use of the systems view will mean that there will be greater emphasis on the systemic causes of errors such as poor procedures, equipment design or training.

The model of human error held by management and the plant culture constitutes the environment in which the data collection system operates. Within this environment, all data collection systems need to address the topics listed in Figure 6.1. These topics, from the types of data collected, to the feedback systems that need to be in place, will be addressed in subsequent sections of this chapter.

Figure 6.1 emphasizes the fact that the outputs from data collection systems, particularly those that address safety and environmental issues, are of critical importance to an organization in that they are used as major indications of the acceptability of a company's operating practices by regulators, shareholders, and the general public. This criticality has both advantages and disadvantages. From the positive perspective, there is considerable pressure on a company to ensure that its policies produce low accident rates. On the negative side, there is equally strong pressure to produce data collection systems that present the operating record of a company in the best possible light. Unfortunately, these considerations can often work against the development of reporting systems that are designed to get at underlying causes of accidents.

Figure 6.1 also indicates that the output from data collection systems is a vital aspect of the Continuous Process Improvement cycle advocated in Total Quality Management. Feedback on the underlying causes of problems is necessary to ensure continuing support for error and accident reduction programs by senior management. Feedback also leads to changes in the model of error causation held by senior management and to changes in plant culture which can further enhance the effectiveness of data collection systems by gaining ownership and commitment from the workforce.

6.3. TYPES OF DATA COLLECTION SYSTEMS

Many data collection systems place the primary emphasis on the technical causes of accidents. There is usually a very detailed description of the chemical process in which the accident occurred, together with an in-depth analysis of the technical failures that are seen as the major causes. The human or system failures that may have contributed to the accident are usually treated in a cursory manner. Technically oriented reporting systems are very common in the CPI, where engineers who may be unfamiliar with human factors principles will, not unnaturally, tend to focus on the technical causes of accidents unless provided with very clear guidelines or training to allow them to consider the human causes.

Where data collection systems do address human error, they are generally driven by the traditional safety engineering view and focus on the outcomes of errors, which are usually assumed to be due to basic human weaknesses such as inattention or lack of motivation and commitment. The outputs from traditional data collection systems may be both descriptive and statistical. Descriptive information about specific accidents may be used to emphasize the implications of frequently occurring and potentially dangerous behaviors such as entering confined spaces without carrying out checks to test for toxic gases or violating operating instructions. Generally, little attempt is made to identify any systemic causes of these behaviors. The statistical information is in the form of aggregated data such as lost time accidents. These data are often used on a comparative basis to indicate trends over time, or differences among groups of workers or organizations. Another application is to provide inputs to safety campaigns and other motivational approaches to safety improvement.

The systems which are the major focus of this chapter are described below. They emphasize the identification of underlying causes and the use of this information to specify error and accident reduction strategies.

6.3.1. Incident Reporting Systems (IRS)

The main function of an incident reporting system (IRS) is to identify recurring trends from large numbers of incidents with relatively minor outcomes, or from near misses. One of the important characteristics of an IRS is that the time and resources required to evaluate an incident and incorporate it into the database must be minimized. This means that the designers of an IRS have to carefully evaluate the benefits and costs of requiring more comprehensive information from each incident that is to be reported. A requirement for too much information will bring the system into disrepute, and too little information will mean that the results are too general to be of any real value.

Other important considerations in the design of an IRS are the data storage and analysis requirements. These need to be considered early in the design of the system if it is to be used to research and display trends effectively. For example, in addition to the answers to specific questions, the accident data analyst may wish to make use of free text descriptions of the circumstances of the accident. This implies that a text-based retrieval system will be required.

6.3.2. Near Miss Reporting Systems (NMRS)

The value of near miss reporting has been emphasized at a number of points in this book. Near misses represent an inexpensive way to learn lessons from operational experience, since they have the potential for providing as much information about the systemic causes of accidents as events with serious consequences. However, it is unusual to find an effective NMRS in the process industry. This is because the success of a NMRS depends critically on the voluntary reporting of events which would otherwise have gone unnoticed. This requires a culture which is highly supportive in terms of emphasizing the value of this type of information and minimizing the use of blame and punishment as a method of error control. Although such an approach is a fundamental aspect of modern quality assurance approaches such as Total Quality Management (TQM), it is still rare in many parts of the industry. Another factor is the need for a careful definition of exactly what constitutes a near-miss. Unless this is clearly specified, the system may be swamped with large numbers of reports which have little value in the context of establishing the underlying causes of accidents. Van der Schaaf et al. (1991) provide a comprehensive discussion of near-miss reporting systems and data collection issues in general.

6.3.3. Root Cause Analysis Systems (RCAS)

The term *root cause analysis system* is used to denote systems that are concerned with the detailed investigations of accidents with major consequences such as loss of life, or severe financial or environmental implications. These systems are characterized by the use of comprehensive, resource-intensive techniques designed to evaluate both the direct and indirect root causes. Although resource limitations are less important with RCAS, a clearly structured methodology is nevertheless needed in order to ensure that investigations are both comprehensive and consistent. The requirement for consistency is particularly important if the lessons learned from accident analyses are to be useful from a comparative basis and for evaluating trends in underlying patterns of causes over time. As with IRS, an investigation procedure based on a model of accident causation such as the systems approach (see Chapters 1 and 2) will provide a systematic framework to ensure that the right questions are asked during the investigation. Comprehensive methodologies have been developed to support RCAS, and these are explained in detail in Section 6.8.

6.3.4. Quantitative Human Reliability Data Collection Systems

There is considerable interest in developing a database on human error probabilities for use in chemical process quantitative risk assessment (CPQRA). Nevertheless, there have been very few attempts to develop such a database for the CPI compared, for example, with the nuclear industry. Some of the reasons for this are obvious. The nuclear industry is much more highly integrated than the CPI, with a much greater similarity of plant equipment and less direct competition among companies. This, at least in theory, makes it more feasible to develop shared databases of error probabilities for standardized human actions in response to specific emergency scenarios. Also, probabilistic safety analysis has been applied to a much greater extent in the nuclear industry via the regulatory process, and hence there has been a greater requirement over a long period of time for data and techniques to support these analyses. Although human reliability analyses have been performed (primarily in the offshore sector), these have mainly used extrapolated data from sources such as the THERP (Technique for Human Error Rate Prediction) database (see Chapter 5) which was largely developed in a nuclear context.

The requirements for the development of a CPI-specific quantitative human reliability data collection system are as follows:

- The users of quantitative human reliability data need to specify their needs for such data in the context of CPQRA, in terms of the types of human operations for which data are required, analytical data techniques to be used, etc..
- The PIFs that determine human reliability in these situations need to be defined.
- An industry-wide data collection effort needs to be organized that would use a common classification for human error data. This would allow a large number of errors in each category to be collected. This, together with information on the number of opportunities for errors, would allow probabilities to be estimated from the frequency of errors in each category.
- Methods for extrapolating these probabilities to specific situations, on the basis of differences among PIFs, would need to be developed (see Chapter 5).
- Where field data were unavailable, a program of experimental work (for example, based on the use of simulators for control room operations) could be implemented to generate appropriate generic data.

Although the steps outlined above would in theory be capable of generating a quantitative database, it seems unrealistic to expect the degree of cooperation that would be required across the industry to develop such a resource. A more likely possibility is that large multinationals will support the development of in-house databases, possibly using the same approach as advocated here.

6.3.5. Conclusions on Data Collection System Types

The discussion of alternative types of data collection systems serves to emphasize the fact that the design of such systems needs to have very clear objectives. Although a range of data collection systems have been described as if they were independent, in fact many systems will be combinations of these types. For example, root cause analysis systems will need to consider both the technical and human causes of major accidents. A comprehensive Incident Reporting and Investigation System would probably include near misses as well as actual incident reporting.

In subsequent sections the emphasis will be on the human factors aspects of these systems. In general, the design principles which will be set out will apply to both types of system. However, distinctions will be made where appropriate.

6.4. ORGANIZATIONAL AND CULTURAL ASPECTS OF DATA COLLECTION

The first area focuses on the cultural and organizational factors that will have a major influence on the effectiveness of a human error data collection system and how well the information derived from such a system is translated into successful error reduction strategies. Regardless of how effectively the technical issues are dealt with, the system will not be successful unless there is a culture in the organization which provides support for the data gathering process. No data collection system aimed at identifying human error causes of accidents will be workable without the active cooperation of the workforce.

6.4.1. Model of Accident Causation Held by the Organization

The type of data collected on human error and the ways in which these data are used for accident prevention will vary depending upon the model of error and accident causation held by the management of an organization. This model will also influence the culture in the plant and the willingness of personnel to participate in data collection activities. In Chapters 1 and 2 a number of alternative viewpoints or models of human error were described. These models will now be briefly reviewed and their implications for the treatment of human error in the process industry will be discussed.

6.4.1.1. The Traditional Safety Engineering (TSE) View
The traditional safety engineering view is the most commonly held of these models in the CPI (and most other industries). As discussed in Chapter 1, this view assumes that human error is primarily controllable by the individual, in that people can choose to behave safely or otherwise. Unsafe behavior is assumed to be due to carelessness, negligence, and to the deliberate breaking of operating rules and procedures designed to protect the individual and the system from known risks.

The responsibility of management from the TSE perspective is to provide a safe system of work to minimize the exposure of the individual and the process system to these risks. This is achieved by technical approaches such as barriers and interlocks, and through the provision of personal protective equipment. Management also has the responsibility to inform workers of these risks and to ensure that safe methods of work are adopted by providing appropriate training. Given that management carries out these functions adequately, the main strategy for maximizing safety from this perspective is to motivate the workforce so that they do not commit deliberately unsafe acts

6.4.1.2. Implications of the TSE View for Data Collection

The implications of this approach for the data collection philosophy will be as follows:

Causal Analysis

There will be comparatively little interest in the underlying causes of errors leading to accidents. This is because the TSE view assigns virtually all errors to unsafe acts that are preventable by the individual workers concerned. There is therefore little incentive to delve into other causes.

Prevention Strategies

Emphasis for prevention will be on changing individual behavior by symbolic or tangible rewards based on statistical evidence from the data collection system. "Hard" performance indicators such as lost time incidents will therefore be preferred to "softer" data such as near-miss reports. Accident prevention will also emphasize motivational campaigns designed to enhance the awareness of hazards and adherence to rules. If a severe accident occurs, it is likely that disciplinary sanctions will be applied.

Changes in Data Collection Strategies

The TSE model of causation that accidents are primarily due to individually controllable unsafe acts is unlikely to be modified over time. This is because very little evidence on the background and conditions which led up to an accident will be collected. The data collection strategy is therefore likely to remain static, since the data collected will, by definition, not contradict the underlying assumptions.

6.4.1.3. The System-Induced Error Approach

As described in Chapters 1 and 2 the system-induced error approach comprises the following elements:

Error Tendencies and Error-Inducing Environments

Human errors occur as a result of a combination of inherent human error tendencies, and error-inducing conditions. Errors then combine with unforgiving situations (lack of recovery and the presence of hazards) to produce an accident, as illustrated in Figure 6.2. The error-inducing conditions consist of two aspects. The first of these is the presence of factors such as poor procedures, inadequate training and time stress, which mean that the worker is unlikely to have the mental or physical resources available to meet the demands arising from the job. This mismatch creates a situation of high error potential. The other aspect of error-inducing conditions is the presence of specific triggering events such as unexpected fluctuations in demand, distractions, or other additional pressures

Multiple Causation
Accidents do not arise from a single cause but from a combination of conditions which may be human caused (active or latent failures), characteristics of the environment, or operating states of the plant (see Chapter 2).

Role of Latent Failures
The systems approach emphasizes the effects of organizational and managerial policies in creating the preconditions for errors described above. In addition to the direct effects of these policies, management is also responsible for determining the culture in the organization. This may, for example, influence the choices made among profitable but possibly risky ways of working and adherence to stated safety practices (see Chapter 2, Section 2.7).

Emphasis on the Modification of System Factors as a Major Error Reduction Strategy
This emphasis replaces the reliance on rewards and punishment as a means of error control which characterizes the TSE approach.

6.4.1.4. Implications of the System-Induced Error Approach for Data Collection

Causal Emphasis
There will be strong emphasis on the collection of data on possible causal factors that could have contributed to an accident. The specific data that are collected may be based on an error model such as that shown in Figure 6.2. However, this model will usually be modified depending upon the extent to which it fits the data collected over a period of time. The systems approach is therefore dynamic rather than static.

Organizational Perspective
Monitoring and detailed accident investigation systems will attempt to address the organizational and work culture factors that influence accident causation. This will encourage the investigation of the global effects of organizational policies in creating the precursors for accidents.

Use of Near-Miss Data
The Systems Approach emphasizes the value of near-misses as a rich source of information about accident causes. This is based on the concept of accidents

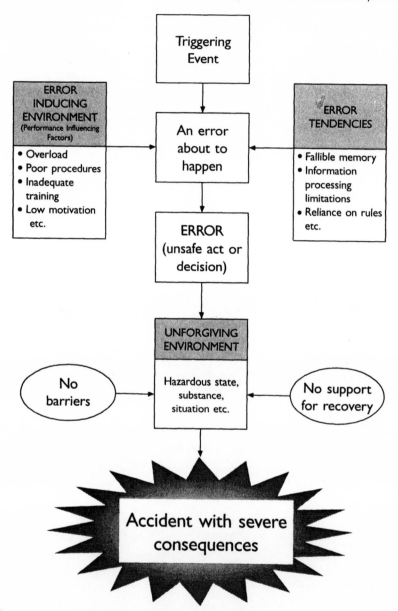

FIGURE 6.2. **Accident Causation Model (From Chapter 2).**

as resulting from combinations of conditions such as a poor safety culture, inadequate training and poor procedures, together with a triggering event (see Figure 6.2). Near-miss reporting systems are therefore important to provide early warnings of these conditions before they lead to an accident.

Changes in Data Collection Strategies

Because of the emphasis on modeling accident causation, data collection systems based on the system-induced error approach are likely to modify their data collection strategies over time. Thus, as evidence accumulates that the existing causal categories are inadequate to account for the accidents and near misses that are reported, the data collection philosophy will be modified, and a new accident causation model developed. This, in turn, will be modified on the basis of subsequent evidence.

6.4.2. Cultural Aspects of Data Collection System Design

A company's culture can make or break even a well-designed data collection system. Essential requirements are minimal use of blame, freedom from fear of reprisals, and feedback which indicates that the information being generated is being used to make changes that will be beneficial to everybody. All three factors are vital for the success of a data collection system and are all, to a certain extent, under the control of management. To illustrate the effect of the absence of such factors, here is an extract from the report into the *Challenger* space shuttle disaster:

> *Accidental Damage Reporting.* While not specifically related to the *Challenger* accident, a serious problem was identified during interviews of technicians who work on the Orbiter. It had been their understanding at one time that employees would not be disciplined for accidental damage done to the Orbiter, providing the damage was fully reported when it occurred. It was their opinion that this forgiveness policy was no longer being followed by the Shuttle Processing Contractor. They cited examples of employees being punished after acknowledging they had accidentally caused damage. The technicians said that accidental damage is not consistently reported when it occurs, because of lack of confidence in management's forgiveness policy and technicians' consequent fear of losing their jobs. This situation has obvious severe implications if left uncorrected. (Report of the Presidential Commission on the Space Shuttle Challenger Accident, 1986, page 194).

Such examples illustrate the fundamental need to provide guarantees of anonymity and freedom from sanctions in any data collection system which relies on voluntary reporting. Such guarantees will not be forthcoming in organizations which hold a traditional view of accident causation.

Feedback is a critical aspect of voluntary reporting data collection systems. If personnel are to continue providing information they must see the results of their input, ideally in the form of implemented error control strategies. A method for providing feedback which aims to share any insights gained from a scheme will indicate to all personnel that the system has a useful purpose. One example of an incident reporting scheme with an effective feedback channel is the Institute of Nuclear Power Operations human performance evaluation system (HPES) (Bishop and Larhette, 1988). Here a newsletter called "Lifted Leads" is used to publicize anonymous reports of incidents together with any error control strategies implemented. The newsletter is circulated to all plants participating in the HPES program. In addition, humorous posters have been developed from certain reported incidents and these are also circulated freely.

As well as a nonpunitive culture with guarantees of anonymity and feedback there are three other necessary conditions for an effective data collection system. First, it is important that the future users of the system are involved in its design and implementation. Second it is essential that those who use the system should eventually own it. Such owners should be willing to view the information in any database as a neutral commodity for all to use. Finally, it is crucial that effective training is given. This includes training in communication skills and analysis methods for the investigators of incidents, and an awareness training program for all levels of staff who will be involved.

6.5. TYPES OF DATA COLLECTED

The types of data collected in both incident reporting and root cause analysis systems are obviously very closely linked to the particular model of accident causation which exists in the company. If, for example, the emphasis is on the underlying causes of errors, this will mean that information necessary to distinguish among different underlying causes will need to be collected (see 6.5.2.1 below). In the case of root cause analysis systems, more detailed data on indirect causes such as organizational policies will need be required (see 6.5.2.3). In both systems, information on key performance influencing factors (PIFs) will be needed. With incident reporting systems, because of the limited time available for investigations only a critical subset of the PIFs will be recorded. In the case of root cause analysis systems, a much more comprehensive PIF evaluation tool similar to the human factor assessment methodology (HFAM) tool described in Chapter 2 can be employed.

In the first of the following subsections, the data collection approaches adopted in most CPI incident reporting systems will be described. The fact that these systems provide little support for systematically gathering data on underlying causes will provide an introduction to the later sections which emphasize causal analysis techniques.

6.5.1. Data Collection Practices in the Chemical Processing Industry

The following types of information are collected in most CPI safety-related data collection systems:

Severity of the Incident

This typically considers injuries to personnel and damage to plant and equipment. In a few more highly developed systems, information on **potential** consequences is also collected. Normally the severity of the incident consequences (or in some cases its potential consequences) will determine the resources that are put into its investigation.

General Descriptive Data

This typically includes the following areas:

- Brief description of the incident in terms of when and where it occurred, etc.
- Details of any injury
- A more complete narrative description of the incident

Work Control Aspects

This describes any work permits associated with the work relevant to the incident.

Technical Information

This is mainly applicable to equipment and other technical failures. It also considers areas such as loss of containment, environmental impact, fires, and explosions.

Causal Aspects

In the majority of reporting systems this area receives relatively little attention. The user of the reporting form is asked to provide a single evaluation of the cause, with little guidance being available to assist in the process. In a few large companies, the causal aspect is addressed more comprehensively. In one multinational, for example, the form asks the investigator to evaluate both immediate and underlying causes. Guidance is also provided by providing pre-specified categories in these areas. However, information on systemic causes such as incorrect policies or discrepancies between policies and practices is rarely included in these categories.

Remedial Actions

The section on remedial actions is usually directed at preventing a recurrence of the specific accident which is the focus of the investigation. It often consists of a sequence of recommended actions linked to the causal categories identified in the previous section. Again, remedial actions directed at more fundamental systemic causes are rarely addressed.

Management of the Investigation
In some cases, the final part of the form provides a checklist which tracks the management permissions and endorsements associated with the conduct of the investigation.

Conclusions on CPI Data Collection Systems
The overall conclusion that can be drawn from a survey of CPI data collection systems is that the better systems do attempt to address the causes of human error. However, because of the lack of knowledge about the factors which influence errors, the causal information that is collected may not be very useful in developing remedial strategies. General information in areas such as severity, work control aspects and the technical details of the incident will be required in all data collection systems. However, in almost all cases a structured process for causal analysis is lacking. Some of the requirements for causal analysis are set out in the following sections.

6.5.2. Causal Data Collection

All causal data collection processes require information in the following areas:

- What happened
- How it happened
- Why it happened

As discussed earlier, most data collection systems in the CPI place considerable emphasis on the "what," but provide little support for the "how" or "why." Causal analysis methods can be broadly divided into techniques which emphasize the structure of an accident and those which focus on causes. Structural techniques provide information on the "what" and "how," and the causal techniques enable the "why" to be investigated.

The areas that need to be addressed in causal analysis can be specified by considering the contributing causal factors as a series of layers. Accident investigation can be likened to the process of peeling an onion. The onion model of accident investigation is shown in Figure 6.3. The onion analogy is not quite correct in the sense that accident investigation (peeling the onion) usually proceeds from the middle outward. However, it does provide a useful metaphor for the accident causation process.

Typically, the first phase of a comprehensive accident investigation process will involve describing the way in which the hardware, the chemical process, individual operators and operating teams are involved in the accident process. This is the domain of the structural analysis techniques and the technical analysis of the chemical process which gave rise to the accident. Analyses of human error will primarily address the interactions between hardware systems and individuals or operating teams (the first two layers

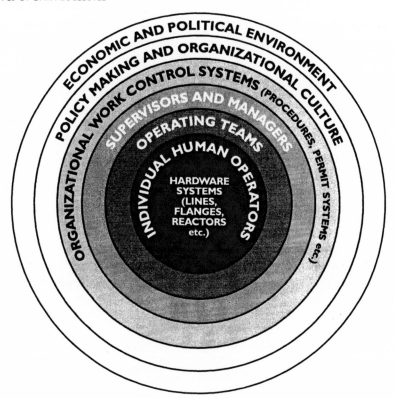

FIGURE 6.3. **Onion Model of Accident Causation.**

starting from the center of the onion). The next level of analysis is likely to address interactions between the plant operator level and the supervisory and management levels (the second and third layers). It is at this interface where communication failures or violations due to incorrect interpretations of policy are likely to occur.

The next layer contains the work control systems such as operating instructions and permit systems. Although such systems are generally executed at the operational level, the extent to which they are adhered to is strongly influenced by how they are regarded by managers and supervisors. These are in turn influenced by the general policies and safety culture that exists in the plant and the organization. Finally, these policies will be influenced by the economic and political climate in which the company operates.

6.5.2.1. Data on Event Sequence and Structure
The amount of time available for the recording of data in incident reporting systems is limited, and hence the information collected is usually confined to short descriptions of the event, its actual and potential consequences and

immediate and indirect causal factors such as those discussed in the preceding section.

However, in the case of a root cause analysis system, a much more comprehensive evaluation of the structure of the accident is required. This is necessary to unravel the often complex chain of events and contributing causes that led to the accident occurring. A number of techniques are available to describe complex accidents. Some of these, such as STEP (Sequential Timed Event Plotting) involve the use of charting methods to track the ways in which process and human events combine to give rise to accidents. CCPS (1992d) describes many of these techniques. A case study involving a hydrocarbon leak is used to illustrate the STEP technique in Chapter 7 of this book. The STEP method and related techniques will be described in Section 6.8.3.

6.5.2.2. Data on Human Error Tendencies

In order to establish the psychological causes of errors, data from accidents or near misses which are relevant to the human error tendencies discussed in Chapter 2 should be collected. These include information on the following:

- Distractions or competing activities (demand/resource mismatch). Experience or familiarity with the task (useful for identifying slips and mistakes)
- Ambiguous or difficult-to-detect information which is necessary to perform the task (possible failure to acquire critical information, see the flow diagram in Chapter 2, Appendix 2B)
- Aspects of the job requiring excessive demands on memory (task steps omitted or misordered)
- Sequences of operations in different tasks which are very similar, apart from the critical steps which could have critical consequences when performed in the wrong situation (strong habit intrusions)
- Evidence that a misdiagnosis was involved (distinction between slips and mistakes)
- Identification of possible violations

6.5.2.3. Data on Performance Influencing Factors

Another group of factors is relevant to establishing **situational causes**. Some of these will overlap with the previous group, but will include some of the PIFs considered in Chapter 3, such as:

- Quality of procedures
- Adequacy of human-machine interface or task design
- Time available
- Time of day and fatigue effects
- Environmental conditions

It should be emphasized that it is usually necessary to develop the data collection specification on an incremental basis and to utilize feedback from the system to modify the initial model relating causal factors to error types. This dynamic approach provides the best answer to the problem that no predefined error model will be applicable to every situation.

This is in contrast to many data collection systems, in which considerable efforts are expended in developing a "definitive" data collection philosophy. However, once the system is in place, there is little attempt to modify this on the basis of operational feedback.

The fact that the model connecting error types with their causes may change as a result of gaining further experience with the data collection system means that the information gathered on the PIFs in a situation may also change. For example, if incident data indicates the neglect of safety procedures because of production pressures, then the questions relating to this area will need to be extended.

In the case of root cause analysis systems, more comprehensive evaluations of PIFs will normally be carried out as part of a full-scale human factors audit. This could make use of the types of comprehensive PIF evaluation methods described in Chapter 2 (see Section 2.7.7 and Figure 2.12).

6.5.2.4. Data on Organizational Causes

The two categories of data described above relate to immediate causes of error. However, the question of how these factors came to be as they are, involves a consideration of the effects of organizational, and management and cultural issues.

An evaluation system currently being developed for process industry operations (the HFAM technique described in Chapter 2, Section 2.7) addresses organizational and work culture factors such as:

- Possible conflicts between well-established work practices and those specified in safety policy
- Policies for procedures and training
- Communications and feedback systems
- Clarity of roles and responsibilities
- Reward system
- Perceived credibility of organizational commitment to safety policy

Information on these factors is critical in establishing more general influences that impact indirectly but powerfully on the probability of an accident occurrence.

6.6. METHODS OF DATA COLLECTION, STORAGE, AND RETRIEVAL

6.6.1. Personnel Involved in Data Collection

The personnel responsible for the collection and analysis of incident data vary in different organizations. One common practice is to assign the responsibility to an investigation team which includes the first line supervisor, a safety specialist and a plant worker or staff representative. Depending on the severity of an incident, other management or corporate level investigation teams may become involved.

In some organizations, designated individuals have specific responsibility for eliciting detailed information from operational staff on the immediate and underlying causes of incidents. An example is the Human Performance Evaluation System (HPES) developed for the nuclear industry, which is described in Bishop and Larhette (1988). These coordinators provide a certain level of guaranteed immunity from sanctions which allows individuals to be frank about the contributory causes that they may not be willing to discuss in an open forum. As discussed earlier, the need for this approach is a consequence of the fact that in many organizations a blame culture exists which is likely to inhibit a free flow of information about the causes of accidents.

6.6.2. Design of Reporting Forms

The information gathered from the interviews conducted as part of the human error data collection process is entered on paper forms. In order to facilitate the ease and accuracy of data collection, the forms should be designed using human factors guidelines for written materials (e.g., Wright, 1987; Wright and Barnard, 1975).

In a data collection system that was developed in the transportation sector, the application of these principles generated the following format for a data collection form:

- Each part contains distinct sections in which related questions are grouped.
- Two types of questions predominate: simple yes/no options and multiple choice questions. For each the user is asked to tick the appropriate box.
- An indication of who is to fill in the questions is made by the use of symbols.
- For certain questions the user is provided with a "maybe" option, that is, yes/maybe/no. Much valuable information will be lost without this option.

- Questions are made as short as possible and each question only asks about one aspect of the incident.
- Notes are provided to assist the user. These are compiled as a separate document and brief cross-references are given in each part of the form.

6.6.3. Data Collection Procedures for Major Incident Analysis

For a major incident investigation using a comprehensive root cause analysis system, teams will be formed to acquire information relevant to determine the structure and analyze the causes in depth. In addition to evaluations of the immediate causes, underlying causes are likely to be evaluated by investigations in areas such as safety and quality management. Both paper- and computer-based systems will be used to acquire and record information for subsequent detailed analyses.

The systems for handling the large amounts of data generated by major incident investigations need to be in place before they are called upon to be used under pressure. It is well known that the data necessary for establishing causes becomes more difficult to obtain the longer the period that elapses after the incident. There is a strong case for ensuring that any emergency response system has a built-in facility for acquiring important status information while an incident is still in progress. The robustness of data collection systems that are required to operate under conditions of high stress needs to be tested regularly, by means of frequent exercises and simulations.

6.6.4. Storage and Retrieval of Information

With the advent of notebook computers, it is feasible to use interactive software to structure the data collection process at the workplace. There are many potential advantages with this approach.

- The capability to modify the sequence of questions interactively means that the information elicitation process can home-in on particularly useful areas for establishing causes.
- The data collected on site can easily be downloaded to a central data base, thus ensuring that any significant trends in error causation could be rapidly identified and remedied.
- Individuals involved in accidents where error was a possible factor can have access to a computer which will allow them to provide information on a confidential basis. Although portable computers have not yet made a significant impact on incident data collection, there is clearly considerable potential in this area.

The databases that exist in most large companies for the accident data are usually oriented toward coded information. Each of the items on the form is keyed into a field and stored in a standard data base, where it can be interrogated to produce the familiar range of numerical and descriptive data such as bar charts and graphs. The disadvantage of the standard format for human error data is that there is usually insufficient space to allow free text descriptions of accidents to be entered in full. These descriptions are a rich source of data for the human error analyst. It is therefore recommended that the collection and storage systems for human error data provide these facilities. In order to search these free text descriptions, a database system which is capable of storing variable length records and performing text searches is desirable. Examples of database and text retrieval software which can be used for this purpose are Pagefinder® by Caere Systems (USA) and Idealist by Blackwell Software (Europe).

6.7. DATA INTERPRETATION

There is considerable overlap between the processes of data collection and interpretation as discussed in earlier sections of this chapter. The nature of the data collected will be strongly influenced by the assumed relationship between the observable characteristics of errors and their underlying causes. Similarly, the interpretation process will also be driven by the causal model.

The overall process of data interpretation and the development of suitable remedial strategies once a set of causes has been identified, is set out in Figure 6.4. The two-stage process of confirming the initial causal hypothesis is recommended to overcome the tendency to jump to a premature conclusion and to interpret all subsequent information on the basis of this conclusion.

In the following sections, a number of methodologies for accident analysis will be presented. These focus primarily on the sequence and structure of an accident and the external causal factors involved. These methods provide valuable information for the interpretation process and the development of remedial measures. Because most of these techniques include a procedure for delineating the structure of an incident, and are therefore likely to be time consuming, they will usually be applied in the root cause analysis of incidents with severe consequences.

In the case of incident reporting systems, the data interpretation process will be more concerned with identifying trends in recurrent causes for a large number of incidents than a detailed investigation of specific situations. These analyses could identify the repeated occurrence of failures arising, for example, from inadequate procedures, work systems, training, and equipment design. In addition, classifying errors using some of the concepts from Chapter 2, such as slips, mistakes, and violations, can be useful. Essentially, the interpretation process should be based upon an explicit causal model, which should specify the types of data to be collected by the incident reporting system. This causal

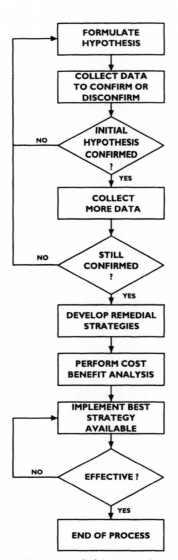

FIGURE 6.4. **Data Interpretation, Remedial Strategy Generation, and Implementation.**

model must not, however, be cast in concrete. If there is evidence that important causes are not being addressed by the existing causal model, then this must be updated and the new information generated by the revised model must be collected and incorporated in the interpretation process.

A specific example of a causal model is the root cause tree described in Section 6.8.4 and Figure 6.8. This is a very elaborate model which includes several levels of detail for both equipment and human causes of incidents. The root causes tree is a generic causal model, and may require tailoring for application to specific plants and processes (e.g., in the offshore sector) where other error causes may need to be considered.

6.8. ROOT CAUSE ANALYSIS TECHNIQUES

Root cause analysis techniques are formalized methodologies that are usually applied to incidents with severe consequences, for example, major financial loss or injuries to personnel. The ideal root cause analysis technique would include all the dimensions discussed in Section 6.5.2—event sequence and structure, human error tendencies, PIFs, and organizational causes. Unfortunately no incident analysis technique currently exists that comprehensively addresses all of these areas. However, several of the available techniques provide a highly structured approach for performing an investigation which will provide insights into incident root causes. These techniques are described in subsequent sections. The description of techniques is necessarily selective, since a large number are available. (See Ferry, 1988, and CCPS, 1992d, for an extended analysis of these techniques.)

6.8.1. Tree of Causes/Variation Diagram

The Tree of Causes investigative method was developed by the Institute National de Recherché et de Sécurité (Leplat, 1982). The underlying principle of the method is that an accident results from changes or variations in the normal process. These antecedent variations must be identified listed, and finally organized into a diagram in order to define their interrelationship. Unlike a fault tree, the method starts with a real accident and results in a representation which only includes the branches actually leading to the accident. Thus, no OR gates are represented. The construction of the diagram is guided by simple rules which specify event chains and confluence relationships. These correspond to AND gates in fault trees, in other words, event C would only have occurred if events A and B also occurred. Suokas (1989) used the tree of causes to describe the Spanish campsite disaster (see Example 6.1 and Figure 6.5).

Example 6.1. The Spanish Campsite Disaster
(based on a description in Mill, 1992)

A tank truck was delivering propylene from Tarragon to Puertotollano, a road journey of 270 miles. Prior to the journey, the tank truck had frequently carried anhydrous liquid ammonia, which probably affected the strength of the high tensile steel storage tank. Another contributory factor to the accident was the fact that no pressure relief valve was fitted. At the loading bay, the tanker was filled with propylene. No metering facilities or overload cut-out devices were provided. The driver measured the weight of propylene at a scale at the exit to the site. The weight of propylene in the tank was 23 tons, which was 4 tons over the maximum allowed weight.

The driver of the tank truck decided to take the coastal route to Puerto-tollano, which passed close to several campsites. During the journey, the pressure in the tank built up and, because of the absence of a pressure relief valve, the weakened tank cracked. The propylene that was released ignited, and a flash fire burned near the tank. Eventually this ruptured and an explosion occurred close to a campsite, killing 210 people.

It should be noted that the completed diagram is not a diagram of causes as such, since variations are the active factors necessary to generate an accident in conjunction with other latent factors already present in the system. The method recognizes that there may be permanent factors in a system which need to be represented in order to improve the comprehensiveness of the diagram, and it is by representing these "state antecedents" that one moves toward a comprehensive description of causes. For example, in Figure 6.6 the situation "no relief valve in tank" could have arisen from "design error" as an antecedent cause. The goal of the method is to identify those changes which can be introduced to break the flow of events in the diagram.

The finished diagram is used to identify nodes representing inappropriate acts and abnormal physical states in the system, and to extract a list of factors involved in the accident with a view to improving the conditions for human decision-making and action, hence improving the work environment and system design. Also, the sequence of events is analyzed with the objective of breaking the causal relations among nodes by either making physical changes or providing operator feedback concerning a risky course of events. Both of these interventions act as barriers against the flow of events which could lead to an accident

Although the diagram is easy to construct and represents the incident process in an accessible manner, the method provides little guidance on how to collect all the relevant information or identify the main events involved. The method also relies heavily on the analyst's knowledge of the system conditions. Without this knowledge, it is necessary to perform a task analysis of the system in order to identify all the deviations. The root causes may remain undiscovered if the analyst is not experienced in incident investigation, as the method deals mainly with identifying direct causes, trigger events and prevailing conditions, but not the underlying causes which lead to these.

An extension of the tree of causes, called **variation diagrams** (Leplat and Rasmussen, 1984) was developed to answer some of these criticisms. In this method, the Rasmussen stepladder model of human error (see Chapter 2) is applied to analyze causal factors at each node of the tree. A detailed example of the use of this technique is provided in Chapter 7 (Case Study 1).

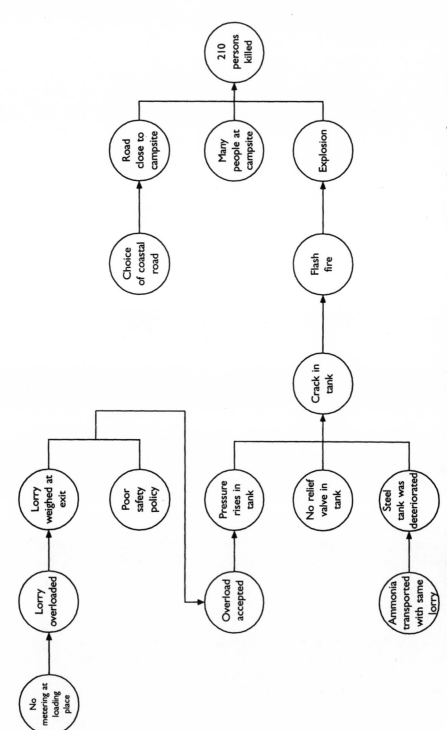

FIGURE 6.5: **The Spanish Campsite Disaster Described Using the Tree of Causes Diagram (from Suokas, 1989).**

6.8.2. The Management Oversight and Risk Tree (MORT)

The development of MORT was initiated by the U.S. Atomic Energy Commission, and is described in Johnson (1980). MORT is a comprehensive analytical procedure that provides a disciplined method for determining the causes and contributing factors of major accidents. It also serves as a tool to evaluate the quality of an existing safety program.

Management Oversight and Risk Tree is designed as an investigative tool with which to focus on the many factors contributing to an accident. A unique feature of the method is a logic diagram (see Figure 6.6) which represents an idealized safety system based upon the fault tree method of system safety analysis. The diagram comprises specific control factors and general management factors. Detailed consideration of the former is accomplished by reasoning backward in time through several sequences of contributing factors. This analysis ends when the question posed by the MORT statements is answered "yes" or "no." The analyst must focus upon the accident sequence when evaluating the specific control factors and, when evaluating the management factors, must consider the more global or total management controls. The diagram is supplemented by the MORT text which is a commentary on best concepts and practices found in the safety literature. It contains criteria to assist the analyst in judging when a factor is adequate or less than adequate. In summary, MORT provides decision points in an accident analysis which help an analyst detect omissions, oversights, or defects. Johnson (1980) claims that MORT considerably enhances the capability of the analyst to identify underlying causes in accident analyses.

However, MORT does not aid in the representation of the accident sequence which must first be determined before the method can be effectively used. Although MORT provides a comprehensive set of factors which may be considered when investigating an incident, it can easily turn an investigation into a safety program review as no guidance is provided on the initial investigative process.

MORT excels in terms of organizational root cause identification, as factors such as functional responsibilities, management systems and policies are well covered, but this strength of the method requires an accurate description of the incident process, and an experienced MORT analyst who is knowledgeable and well-practiced in the methodology.

6.8.3. Sequentially Timed Ever.ts Plotting Procedure (STEP)

The STEP procedure, described by Hendrick and Benner (1987), was developed from a research program on incident investigation methods. STEP is based on the multiple events sequence method and is an investigative process which structures data collection, representation, and analysis.

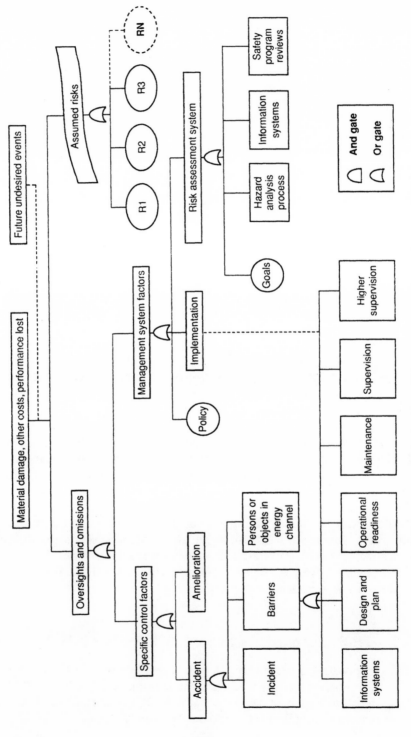

FIGURE 6.6. **Management Oversight and Risk Tree (Johnson, 1980).**

274

The method distinguishes among actors (referred to as "agents" in this book), actions, and events. Agents can be people, equipment, substances, etc., whereas actions are anything brought about by an agent. Events are the unique combination of one agent plus one action during the main incident process. The method's primary aim is to help the analyst identify the main agents and their actions and map the relations among these events along a flexible time line.

The main agents are identified based on a description of the incident and its end state. The initial state is determined by identifying the first event in the incident which is an unplanned change by an agent within the planned process. The method proceeds by developing an event sequence diagram which involves listing the agents down a vertical axis and establishing a time line on the horizontal axis. It should be noted that the time axis is not necessarily linear. Nevertheless, the actual time that events occur needs to be recorded. Each agent's actions are traced from the start of the incident to the finish. Agents which initiate changes of state in other agents are also identified. This reveals new agents not previously implicated in the incident. Events are positioned relative to one another along the time line and causal links are represented. Figure 6.7 provides an example of the structure of the STEP work sheet using the Spanish campsite disaster described in Section 6.8.1 and Figure 6.5. Case Study 1 in Chapter 7 provides a detailed example of a STEP analysis.

As the diagram develops, a necessary and sufficient test is applied to pairs of events, and checks for completeness and sequencing are made. One-to-many and many-to-one relations can be represented in the diagram. If data cannot be found to verify the relation between an event pair, then a technique called back-STEP can be used to explore gaps in understanding. Essentially back-STEP is a fault tree which uses the event with no other events leading to it as the top node. The analyst then develops possible event flows which could describe what happened during the gap in events in order to cause the top node.

When the diagram is complete, the analyst proceeds through it to identify sets of events that were critical in the accident sequence. These critical events are then subjected to a further causal analysis using other techniques such as **root cause coding**, described below in Section 6.8.4.

The method is well-structured and provides clear, standardized procedures on how to conduct an investigation and represent the incident process. Also it is relatively easy to learn and does not require the analyst to have a detailed knowledge of the system under investigation. However, the method alone does not aid the analyst in identifying root causes of the incident, but rather emphasizes the identificaticn of the propagation of event sequences. This is an important aspect of developing a preventive strategy.

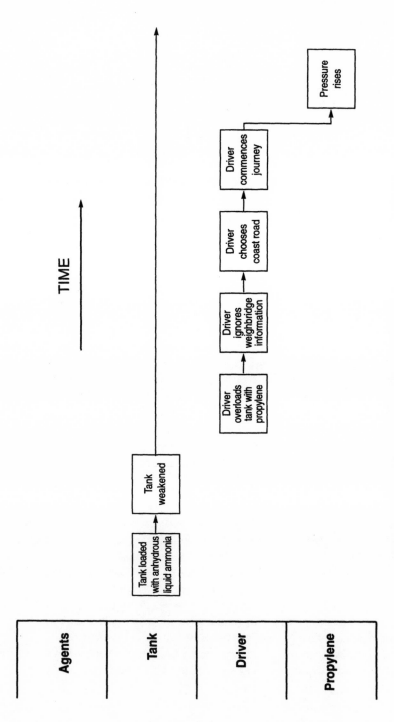

FIGURE 6.7. STEP Diagram for the Spanish Campsite Disaster (page 1).

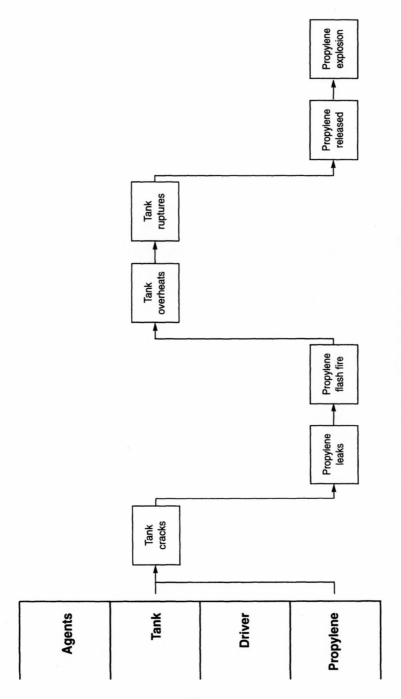

FIGURE 6.7. STEP Diagram for the Spanish Campsite Disaster (page 2).

277

6.8.4. Root Cause Coding

As discussed in the previous section, STEP, although an excellent method for representing the accident sequence, does not in itself provide direct insights into the causal factors underlying an incident. However, it can be used in conjunction with a technique called root cause coding to produce a comprehensive accident investigation framework. The most important aspect of root cause coding is a root cause tree (see Figure 6.8). This is a decision aid to assist the analyst in identifying the underlying causes of accidents at a number of different levels.

The root cause tree was originally developed in the U.S. nuclear industry and has close links with MORT. Armstrong (1989), and Armstrong et al. (1988) provide full descriptions of its development and construction. It consists of six levels covering equipment failures, quality failures, management systems failure and human error. The decision tree codes critical actions and events. By entering the top level of the tree, the analyst has to determine whether the critical event involved an equipment difficulty, operations difficulty or technical difficulty. Based on the answers to these general questions, the investigator branches down to more specific levels of the tree. These relate to: functional area, equipment problem category, major root cause (such as training and management system), near root cause (such as incorrect procedure, and training methods less than adequate), and finally root causes themselves (such as procedures design and training inadequate). This root cause coding allows the investigator to specify the underlying reason for a given critical event. Critical events in the STEP analysis are those which lead directly to critical outcomes or which influenced the course of subsequent events in a critical manner. The use of root cause coding in conjunction with STEP is illustrated in Chapter 7, case study 5.

6.8.5. Human Performance Investigation Process (HPIP)

The HPIP process is a hybrid methodology which combines a number of the techniques discussed earlier. The development of HPIP was supported by the U.S. Nuclear Regulatory Commission, and most of its early applications have been in the nuclear industry. A description of the approach is provided by Paradies et al. (1992). The structure of the incident investigation process is represented in Figure 6.9. The HPIP method was originally developed for use by investigators external to the plant (specifically NRC inspectors), and hence some steps would be modified for use by an in-plant investigation team in the CPI. The stages in the investigation process and the tools used at each of these stages are discussed below.

ROOT CAUSE TREE

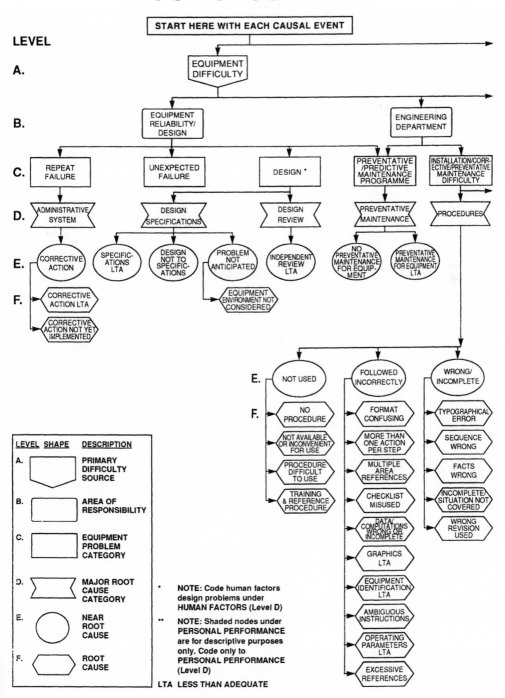

FIGURE 6.8. **Root Cause Tree** *(continues on next two pages).*

ROOT CAUSE TREE

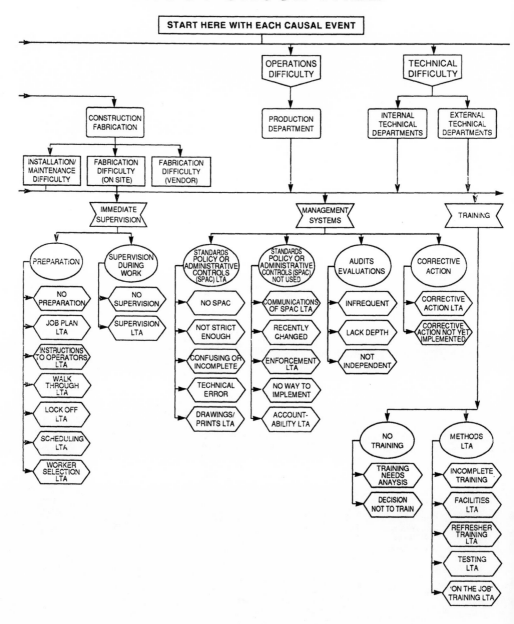

FIGURE 6.8. **Root Cause Tree** *(continues on next page).*

ROOT CAUSE TREE

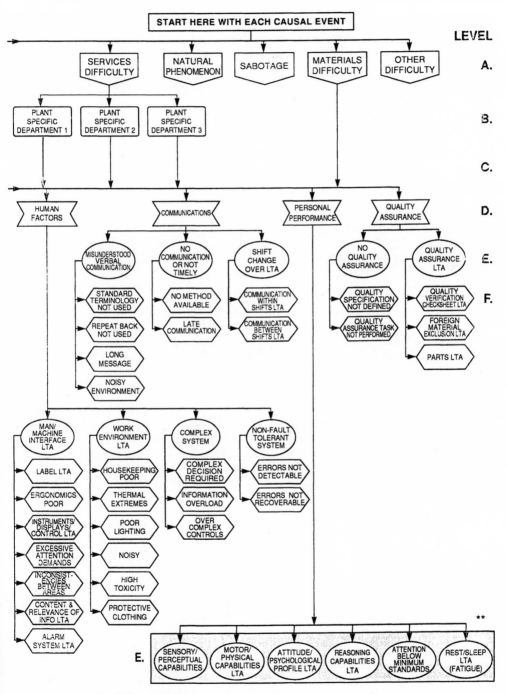

FIGURE 6.8. **Root Cause Tree.**

Plant Investigation

This involves collecting physical and documentary evidence and interviewing key witnesses. A preliminary representation of the accident sequence is developed using the Events and Causal Factors Charting (ECFC) method. This is an event sequence representation method similar to the Tree of Causes and related techniques, and was originally developed for use with the root cause tree described in the last section (see Armstrong et al., 1988). A worked example of the ECFC method is provided in Chapter 7, case study 1. Stimulus operation response team performance (SORTM) is a structured series of questions addressing the following aspects of the causal factors involved in the incident:

- Stimulus—the initiating events for the actions involved in the accident
- Operation—the mental and physical skills and information requirements
- Response—the nature of the individual actions
- Team performance—the team performance aspects of the incident
- Management factors

Develop Event Sequence

This is accomplished using the ECFC and the Critical Human Action Profile (CHAP), a task analysis-based method used to identify the most critical actions necessary for the performance of the task. Change Analysis is a technique for investigating the role of change in accident causation. It will be described in Section 6.8.6.

Analyze Barriers and Potential Human Performance Difficulties

During this phase of the analysis process, the barriers that have been breached by the accident are identified. These barriers could include existing safety systems, guards, containment, etc. This analysis is called barrier analysis. The causal factors from SORTM are also applied in more detail.

Analyze Root Causes

Using the ECFC representation of the incident, a series of detailed questions which address specific causal factors (e.g., poor procedures), are applied to evaluate direct and indirect root causes. These detailed questions are contained in a series of HPIP modules.

Analyze Programmatic Causes

This stage is used to evaluate indirect generic causes such as inadequate human factors policies in the plant or the company.

Evaluate Plant's Corrective Actions and Identify Violations

This stage is used to develop remedial strategies, based on the findings of previous stages. In the original HPIP framework, this stage is simply a check

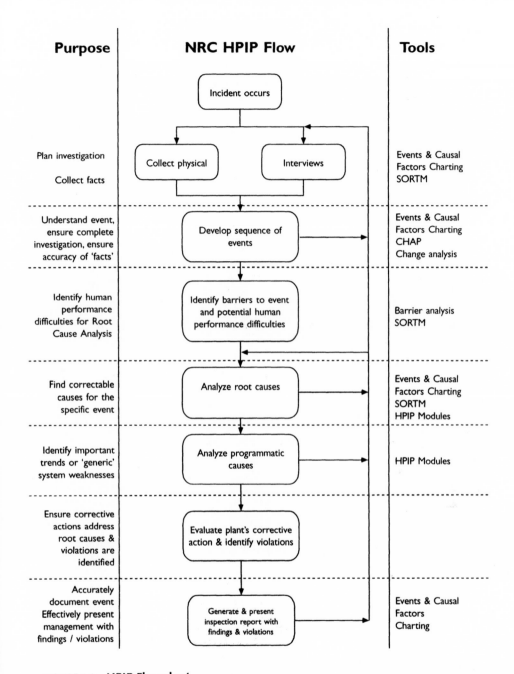

Purpose	NRC HPIP Flow	Tools
	Incident occurs	
Plan investigation Collect facts	Collect physical Interviews	Events & Causal Factors Charting SORTM
Understand event, ensure complete investigation, ensure accuracy of 'facts'	Develop sequence of events	Events & Causal Factors Charting CHAP Change analysis
Identify human performance difficulties for Root Cause Analysis	Identify barriers to event and potential human performance difficulties	Barrier analysis SORTM
Find correctable causes for the specific event	Analyze root causes	Events & Causal Factors Charting SORTM HPIP Modules
Identify important trends or 'generic' system weaknesses	Analyze programmatic causes	HPIP Modules
Ensure corrective actions address root causes & violations are identified	Evaluate plant's corrective action & identify violations	
Accurately document event Effectively present management with findings / violations	Generate & present inspection report with findings & violations	Events & Causal Factors Charting

FIGURE 6.9. **HPIP Flow chart.**

performed by the regulatory authority to ensure that the plant response is adequate. This stage was included because each nuclear power plant has a resident inspector who would be expected to evaluate the response of this plant to the incident.

Generate and Present Inspection Report
The results of the investigation would be presented to management at this stage.

6.8.6. Change Analysis

In many accidents an important contributory cause is the fact that some change has occurred in what would otherwise be a stable system. The importance of change as an antecedent to accidents has lead to the development of a formal process to evaluate its effects.

The technique of change analysis was originally developed by Kepner and Tregoe (1981) as part of research sponsored by the Air Force. It was subsequently incorporated in the MORE technique described earlier. A comprehensive description of the process is provided in Ferry (1988). The main stages of the process are shown in Figure 6.10. The MORT process indicates that the

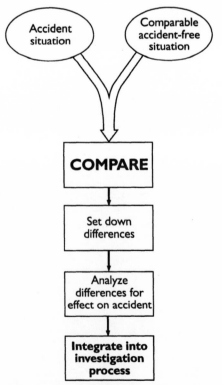

FIGURE 6.10. **The Six Steps of Change Analysis (Ferry, 1988).**

following types of change should be considered (these have been interpreted in the context of process safety):

Planned versus Unplanned Changes
Planned changes should be documented as part of a formal change monitoring process (for example via a quality assurance system). Unplanned changes should be identified during the accident investigation process.

Actual versus Potential or Possible Changes
Actual changes are those identified by a surface analysis of the incident sequence. Potential changes would be revealed by a more in-depth analysis.

Time Changes
These are changes in the system over time due to factors such as wear and deterioration of hardware, and also the erosion of human systems such as supervision and permits to work.

Technological Changes
Changes resulting from the implementation of new processes and plant.

Personnel Changes
These are changes in individuals or teams which may mean that normally assumed unwritten knowledge (e.g., about the particular operational characteristics of the plant) is not available.

Sociological Changes
These can arise from changes in values of process workers (e.g., an increased focus on production rather than safety, because of a fear of reduced pay or job losses) due to larger changes in society (e.g., reduced job security because of an economic depression).

Organizational Changes
These may give rise to lack of clarity with regard to who is responsible within an operating team.

Operational Changes
These are defined in MORT as changes in procedures without an associated safety review.

6.8.7. Evaluation of Root Cause Analysis Techniques

On the basis of the descriptions of incident analysis techniques in the previous section and the comprehensive reviews available in Ferry (1988) and CCPS

(1992d) it is clear that there is no shortage of methods to provide a framework for the detailed incident analysis that would form part of a root cause analysis system. However, despite the variety of techniques which are available, very few of these appear to effectively address the following areas:

- Incorporation of psychological models of human error into the investigation process
- Evaluation of effects of management influences and policy factors on error causation
- Consideration of how formal data collection incident investigation methods are to be introduced into a plant in order to ensure acceptance and long-term support by the workforce

Use of Psychological Models

With regard to this issue, it can be argued that a knowledge of the psychological processes underlying error may not be necessary in order to carry out effective incident analyses. If the direct causes of errors are identified, in the form of the PIFs that were present when the error occurred, then it may appear to be unnecessary to try to evaluate the actual mental processes that occurred.

However, a knowledge of the psychology of error from the cognitive engineering perspective provides unique insights. In the case study of a reactor explosion quoted in Section 1.8, one of the error mechanisms identified was the reversion, under stress, to a familiar set of operating instructions which were similar to those which should have been used, but which omitted a critical step. This "strong stereotype take-over" error mechanism (see Chapter 2) would not have occurred to the analysts without some knowledge of cognitive psychology. This would mean that an important error reduction strategy, the use of less confusing procedures, would have been neglected.

In general, the value of a psychological perspective in incident analysis is that it directs the analyst to search for causes that would not otherwise have been considered. This means that the development of preventative strategies will be better informed. In addition, an evaluation of causes from a psychological perspective can be useful when the "root cause" appears to be an otherwise incomprehensible failure on the part of an individual. A psychological analysis can break the "causal log jam" by providing an explanation.

Management and Policy Influences on Error and Accident Causation

As has been emphasized in Chapters 1, 2, and 3, the system-induced error view states that it is insufficient to consider only the direct causes of errors. The underlying organizational influences also need to be taken into account. However, most of the available techniques stop when an immediate cause has been identified, such as less than adequate procedures or poor equipment design. The questions of why the procedures were poor, or why the equipment was badly designed, are rarely addressed at the level of policy. Kletz (1994a) has described the importance of identifying these policy level factors, and including them in recommendations for corrective action.

With regard to evaluating these factors, it is recommended that structured checklists be used, such as those provided by the HFAM method described in Chapter 2. These checklists provide an explicit link between the direct causal factors and management policies. Figure 2.12 shows how these checklists could be used to investigate possible procedures deficiencies, and the policies that led to the deficiencies, as part of the incident investigation. Similar checklists can be used to investigate possible culture problems (e.g., inappropriate trade-offs between safety and production) that could have been implicated in an accident.

Workforce Support for Data Collection and Incident Analysis Systems
Few of the incident investigation and data collection systems reviewed provide any guidelines with regard to how these systems are to be introduced into an organization. Section 6.10 addresses this issue primarily from the perspective of incident reporting systems. However, gaining the support and ownership of the workforce is equally important for root cause analysis systems. Unless the culture and climate in a plant is such that personnel can be frank about the errors that may have contributed to an incident, and the factors which influenced these errors, then it is unlikely that the investigation will be very effective.

6.9. IMPLEMENTING AND MONITORING THE EFFECTIVENESS OF ERROR REDUCTION MEASURES

As shown in the flow diagram in Figure 6.1, the process of identifying underlying causes leads naturally to the development of error reduction strategies that will address these causes. Remedial strategies can be formulated to prevent a recurrence of the specific type of accident under investigation and/or can address more fundamental systemic causes possibly at the level of management or even policy systems. Although there is often pressure to be seen to be implementing measures which address the direct causes of accidents, it is obviously important, and in the long run highly cost effective, to remedy problems at a fundamental level. If these underlying causes are not addressed, it is likely that an accident of the same type will recur in the future.

Establishing the effectiveness of error reduction measures is difficult in an environment where there are a number of other changes occurring. Nevertheless, a properly designed reporting system should be able to detect changes in the incidence of particular types of error as a result of the effectiveness of the preventive strategy. One of the advantages of near-miss reporting systems is that they can provide a greater volume of evidence to allow the effectiveness of preventive measures to be evaluated.

6.10. SETTING UP A DATA COLLECTION SYSTEM IN A CHEMICAL PLANT

In previous sections of this chapter, the required characteristics of effective causally based data collection systems to reduce human errors and accidents have been described. In this final section, the stages of setting up such a system in a plant will be described.

Specify Objectives

The first stage of the process will need to specify the overall boundaries and objectives of the proposed system. For example, will the system perform both incident reporting and root cause analyses, what types of data will be stored, who will be involved in setting up and operating the system? In order to ensure that the system engenders ownership from its inception, a data collection group should be set up including representatives from all levels of the organization. This group should be provided with visible management support and adequate resources. The purpose of this group is to provide a stewardship function to ensure that the data collection systems are implemented and maintained.

Evaluate Potential Cultural Barriers to Data Collection

It is advisable at an early stage in the development process to determine if problem areas such as a negative view of human error or a blame and punishment culture exist in the organization. If these are identified as a problem, then appropriate measures such as culture change programs can be implemented. If these problems are not addressed at the design stage, then it is unlikely that any human error data collection initiative will be successful. Since cultural problems often have their long-standing origins in senior or middle management beliefs (see Section 6.5.1), the development of a supportive culture may not be an easy task. Nevertheless, it is wise to evaluate the extent of these problems and the possibility of developing a supportive culture within which the data collection process can operate. Otherwise, resources may be wasted in implementing a system which is technically adequate but which fails because of lack of support.

Specify Data Collection Methods and Responsibilities

Several types of data collection have been specified in earlier sections. It is important that the responsibilities for operating the various aspects of the system are unambiguously defined.

Specify the Analysis, Interpretation Framework, and the Type of Input Data Required

The purpose of this stage is to specify how underlying causes will be derived from plant data and the type and level of detail required to perform these analyses. During this stage data reporting and incident analysis forms will be developed, based on the underlying causal model together with a consideration of the practicalities of data collection (e.g., amount of time available, other competing priorities).

Develop Procedure for Identifying and Implementing Remedial Measures

This process will specify the methods for deriving error reduction strategies from the data collected, and the responsibilities for implementing these measures and monitoring their effectiveness.

Specify Feedback Channels

This phase of the program is designed to ensure that the information produced by the system is fed back to all levels of the workforce, including process operators, managers, supervisors, engineers, and senior policy makers.

Develop Training Programs

This phase will proceed in parallel to some of the earlier phases. In addition to launching the system, and raising the general level of awareness at all levels of the organization regarding the importance of human performance, the training program will provide the specific skills necessary to operate the system.

Implement Pilot Data Collection Exercise in Supportive Culture

In order to ensure that the data collection system has been thoroughly checked and tested prior to its launch, it is advisable to test it in a plant or plant area where there is likely to be a supportive culture. This will allow the effectiveness of the system to be addressed prior to a larger-scale implementation in a less controlled environment.

Evaluate Effectiveness on the Basis of Outputs and Acceptance

Once the system has been implemented on its chosen site, its effectiveness needs to be evaluated at frequent intervals so that corrective action can be taken in the event of problems. The first criterion for success is that the system must generate unique insights into the causes of errors and accidents, which would not otherwise have been apparent. Second, the system must demonstrate a capability to specify remedial strategies that, in the long term, lead to enhanced safety, environmental impact and plant losses. Finally, the system must be owned by the workforce to the extent that its value is accepted and it demonstrates its capability to be self-sustaining.

Maintain the Momentum

In order to maintain motivation to participate in the data collection process, the providers of information need to see that their efforts produce tangible benefits in terms of increased safety. This is particularly important in the case of near miss reporting systems, where the benefits of participation may be less obvious than with accident reporting systems. This type of feedback can be provided via regular newsletters or feedback meetings. Even if tangible improvements cannot be demonstrated in the short term, then it is at least necessary to show that participation has some effects in terms of influencing the choice of safety strategies. As with the data providers, data analysts also need to be motivated by seeing that their work is recognized and used effectively and that recommendations are implemented.

Since the resources for data collection systems will be provided by senior management it is essential that information from the system is fed back to policy makers at this level. It is also important that the system indicates the problem areas as well as the successes. Many organizations have drifted to a state where safety standards have fallen to below acceptable levels over time as a result of suppression of information feedback to senior managers. This may be carried out with good intentions, but its long-term effect can be disastrous.

6.11. SUMMARY

This chapter has adopted a broad perspective on data collection and incident analysis methods. Both qualitative and quantitative aspects of data collection have been addressed, and data collection approaches have been described for use with large numbers of relatively low-cost incidents or infrequently occurring major accidents.

Three major themes have been emphasized in this chapter. The first is that an effective data collection system is one of the most powerful tools available to minimize human error. Second, data collection systems must adequately address underlying causes. Merely tabulating accidents in terms of their surface similarities, or using inadequate causal descriptions such as "process worker failed to follow procedures" is not sufficient to develop effective remedial strategies. Finally, a successful data collection and incident investigation system requires an enlightened, systems oriented view of human error to be held by management, and participation and commitment from the workforce.

7

Case Studies

7.1. INTRODUCTION

The purpose of this chapter is to show that improvements in safety, quality, and productivity are possible by applying some of the ideas and techniques described in this book. The fact that error reduction approaches have not yet been widely adopted in the CPI, together with questions of confidentiality, has meant that it has not been possible to provide examples of all the techniques described in the book. However, the examples provided in this chapter illustrate some of the most generally useful qualitative techniques. Case studies of quantitative techniques are provided separately in the quantification section (Chapter 5). The first two case studies illustrate the use of incident analysis techniques (Chapter 6).

The first case study describes the application of the sequentially timed event plotting (STEP) technique to the incident investigation of a hydrocarbon leak accident. Following the analysis of the event sequence using STEP, the critical event causes are then analyzed using the root cause tree.

In the second case study, variation tree analysis and the events and causal factors chart/root cause analysis method are applied to an incident in a resin plant. This case study illustrates the application of retrospective analysis methods to identify the underlying causes of an incident and to prescribe remedial actions. This approach is one of the recommended strategies in the overall error management framework described in Chapter 8.

Case study 3 illustrates the use of proactive techniques to analyze operator tasks, predict errors and develop methods to prevent an error occurring. Methods for the development of operating instructions and checklists are shown using the same chemical plant as in case study 2.

Case study 4 is based on the updating of information displays for refinery furnace control from traditional pneumatic panels to modern VDU-based display systems. In addition to illustrating the need for worker participation in the introduction of new technology, the case study also shows how task

291

analysis and error analysis techniques (Chapter 4) can be used in human–machine interface design.

Case study 5 provides an example from the offshore oil and gas production industry, and illustrates the fact that in solving a specific practical problem, a practitioner will utilize a wide variety of formal and informal methods. Table 7.1, which describes some of the methods used in the study, includes several techniques discussed in Chapter 4, including interviews, critical incident techniques, walk-throughs and task analysis.

7.2. CASE STUDY 1: INCIDENT ANALYSIS OF HYDROCARBON LEAK FROM PIPE

7.2.1. Introduction

This case study concerns the events leading up to the hydrocarbon explosion which was the starting point for the Piper Alpha offshore disaster. It describes the investigation of the incident using the sequentially timed events plotting (STEP) technique. Based on the STEP work sheet developed, the critical events involved in the incident are identified and analyzed in order to identify their root causes.

The following description is taken from Appendix D of CCPS (1992a). (The results of the public inquiry on the disaster are in Cullen, 1990.)

> An initial explosion occurred on the production deck of the Piper Alpha Offshore Platform in the North Sea at about 1:00 PM on July 6, 1988. The incident escalated into a tragedy that cost the lives of 165 of the 225 persons on the platform. Two additional fatalities occurred on a rescue boat. The Piper Alpha Platform was totally devastated.
>
> Immediately after this blast, a fire originated at the west end of B Module and erupted into a fireball along the west face. The fire spread quickly to neighboring portions of the platform. Approximately 20 minutes later, a major explosion happened due to the rupture of the Tartan gas riser. This occurrence caused a massive and prolonged high pressure jet of flames that generated intense heat. At about 10:50 PM, another immense blast occurred that was believed to be a result of the rupture of the MCP-01 gas riser. Debris from this explosion was projected up to 800 m. away from the platform. Structural deterioration at the level below Module B had begun. This failure was accelerated by a series of additional explosions. One of these eruptions was caused by the fracture of the Claymore gas riser. Eventually, the vast majority of the platform collapsed.

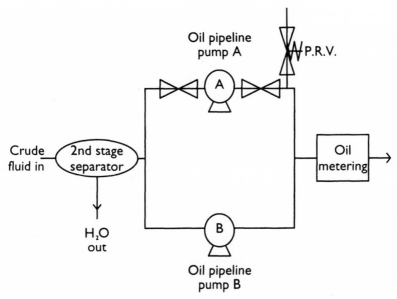

FIGURE 7.1. **Simplified Process Diagram: Hydrocarbon Leak From Pipe (from Cullen, 1990).**

7.2.2. Process Description

The process involved in the incident is concerned with the separation of crude into three phases. The crude is pumped into a two stage separation process where it is divided into three phases; oil, gas, and water. The water is cleaned up and dumped to drain. The remaining mixture of oil and gas is then pumped into the main oil line where it is metered and sent on for further processing. A simplified process diagram is shown in Figure 7.1. The case study described here is centered on a flange leak in one of the oil pipeline pumps (pump A) and its associated pressure relief valve piping.

7.2.3. Events Leading to the Explosion

The separation plant had been running smoothly for several weeks and the planned shutdown was some time away. On the day of the incident a number of unusual events occurred which contributed to its development. Shortly after the start of his day shift, the control room operator received a high vibration alarm from booster pump A in the crude fluid separation building. Following normal procedure, he switched over to the standby pump (pump B), switched off pump A, and told his supervisor of the alarm. The supervisor subsequently organized a work permit for work to be carried out on pump A by the day shift maintenance team. The permit was issued and repair work started. Since pump A and its associated pipework was off-line, the supervisor took the opportunity to carry out scheduled maintenance on the pressure relief valve (PRV) downstream of pump A. The valve had been malfunctioning, and although the work was not scheduled to be done for some weeks, the specialist contractor team who maintain the PRVs had a team available to carry out the work immediately. The supervisor therefore now had two teams working on the pump A systems: the shift maintenance team working on the pump itself, and a two-man contractor team working on the PRV and its associated pipework. The PRV for pump A is not located immediately adjacent to the pump, and is above floor level, close to a number of other pipe runs. The following description represents a hypothetical sequence of events based on the inquiry findings, but embellished for the purposes of the case study.

During the course of the day, the shift maintenance team identified the cause of the vibration and rectified it. They rebuilt the pump and completed the work at about 17:30, before their shift ended. The permit was returned to the operations supervisor who duly signed it off. The contractor's work, however, did not go as smoothly. The team removed the PRV and the team leader took it to their workshop for maintenance and pressure testing. His partner remained behind in order to fit a blank to the pipeline, as required by site procedure. The contractor fitted the blank to the pipe, although the job was made difficult by its awkward position, and he returned to the workshop to help with the maintenance on the valve itself. Unknown to the workers, the blank had not been fitted correctly and did not seal the pipe. The PRV required a complete strip and overhaul but the contractors were unable to complete the work by the end of the day. They did not inform the operations supervisor, as they thought that the pump had been signed off for more than one day and that they would be able to complete the work the following morning. The day shift supervisor, having had no contact with the contractor team since signing on their permit, made the assumption that the contractors, as they were no longer on the job, would be working overtime to complete the job during the night shift.

At shift handover at 18:00, the incoming operations supervisor was briefed by the day supervisor. The conversation centered on the vibration fault and subsequent repair work carried out. However, no mention was made of the work on the PRV, so consequently none of the incoming shift were aware of it. The night shift supervisor, wanting to return pump A to standby as soon as possible, asked the plant operator to check the status of the pump, and together with the shift electrician, to reset it and put it back on stand-by. The operator, unaware of the work being done on the PRV, did not check this part of the system and, following inspection of the pump, returned it to stand-by.

Later in the night shift the control room operator received a pump trip alarm from pump B. Soon after, the second stage separator high oil level alarm sounded in the control room. The operator, needing to reduce the level, switched on to the standby pump A. Monitoring the oil level in the separator, the operator saw the level fall. Unknown to the control room operator, switching to pump A resulted in high pressure oil and gas leaking from the incorrectly fitting blank. The control room operator's monitoring of the oil level was interrupted by the gas monitoring system giving an alarm. The operator accepted the alarm but was not unduly worried, thinking it was a false alarm, as often happens after work has been done on a pump. He decided to radio the plant operator and asked him to check it out. The oil level continued to fall in the separator, and the leaking flange continued to release oil and gas into the separation building. The plant operator, responding to the control room operator's request, went to investigate the low level alarm in the separation building. While the control room operator was waiting for the operator to report back, the high gas alarm sounded. He immediately started safety shutdown procedures. It was at this point that the oil and gas mixture ignited and exploded, and the next phase of the Piper Alpha disaster began.

7.2.4. Investigation Using the STEP Technique

A number of stages are used when applying the STEP technique. These will be illustrated with respect to the investigation carried out for the above incident.

The first stage involves documenting the beginning and the end state of the incident. This bounds the scope of the investigation from the first event that deviated from the planned process to the last harmful event in the incident. In this case study, these are the faulty PRV and pump vibration through to the explosion as a result of the ignition of the leaking hydrocarbon mixture. In documenting end states the intention is to identify the main agents (people and things) involved in the incident. This is achieved by recording, measuring, sketching, photographing, and videotaping the incident scene. For example, who were the last people present? What were they doing? Where were they? What plant was involved? Was it in operation? The start state indicates the state of the agents at the beginning of the incident, which shift personnel were on-site, the plant status, and how it was being run. Drawings, procedures, records, and personnel are the typical sources of such information.

The STEP work sheet shown in Figure 7.2 is developed during the analysis. It will obviously be simple and patchy at first, but serves as an important aid in guiding and structuring the data collection and representation. It is important to use a form of work sheet which is easy to construct and modify. Flip charts and add-on stickers are an ideal basis for the work sheet, and are also easy to change. The agents are placed on the vertical axis of the work sheet. This is the start state, the point at which the first deviation in the planned process which led to the incident is identified. In this incident this was found to be the vibrating pump A. The horizontal axis represents the time line on which events are fixed for each agent. The aim is to trace each agent's actions from the start state to the end state This will picture the effect of each agent in the incident, the effect it had on other agents, and what influenced the agents. This can lead to new agents being considered which were not initially identified as being involved in the incident.

Taking the process control system (PCS) as an agent, for example, it can be identified as an agent at the start and end state of the incident. This provides objective information about the plant before, during, and after the event, and allows fixed time points on the work sheet to be established. Figure 7.5 contains the information available from the printout of the PCS alarm recorder. This locates a number of events for the PCS agent along the time-line, for example, when the vibrations in pump A were detected, when pump B was activated, and when the high oil level alarm was activated. This means that concrete data are available on events relating to the PCS from the start of the incident to its end. Similar data were also available for the gas monitoring system which indicate when both low and high alarms were activated.

A similar process was carried out for each of the agents identified. For agents that are people, however, the process can present problems. For example, in this case study the time period for the development of the incident crossed a shift boundary, another fixed point on the work sheet, and therefore involved different people. Each of these needs to be interviewed to establish their role in the incident. It is important to focus on the events involving the agents and to avoid introducing bias into the work sheet. In this case it was possible to use objective data to validate interview data. The PCS data confirms actions and indicates initiation times for the action taken by the control room operators on both the day and night shift. The interview data used to develop the STEP work sheet for this incident are contained in Figure 7.3 (note that these are hypothetical interview data generated for this case study). Data from the PCS gas monitoring systems was used to verify and help locate data gathered from interviews. Particular focus was paid to agent's actions which initiate changes in the other agents. For example, the control room's request for the plant operator to check out the low gas alarm, or the high oil level alarm leading to the control room operator switching to pump A and directing his attention to monitoring the oil level.

The logic tests for placing building blocks on the STEP diagram help to determine whether all the events for an agent were listed, and whether the relevant building block was placed correctly on the time sheet relative to that and other events. It is here that one of the strengths of the work sheet became apparent. Using the events for each agent and the simple logic tests quickly identified gaps in the analyst's knowledge. These gaps were further defined once the event elements were linked.

As the diagram develops, a necessary and sufficient test is applied to event pairs. For example, the event involving the night shift controller switching from pump B to pump A and the tripping of pump B are necessary for the event, but not sufficient to cause this event. The process control system gave the high oil level alarm which reduced the time window for the operator to take other action, for example, investigating the cause of the trip. However, other events were also necessary. These were the confirmations by the plant operator and electrician that pump A had been placed back on standby. If this had not happened the option to switch to pump A would not have been available. The necessary and sufficient test led to both converging arrow links, as above, and also diverging links, for example where the gas/oil leaking from the flange leads to both high and high high gas alarms and is necessary for the ignition of the leak. In this way, the relationships among events were elicited and the investigator was forced to think about causal events one at a time instead of considering the incident as a whole. The process of data collection, with its conversion to events, building block positioning and logic testing, was an iterative one and this diagram went though several revisions.

The STEP procedure provides investigators with a well-structured, logical, and independently verifiable representation of the events involved in the incident. This, however, only represents the first stage in the investigation. The second stage was to identify the critical agents and events in the incident process.

7.2.5. Identification of Critical Agents/Events

This stage involved the identification of critical actions and events in the incident process. Three critical events were identified from the STEP diagram. These were

- Failure to fit the blank correctly
- Changeover between day and night supervisor
- Contractor fails to report status of work

It was these events which significantly influenced the course of events by triggering later problems.

7.2.6. Identification of Root Causes

Root causes for each of the critical events were then determined using the root cause tree (see Figure 6.8 and Chapter 6, Section 6.8.4). This six-level decision tree was used which, based on answers to general questions, leads through successive levels of the tree until the root cause is identified or the data limitations prevent further progress. These root causes specify the underlying reason for a given critical event. The analysis profiles for each of the critical events are presented below.

CRITICAL EVENT 1 **Failure to Fit Blank Correctly**	
ROOT CAUSE 1	**ROOT CAUSE 2**
A. Equipment difficulty B. Engineering department C. Corrective maintenance D. Human factors E. Human–machine interface F. Ergonomics poor	D. Immediate supervision E. Supervision during work F. Supervision less than adequate (LTA)

The problem manifested itself as an equipment problem, namely a leaking flange joint. The department broadly responsible for this area (but not for implementing, monitoring, and subsequent recommendations) is the engineering department, as the specialist contractors work for them. The critical event took place during a corrective maintenance operation. From here, two separate root causes were identified, based on the data from the investigation.

- **Root cause 1:** Supervision was less than adequate. The team leader should have stayed with his colleague and checked the work as he had responsibility for the team
- **Root cause 2:** The cramped, confined space available made it difficult to verify that the blank had been correctly fitted

The problem was an operational difficulty concerning the production department. Two root causes were identified based on the investigation.

- **Root cause 1:** Procedures for shift changeover were inconvenient for use. The prescribed changeover procedure was detailed and elaborate, so it was not used in practice, being seen as too inconvenient for practical purposes. Consequently, one important aspect was omitted. Supervisors are supposed to go through the work permit book at each shift changeover. This was not done.

CRITICAL EVENT 2 **Changeover between Day and Night Supervisor**	
ROOT CAUSE 1	**ROOT CAUSE 2**
A. Operations difficulty B. Production C. Not applicable D. Communication E. Shift changeover LTA F. Communication among shifts LTA	A. Operations difficulty B. Production C. Not applicable D. Procedures E. Not used F. Inconvenient for use

- **Root cause 2:** The informal method of shift changeover used on the plant meant that vital information relating to plant status was not communicated across shifts.

CRITICAL EVENT 3		
Contractor Fails to Report Status of Work		
ROOT CAUSE 1	ROOT CAUSE 2	ROOT CAUSE 3
A. Operations difficulty	A. Services	
	B. Contractor Maintenance	
D. Immediate supervision	D. Training	D. Procedure Not Use
E. Preparation	?	?
F. Instructions to operators LTA	?	F. Inconvenient

The root causes for this critical event both concern the operations department and the service department who ran the contractor maintenance team. The operations department (i.e., the day shift operations supervisor) failed to provide adequate supervision and instructions to the contractor team. Explanations of the nature of the permit-to-work systems (i.e., the need to report back at end of shift) should have been given, and the possibility and implications of work not being completed before the end of the shift should have been considered by both parties.

On the part of the contractor team, two root causes were identified, root cause 1 being insufficient training of the contractor team leader. He was uncertain of permit systems, specifically whether they should report in at the end of shift, and, if so, who should do it. The second root cause relates to the procedure used at the end of the shift for supervisors to sign back permits. Although according to procedure all workers should hand back permits to supervisors in person, in practice this did not occur. If no one is present, or they are busy, it had become common practice to either leave the permits on the supervisor's desk, or to sign them back in the morning.

7.2.7. Conclusion

The case study has documented the investigation and root cause analysis process applied to the hydrocarbon explosion that initiated the Piper Alpha incident. The case study serves to illustrate the use of the STEP technique, which provides a clear graphical representation of the agents and events involved in the incident process. The case study also demonstrates the identification of the critical events in the sequence which significantly influenced the outcome of the incident. Finally the root causes of these critical events were determined. This allows the analyst to evaluate why they occurred and indicated areas to be addressed in developing effective error reduction strategies.

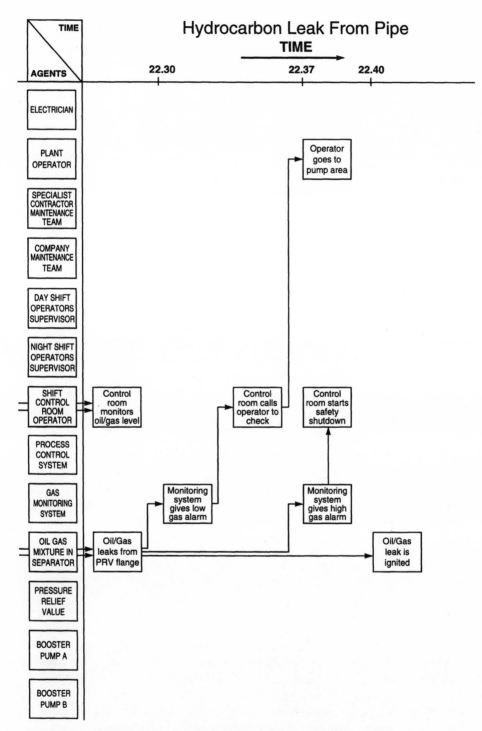

FIGURE 7.2 **STEP Diagram of Hydrocarbon Leak from Pipe, Page 1 of 4.**

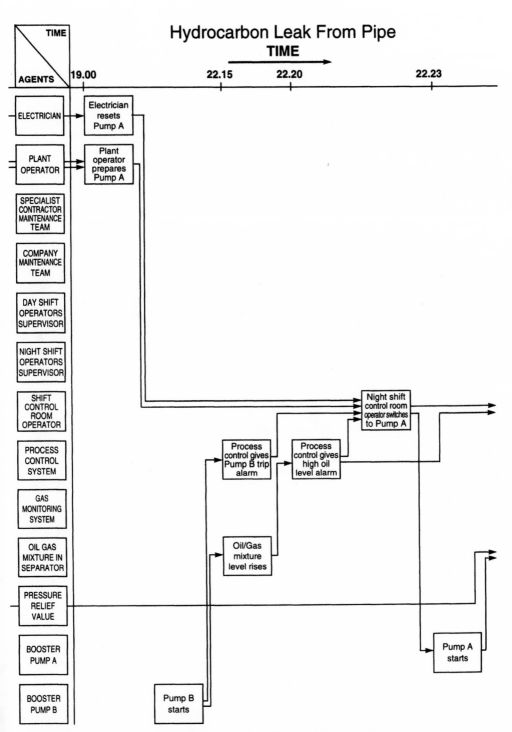

FIGURE 7.2 **STEP Diagram of Hydrocarbon Leak from Pipe, Page 2 of 4.**

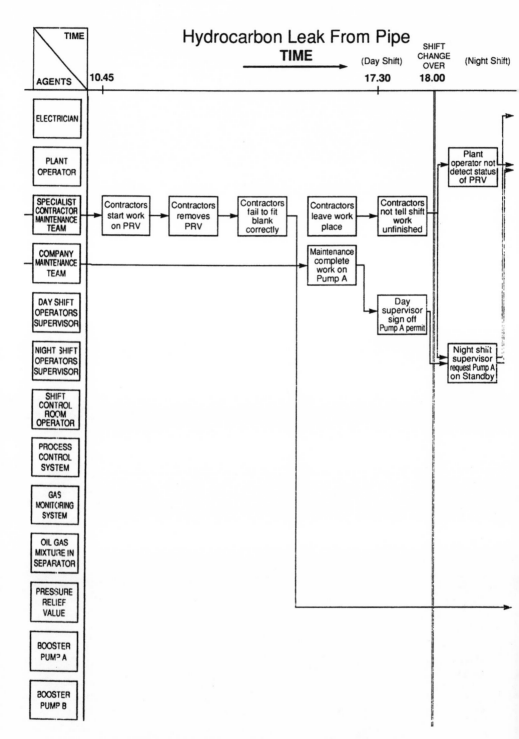

FIGURE 7.2 **STEP Diagram of Hydrocarbon Leak from Pipe, Page 3 of 4.**

302

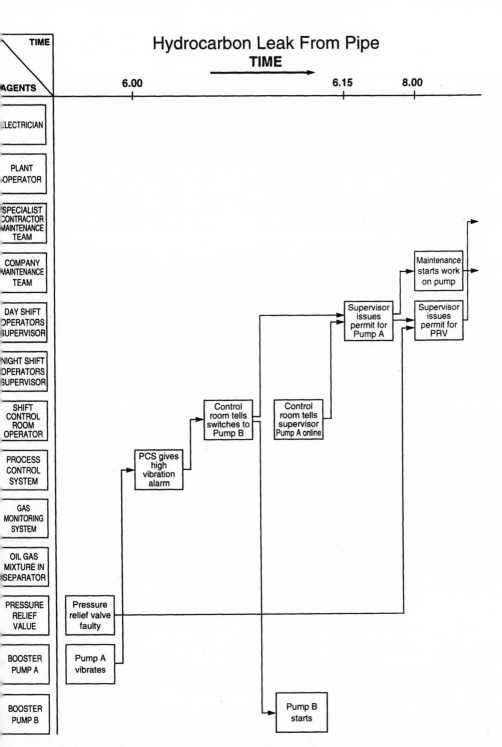

FIGURE 7.2 **STEP Diagram of Hydrocarbon Leak from Pipe, Page 4 of 4.**

303

Plant Operator (night shift)
1. Location: entering oil separation module.
2. Heard an explosion then thrown out of doorway.
3. Ran to phone and rang emergency services.
4. At start of his shift (18:00) had walked around plant, including oil separation module.
5. Supervisor had requested him to prepare pump A for restart. Operator was unaware pump A had been worked on during day shift.
6. Reported pump ready to control room operator. Unsure of time.
7. During break (approx. 22:30) had been called by control room operator to check separation room as gas alarm had gone off. Operator stated that false alarms are common.

Control Room Operator (night shift)
1. Location: Separation plant control room.
2. Heard explosion which was then followed by control room damage, including windows blowing in.
3. Immediately called emergency services followed by radioing plant operator—no reply. Then supervisor came into the control room and sounded plant alarm.
4. Started shift at 18:00, told by night supervisor pump A had been off-line and was to be brought back on standby.
5. Then received notice from plant operator and electrician that pump A was prepared. Unsure of time, approx. 19:15 for operator, 19:25 for electrician.
6. At about 22:10 received pump trip alarm. He then watched level increase in separators, then high oil level alarm sounded. Operator then switched to pump A and started it up.
7. He then monitored oil/gas level display. The oil level dropped.
8. When high gas alarm went off, he thought it was probably a false alarm, as has happened in the past, especially after work done on pump. Then requested plant operator to investigate.
9. While waiting for reply high high gas alarm went off (approx. 22:35) so he immediately started safety shutdown procedure. Explosion occurred while he was doing this.

FIGURE 7.3. **Statements of Witnesses, Page 1 of 2.**

Supervisor (night shift)

1. Location: Supervisor's office.
2. Heard loud explosion. Ran to control room and found that the windows had been blown in. Supervisor then sounded plant alarm. Then went to separation module and found it severely damaged and on fire.
3. Had come on shift at 18:00 and been briefed by day supervisor. Told pump A had been worked on but had just been signed back on by day supervisor. No mention of other work in separation module.
4. Supervisor had requested pump A to be prepared for standby. Did this by asking plant operator to prepare pump and electrician to reset pump. This request was made at about 19:00.

Supervisor (day shift)

1. Contacted at about 23:15 and told of explosion and fire in separation module.
2. During his day shift control room operator had told him pump A had given high vibration alarm and pump B was now in operation. The opportunity was taken to repair/rectify the pressure relief valve on pump A, while the pump was being repaired
3. After signing the permit for the works maintenance team, and seeing the job start (approx. 8:.00), he had contacted the specialist contractors and arranged for them to attend to the PRV. A permit to work had been prepared and the work started about 10:45. He had gone through the procedure for work with the contractor This included instructions for fitting a blank flange on the pipe.
4. Just prior to the end of the shift the works maintenance team reported that the work on pump A was completed, and he had signed off their permit.
5. The supervisor had no further contact with the contractors and had assumed they would be working overtime (after shift change at 18:00) to complete the job.

Contractor Maintenance Leader (days)

1. Had been contacted by oil separation day supervisor and worked to repair and rectify the PRV on pump A. Work started about 11:00 and then PRV had been removed and taken to the contractors workshop to be stripped.
2. One of the two contractors on the job had remained behind to fit a blank to the PRV pipe work.
3. The valve turned out to require a complete strip and overhaul and was unfinished by the end of the work day.
4. They had not informed the plant about this and assumed that the pump was "signed off" for more than one day.
5. The contractor team only work days and currently have no overtime policy in effect.

FIGURE 7.3. **Statements of Witnesses, Page 2 of 2.**

EXPLOSION IN SEPARATION BUILDING
Conclusions:
1. The explosion occurred just after 22:40 hours.
2. The explosion centered around the area of the separation plant holding pumps A and B.
3. The cause of this fire and explosion was ignition of hydrocarbon mixture.
4. The hydrocarbon leak probably resulted from a blank being incorrectly fitted to pump A PRV pipework and subsequent failure to provide a leak-tight seal.
5. The source of ignition is unknown.

FIGURE 7.4 **Investigating Engineer's Report**

GAS MONITORING SYSTEM:	ALARM REPORTS
0Q:00:00	
22:30:45	Low level alarm
22:31:30	Alarm accepted
22:37:50	High level alarm

PROCESS CONTROL SYSTEM:	ALARM REPORTS
00:00:00	
6:00:15	High vibration alarm: Pump A.
6:15:32	Pump B: activated.
22:15:47	Pump B: trip alarm.
22:20:01	High oil alarm: second stage separator.
22:23:17	Pump A: Activated.
22:37:53	Emergency shutdown sequence activated.

FIGURE 7.5 **Data for Process Data Recording System**

7.3. CASE STUDY 2: INCIDENT INVESTIGATION: MISCHARGING OF SOLVENT IN A BATCH PLANT

7.3.1. Introduction

This case study illustrates how the methodologies described in Chapter 6 can be used to analyze plant incidents and identify the root causes of the problems. Based on this information, specific error reduction strategies can be developed to prevent similar incidents from occurring in the future. Also, the findings of such an analysis can provide the basis for more general discussions about the prevalence of similar error inducing conditions in other plant areas.

The incident under consideration occurred in a batch processing plant in which batches of resins produced in various reactors are discharged into blenders to achieve the required viscosity by the addition of solvents. Solvent is charged into the blender via a valved pipeline which originates at the top floor of the building (see Figure 7.6). From the top floor the allocated worker is responsible for the pumping of solvents to the reactors or blenders via a metered solvent bank and a charging manifold.

The solvent bank consists of a number of metered pumps from the storage tanks which, by using flexible pipes (hoses) and quick fit couplings, the top floor man connects to the required pipeline. These pipelines are arranged in a charging manifold with each being labeled and having a valve adjacent to the coupling (see Figure 7.7).

One day, the lead operator gave verbal and written instructions, via a second worker, to the top floor man to pump solvent to 12A blender. The top floor man actually pumped the solvent to 21A blender, as a result of connecting the hose to the 21A blender pipe and not the 12A blender pipe. Consequently the solvent pumped to 21A blender was charged on top of another batch already in 21A blender. The contaminated batch had to be pumped back into a reactor where the mischarged solvent was removed by the application of vacuum.

Mischarging of solvents and oils was a recurrent problem in the plant and on many occasions batches were irrevocably contaminated because incorrect reactions had taken place. Additional problems were related to the unavailability of reactors due to reprocessing of contaminated batches, resulting in disruption of the process schedules. Management response to this problem had involved a number of actions including stressing the importance of checking communication; the issuing of standard operating procedures; and disciplinary action against the operators involved. The analysis of this incident revealed many error-inducing conditions not hitherto recognized by management.

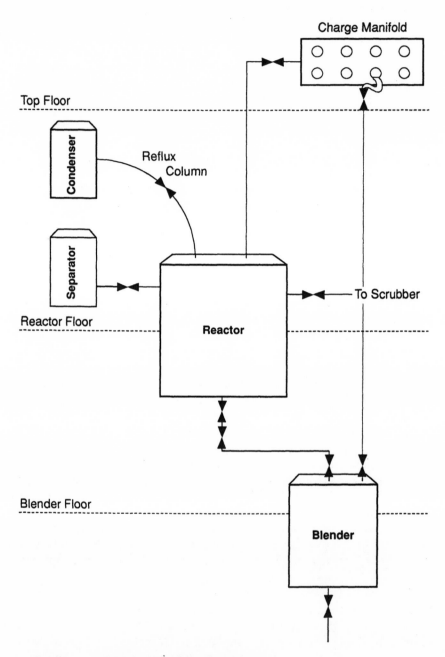

FIGURE 7.6 **Simplified Schematic Plant Diagram.**

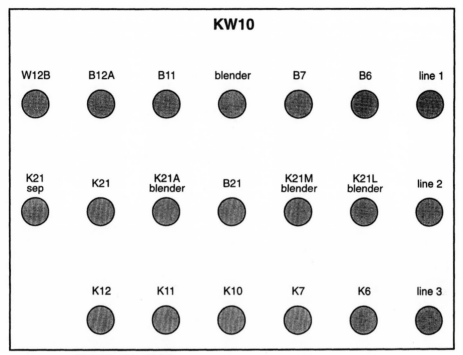

FIGURE 7.7   **Charging Manifold.**

7.3.2. The Application of Accident Investigation Techniques

To gather information about the factors which contributed to the above incident, interviews were held with the workers and their management. Relevant documentation such as standard operating procedures and documentation relating to the incident was also collected. A task analysis (see Case Study 3) of the job of the top floor person was carried out in order to examine the operations involved and the factors which could affect job performance. Two techniques were used for the analysis of this incident, namely variation tree analysis and root cause analysis.

7.3.3. Variation Tree Analysis

The information gathered relating to the incident was used to identify the sequence of causes and consequences which led to the mischarge. From the resulting variation tree (see Figure 7.8) two critical points and their contributory factors can be identified.

First, the selection of the wrong pipeline was influenced by a number of factors such as poor labeling and layout. A "spaghetti" of confusing pipework was already on the charging manifold as a result of a number of concurrent

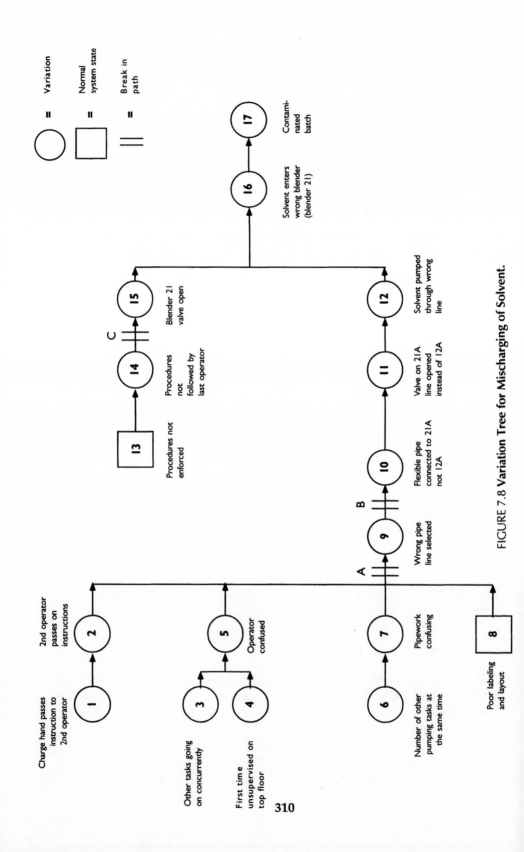

FIGURE 7.8 Variation Tree for Mischarging of Solvent.

pumping tasks. Also the worker was confused and possibly stressed due to inadequate training and the number of tasks being carried out at that time. Second, the charging of the solvent into the wrong blender could still have been avoided if the blender valve had been closed, as specified in the procedures. It was found that it was standard practice to leave this valve open in order to save the staff from having to open and close the valves with each charging. Also, there was no feedback to the worker with regard to the status of the valves.

To generate error reduction strategies, the incident sequence can be altered either by **canceling the variations** (the numbered nodes in Figure 7.8) or by **breaking the path** (shown by numbered vertical lines across the path). This is achieved by considering the cognitive processing level of the operation required at each stage, using Rasmussen's stepladder model (see Chapter 2 and Figure 4.11, Chapter 4), and addressing error reduction strategies at the levels which are specified in the model. A selection of possibilities follows.

Canceling the Variations (Nodes)

- **nodes 3, 4, 5, 6** (relate to the training and workload of the operator)
  Changes to remove some of these nodes could include:
  Identification level: too many jobs being done at once; limit work operator undertakes at one time.
  Observation level: some form of queuing system for jobs required.
  Evaluation level: management must outline explicit criteria for running concurrent tasks; train staff not only in normal work situations, but also in heavy work load situations; use valid training methods and provide procedures.
- **node 15**
  Activation level: install alarm system for indicating valve left open.
  Observation level: provide checklist to operators; make valve more accessible or remotely operable.
  Interpretation level: increase staff awareness of the consequences for safety, etc.
  Evaluation level: change operators' criteria for performance—enforce meaningful procedures—address work conditions and social content. Also it is necessary to change management's perception of the need for supervision and the importance of solvent charging. Obviously it is a failure of supervision to be unaware or to tolerate that valves were not reset after use.
  Execution Level: Enhance accessibility of valve or introduce remote operation.
- **nodes 5, 7, and 8** (relate to the design and ergonomic considerations of the situation). Designers need to acknowledge and address human–machine interface issues:

Observation and identification levels: ensure labeling is clear, consistent, and easily distinguished using color-coded pipework; improve work environment (e.g., lighting and general housekeeping).

Evaluation level: create a system, in accordance with ergonomic criteria, that is error tolerant and supports error recovery; redesign charging manifold (see Figure 7.7) using functional grouping corresponding to the actual layout of system.

Breaking the Paths

- At A, develop a method for facilitating selection of correct pipework such as a light system activated by the person giving the instruction, or a key system that requires a key from a specific valve to operate the coupling on the pipeline related to this valve
- At B, develop an alarm system indicating when a wrong connection is made
- At C, change procedures to reflect actual performance, provide check-lists, alter worker's perception of the importance and consequences of leaving valves open, and improve the supervision of procedures. Some form of lock system would be of benefit

This selection gives an indication of the variety of error reduction strategies, suggested from a combination of the Variation Diagram and a consideration of the cognitive processes underlying either the worker's performance at a particular stage or those of the design and management of the system.

7.3.4 Root Cause Analysis

The events and causal factors chart for this incident is shown in Figure 7.9. The primary sequence of events is shown horizontally in bold boxes. Secondary events are shown in the other boxes, and conditions are in ovals. From the diagram three causal factors were identified and carried forward to the Root Cause Coding to establish the root causes of the causal factors.

Causal Factor 1: Operator A Connects Pump to 21A Pipe Not 12A Pipe
Root cause coding identified the following root causes:

- **Root cause 1.** Scheduling less than adequate; the excessive number of jobs required of the worker had a detrimental effect on his performance.
- **Root cause 2.** Incomplete training; the incident occurred during the worker's first time alone on the top floor. Training had been given but only in low task demand situations.
- **Root cause 3.** Corrective action not yet implemented; management had been aware of problems with the vessel and solvent banks but had done nothing about it.

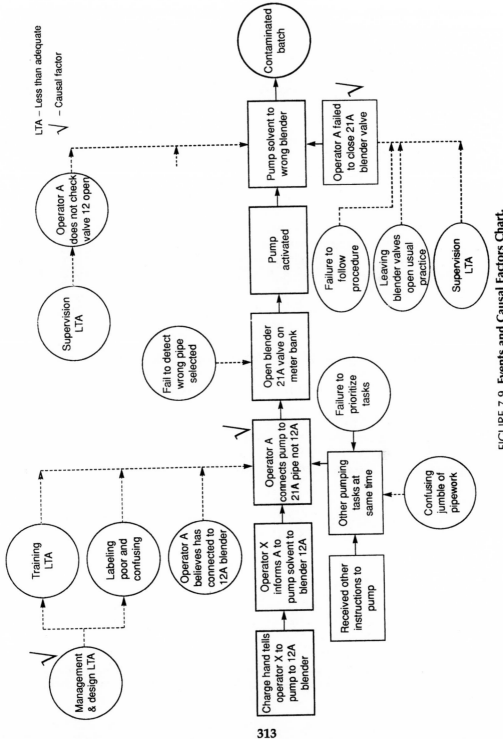

FIGURE 7.9. Events and Causal Factors Chart.

313

- **Root causes 4 and 5**. Human–machine interface less than adequate. The labeling of the pipe was poor and confusing, the general ergonomics of the work situation was poor.
- **Root cause 6**. The system was not error tolerant. The error made was not detectable.
- **Root cause 7**. Personal performance. Attention may have been less than adequate.

Causal Factor 2: Operator A Failed to Close 21A Blender Valve
Root cause coding identified the following root causes:

- **Root causes 1, 2, and 3**. Procedures were not followed. Procedures were not written down and in practice were inconvenient to use. No audit was made to verify the usability of the procedures.
- **Root causes 4 and 5**. There had been no supervision of the worker who should close the blender valves after completion of the job. No audit was made to verify that valves were routinely closed.
- **Root causes 6, 7, and 8**. Human factors aspects were inadequately addressed. Specifically, ergonomics of the plant was poor, there were differences in layout among different areas and the labeling was poor.
- **Root causes 9 and 10**. There may have been a communications problem in telling the worker to close the valve or the personal performance of the operator was less than adequate.

Causal Factor 3: Management and Design Less Than Adequate
This can apply to a number of areas in the sequence. Contributory root causes include: equipment design was poor with no human factors design for the vessel bank; supervision was poor; standards relating to design and training were poor with violations accepted as the norm. Communication among staff members was informal and unstructured.

7.4. CASE STUDY 3: DESIGN OF STANDARD OPERATING PROCEDURES FOR TASK IN CASE STUDY 2

7.4.1. Introduction

Standard operating procedures (SOPs) are step-by-step job instructions which can help workers perform their jobs safely and efficiently. When the end users are involved in their design, SOPs can provide a basis for arriving at a method of work agreed-to by different shifts. In this sense, SOPs can be used to develop training programs and specify measures of competence. Because of the importance of SOPs in the work situation, a systematic framework is needed to enable the design of reliable procedures which are acceptable by the workforce.

This section illustrates how the techniques described in Chapter 4 can be used to develop a procedure for the job of the top floor operator in the batch plant considered earlier. Two techniques are illustrated: (i) a hierarchical task analysis (HTA) of the job, and (ii) a predictive human error analysis (PHEA) of the operations involved. HTA provides a description of how the job is actually done while PHEA identifies critical errors which can have an impact on the system in terms of safety or quality. The basic structure of the procedure is derived from the HTA which specifies in increasing detail the goals to be achieved. To emphasize critical task steps, various warnings and cautions can be issued based on the likely errors and recovery points generated by the PHEA.

The first step in the design of procedures is to identify the required information sources. These can include interviews with the workers and supervisors, reviews of existing documentation (e.g., existing SOPs), actual observation of the job, and reviews of past incidents.

7.4.2. Hierarchical Task Analysis

The main aspect of the job of the top floor person is to pump solvents or oil to various reactors and blenders. Instructions are issued on a job-card or by phone. The instructions are entered in a log book (which is kept by the top floor worker) and on a record card which has to be returned to the laboratory at the end of the shift. To prepare for pumping, protective clothing must be worn. After the required amount of solvent is set on the meter, the worker has to connect the meter and the pipeline with a hose and then open the valve on the pipeline (see Figure 7.10). Before starting the pump, the blender valve

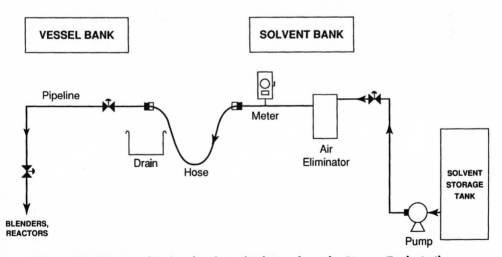

FIGURE 7.10. **Diagram Showing the Flow of Solvents from the Storage Tanks to the Blenders and Reactors**

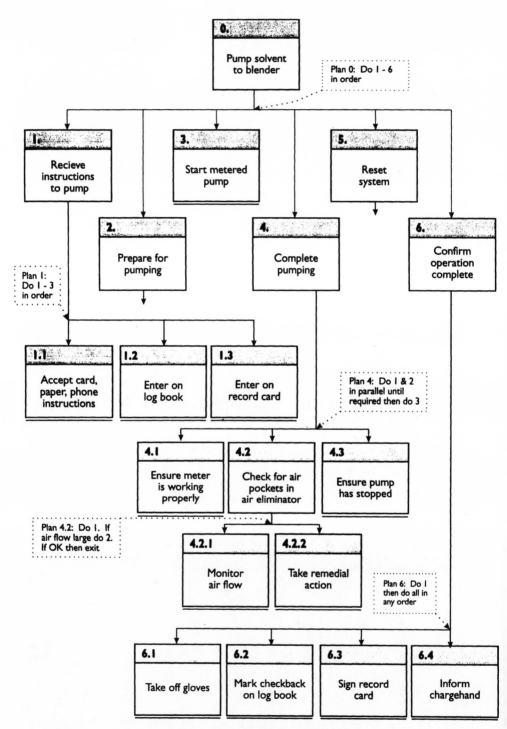

FIGURE 7.11. **HTA of Pumping Solvent to Blender, Page 1 of 2.**

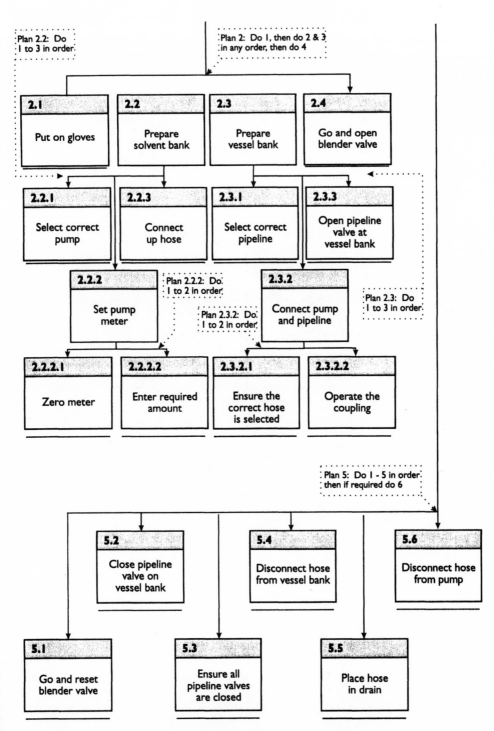

FIGURE 7.11. **HTA of Pumping Solvent to Blender, Page 2 of 2.**

Operation 1: Receive instructions to pump

TASK STEPS	PREDICTED ERRORS
1.1 Accept card, paper, phone instructions to pump	R2 wrong information obtained T2 wrong information transmitted
1.2 Enter in log book	T1 information not transmitted T2 wrong information transmitted
1.3 Enter on record card	T1 information not transmitted T2 wrong information transmitted

Operation 5: Reset System

5.1 Reset blender valve	A4 action too little A6 right action on wrong object A8 action omitted
5.2 Close pipeline valve on vessel bank	A6 right action on wrong object A8 action omitted
5.3 Ensure all pipeline valves are closed	A9 action incomplete
5.4 Disconnect hose from vessel bank	A8 action omitted
5.5 Place hose in drain	A8 action omitted
5.6 Disconnect hose from pump	A6 right action on wrong object

Operation 1: Receive instructions to pump

PREDICTED ERROR DESCRIPTION	PERFORMANCE-INFLUENCING FACTORS
1.1 Wrong instructions received (R2) Wrong instructions transmitted (T2)	High noise levels. Distractions. Unfamiliarity with instructions. Retrieval from memory rather than job card used.
1.2/ Log book/record card not filled in, 1.3 or incorrectly filled in (T1, T2)	Lack of time. Lack of perceived importance. Distractions.

Operation 5: Reset System

5.1 Blender valve not completely closed (A4) Blender valve not closed (A8) Wrong blender valve closed (A6)	Inadequate instructions. No reminder provided. Poor labeling.

FIGURE 7.12. **Extract of PHEA for the "pumping solvent" Task.**

(which is located on the blender floor) must be opened. Reactor valves can be opened by other workers on the reactor floor. While pumping, a number of checks must be made regarding the pump meters and the air eliminators. When pumping has been completed, the hose can be disconnected from the pipeline (reactor bank) and, if required, from the meter (solvent bank). The top

Operation 1: Receive Instructions to Pump	
CONSEQUENCES	REMEDIAL ACTIONS
1.1 Solvent pumped to wrong blender— contaminated batch	Reduce noise levels. Reschedule operator tasks to reduce overload. Incorporate use of procedures in training scheme.
1.2 Laboratory may not have necessary details for testing	Ensure operators spend time in laboratory to understand implications of log book or record card being filled in incorrectly.
Operation 5: Reset System	
5.1 Solvent that is incorrectly routed will contaminate other batches	Clear indications that valve is fully closed. Checklist and training to emphasize importance of closing valve. Improved labeling of valves.

FIGURE 7.12.   **Extract of PHEA for the "pumping solvent" Task (*continued*).**

floor worker has to sign the log book and the record card and reset the blender valve. The operator must also be informed that the operation is complete.

The HTA is shown in Figure 7.11. The overall job is broken down into six subtasks which must be carried out in sequence. The analyst can specify these subtasks as he or she wishes. For instance, "starting the metered pump" and "complete pumping" can be assigned to the same group if this is convenient. The job analysis has been taken to the level of operations where the worker interfaces with the system. This is necessary in order for the PHEA to address possible errors associated with the job.

7.4.3. Predictive Human Error Analysis

During the PHEA stage, the analyst has to identify likely human errors and possible ways of error detection and recovery. The PHEA prompts the analyst to examine the main performance-influencing factors (PIFs) (see Chapter 3) which can contribute to critical errors. All the task steps at the bottom level of the HTA are analyzed in turn to identify likely error modes, their potential for recovery, their safety or quality consequences, and the main performance-influencing factors (PIFs) which can give rise to these errors. In this case study, credible errors were found for the majority of the task steps and each error had multiple causes. An analysis of two operations from the HTA is presented to illustrate the outputs of the PHEA. Figure 7.12 shows a PHEA of the two following tasks: **Receive instructions to pump** and **Reset system.**

Instructions to pump might be received on a job card, by phone or on a scrap of paper. The main errors associated with this task are: wrong instructions obtained or wrong instructions transmitted. This can occur when the wrong information is passed on to the worker, or where he or she misreads or mishears instructions or forgets the instructions because of intervening events. Because of the criticality of this error, a task step "confirm back instructions" was introduced into the procedures developed from this analysis as a possible error detection point. Factors which could contribute to these errors would be: high level of noise at the top floor, inappropriate recording of information (e.g., on a scrap of paper), many pumping tasks performed simultaneously by an inexperienced worker, etc. With respect to the task of recording the specifications on the log book and record card, any omission or commission errors would not have an impact on the actual operation but could cause problems with the processing of information in the laboratory.

Resetting the system entails six operations which must be carried out in sequence. Task step 5.1 is a safety barrier to prevent any mischarges to the wrong blender or reactor and any errors in closing these valves may cancel out this safety function. The PHEA has shown that the worker could fail to close these valves fully (A4) or omit to close them at all (A8) or could close the wrong valve (A6). All these errors were very likely because of the following performance-influencing factors: procedures not followed (because of the absence of a "procedures culture"), many simultaneous pumping tasks, distractions, assumptions that someone else would close the valves etc. Other critical errors were related to tasks 5.2 and 5.3. Omitting to close the pipeline valves (A8, A9) could be seen as unsafe acts because high reactor pressures could cause backflow of solvents or oils to the charging manifold. Also, closing the wrong pipeline valve (A6) could have severe effects in cases where solvent was being pumped to that valve. With respect to task steps 5.4 and 5.5, omission errors were likely but would not have any severe consequences apart from adding to the confusion of pipework caused by a large number of hoses connected to various meters and pipelines. Finally, disconnecting the wrong hose (A6) would be a critical error to make since solvent might be passing through the hose.

All critical errors and recovery points for task steps are recorded in accordance with the conventions of the PHEA and provide a valuable input to the specification of warnings and cautions for the design of procedures. In addition, various performance-influencing factors which contribute to the occurrence of critical errors can be identified which can provide input to the development of other error reduction measures such as training, and control panel design.

7.4.4. Input to the Design of SOPs

To ascertain the appropriate format of a procedure, the analyst must establish how familiar the worker is likely to be with the task steps to be performed. If he or she is very familiar with the task, such that it is usually executed from memory, then the most appropriate form of procedure will be a checklist in which active responses are required at only critical points in the task. For unfamiliar and/or infrequent tasks, a step-by-step form of procedure will be most appropriate.

The top floor operators are usually highly skilled and experienced in many of the steps within the process. Therefore, a checklist may suffice as a job aid. However, new staff members may be sent to the top floor as part of their job-rotation scheme and a step-by-step procedure may be of benefit to them. Detailed procedures can be used either as job aids or as training aids. It is worthwhile, therefore, to develop both a checklist and a step-by-step procedure for the top floor job.

Having established the content of SOPs from the HTA and PHEA, ergonomics guidelines for information presentation can be used to ensure that workers can quickly refer to the instructions and that the possibility of misinterpretation is minimized. An example of such guidelines is United Kingdom Atomic Energy Authority (1991). In this section, the focus will be on the inputs from the HTA and PHEA to the design of SOPs.

Figure 7.13 shows a complete step-by-step procedure for pumping solvents to blenders. The top level goals of the HTA provided the main sections of the procedure, which are also presented as an overview on the first page. HTA provides a means of "chunking" the task steps into meaningful goals or sections so that the task is easily understood and retained by the operator. The structure of the procedure and the order in which the tasks are performed matches the task as it is performed in practice, thus no procedure–task mismatches exist. This is particularly important when workers are required to accurately follow a procedure to ensure production quality. If the procedures are ambiguous, impractical, or unrealistic, then workers will take shortcuts.

Once a draft version of the SOPs has been obtained based on the HTA, a few modifications may be necessary in terms of regrouping task steps. For instance, "starting the pump" and "complete pumping" can be put together in Section 3 (Figure 7.13). On many occasions, the headings of the sections may be quite abstract since they may refer to a number of operations. It is useful to highlight the objective of each section in terms of the final task output. The procedures may be furnished with explanations in the form of notes with regard to why certain operations should be included or performed in the specified order. When operators can understand the underlying logic of a task sequence they may be more willing to apply the procedure and not take shortcuts.

The PHEA is valuable in specifying warnings, checks, and cautions to prevent operators from making mistakes in aspects of the job associated with a high error profile. For instance, "confirm instructions" is underlined and a note is attached at the end of the section to point out the importance of getting

STANDARD OPERATING PROCEDURE (1)

PUMPING SOLVENTS TO BLENDER VESSELS

OBJECTIVE: To pump a specified solvent type to the required blender

PROCEDURE: The following steps are required for the pumping of
solvents to blenders:

1 Receive instructions to pump
2 Prepare for pumping
3 Complete pumping
4 Reset system
5 Confirm operation complete

FOLLOW THE STEPS BELOW IN SEQUENCE

1   Receive instructions to pump

Receive instructions to pump on job card, paper, or phone. **Confirm back**
instructions and then enter specifications on log book and record card.

Note: It is essential that the correct instructions are received.

page 1 of 3.

FIGURE 7.13 **Example of Step-by-Step Procedure for Pumping Solvents, Page 1 of 3.**

2   Prepare for pumping

Objective: To connect the correct pump to the required vessel

Procedure: Do steps 2.1 and 2.2 and then do 2.3.
Correct protective clothing must be worn, i.e. visor, gloves, and
overalls.

Warning: **Solvent can produce intense irritation if it reaches eyes or skin.**

    2.1   Prepare solvent bank by carrying out the steps below in sequence:
    2.1.1   Select correct pump
    2.1.2   Set pump meter to zero and enter the required amount of
solvent
    2.1.3   Connect up hose

    2.2   Prepare charging manifold by carrying out the steps below in
sequence:
    2.2.1   Select correct pipeline
    2.2.2   Connect pump and pipeline by selecting the correct hose
and making the coupling at the pipeline
    2.2.3   Open pipeline valve at vessel bank

    2.3   Go to the blender floor and open blender valve

Warning: **If the correct blender valve is not operated batches in other
blenders may be contaminated.**

3   Initiate and complete pumping

Objective: To feed solvent to the blender, having performed the preparation
checks

Procedure: Start the pump and follow the steps in sequence:

    3.1   Ensure meter is working properly throughout pumping
    3.2   Check for air pockets in air eliminator
    3.3   Ensure pump has stopped after specified amount of solvent has
been pumped
    3.4   Verify that air is completely eliminated by reading 0 pressure
in gauge.

Warning: **If the air in the solvent line is not completely eliminated incorrect
additions may be made.**

page 2 of 3.

FIGURE 7.13   **Example of Step-by-Step Procedure for Pumping Solvents, Page 2 of 3.**

4   Reset system in order to get ready for the next job.

Procedure: Follow the steps in sequence. If required, the hose can also be disconnected from the pump.

   4.1   Go to the blender floor and reset blender valve

Warning: **If valve settings are not returned to the closed position, incorrect additions will occur.**

   4.2   Close pipeline valve at vessel bank
   4.3   Ensure both pipeline valves are closed

Warning: **If pipeline valves are not closed oil or solvent may flow back due to possible mischarges or high reactor pressures.**

   4.4   Disconnect hose from vessel bank
   4.5   Place hose in drain

Note: The hose may contain residual solvent. If this is the case, empty solvent into drain.

5   Confirm operation complete.

Procedure: Take off gloves and follow the steps below:

   5.1   Mark checkback on log book
   5.2   Sign record card
   5.3   Inform chargehand

page 3 of 3.

FIGURE 7.13   **Example of Step-by-Step Procedure for Pumping Solvents, Page 3 of 3.**

the instructions right from the start. In Section 3, a warning is added explaining why workers should monitor the functioning of the air eliminator (large air pockets may give rise to incorrect additions of solvent). Other warnings include reasons why blender valves should be reset after use (Section 4) or why all pipeline valves should be closed after completion of the job (Section 4). All these warnings and notes are based on the errors predicted in the PHEA (Figure 7.12).

On the basis of the information contained in a step-by-step procedure, a checklist can be designed consisting of active **checks** for critical steps which operate as an "aide memoir." Experienced workers who normally carry out a task from memory may forget isolated acts or inadvertently not follow the correct sequence of steps in the procedure.

The function of the checklist is to reduce the likelihood of omission errors and to facilitate recovery of these errors. For those steps which are likely to be omitted (e.g., because no immediate feedback exists to confirm they have been carried out) or which have critical consequences, an active check is specified where the operator has to sign after the step has been executed. The purpose of these checks is to assist the worker in preventing errors rather than keeping a record so that he or she may be held responsible if actions are omitted. If this point is emphasized then it is less likely that personnel will "sign ahead." Other steps, associated with errors of low probability and cost, may be included in the checklist if this helps the operator maintain his or her place within the sequence of the task steps. Warnings play an important role as in the design of step-by-step procedures. An example of a checklist for pumping solvents to blenders is shown in Figure 7.14.

7.5. CASE STUDY 4: DESIGN OF VISUAL DISPLAY UNITS FOR COMPUTER-CONTROLLED PLANT

7.5.1. Introduction

Control room operations in the process industries are notoriously productive of human errors. Typically, this is due to the presence of error inducing conditions such as the presentation of information to the operators which does not allow effective monitoring of the process or diagnosis of abnormal conditions. Unfortunately, the trend of replacing pneumatic panel instrumentation with computer controlled process systems and visual display units (VDUs) for reasons of increased hardware reliability, speed of response, quality, and lower maintenance costs etc., has not always achieved the corresponding benefits for operability. Bellamy and Geyer (1988) analyzed a selection of incidents in computer controlled process systems and concluded that human errors during operations were associated with 59% of the incidents. In a breakdown of the causes of these errors, poor provision of information was predominant. The purpose of this section is to demonstrate how the human factors methodologies discussed in previous chapters can be used to design display systems which support the efficient and reliable performance of process control tasks.

One of the most noticeable differences between pneumatic instrumentation panels and computer-based displays is the way that information is presented to the user. Instrumentation panels allow the workers to obtain a quick overview of the state of the process by simply scanning the panel while computer-based displays present the information in a serial fashion. The worker therefore has only a small "window" on all the data about the present state of the process. Navigating through numerous VDU pages is in itself a secondary task in addition to the primary task of controlling the process. It is apparent, therefore, that computerized displays can increase workload if their configuration is not in accordance with human factors design principles. Application of these design principles can exploit the flexibility of computer-based displays to configure process data in a way that human information processing is enhanced. It is important, therefore, to examine how the methodologies described in previous chapters can take advantage of the flexibility that computer-based displays can offer.

When discussing how to support a worker's task by interface design, it is useful to have a simple framework for the types of task involved, because different design principles are relevant to different types of the overall task. The skill–rule–knowledge (SRK) framework of human performance (see Chapter 2) can be useful in classifying the types of tasks involved in navigating through a computer-based display system and in controlling the chemical process itself. These tasks can be classified as follows:

Skill-Based Tasks
These include identification of process equipment and instruments, interpretation of the meaning of their values and trends, navigation through different VDU pages by means of a selection menu, etc. The common feature of these tasks is handling the display system to search and locate relevant process data. In this respect, "classical" ergonomics checklists (see Chapter 4) are very useful in facilitating performance of such tasks.

Rule-Based Tasks
These refer to the control of a chemical process and include planning for familiar tasks (e.g., change type of fuel-firing in a furnace) or planning for familiar but infrequent tasks (e.g., start-up or shutdown a furnace). Methods of task analysis and error analysis can be used to analyze well-established strategies that operators use to perform procedural tasks and identify the user's information needs. An implication for display design would be that all information needed

☐   Confirm instructions by reading back to lower floor operator

O   Enter specifications in log book and record card

O   Set pump meter

O   Connect up pump and pipeline

O   Ensure hose is coupled firmly to the two edges

O   Open pipeline valve

☐   Open blender valve

O   Start pumping

O   Ensure meter is working properly

O   Check for air pockets in air eliminator

Warning: **Incorrect additions may be made if air in the solvent line is not completely removed.**

O   Verify that the right amount of solvent is charged

☐   Reset blender valve

Warning: **Product will be contaminated if valves are not returned to the closed position.**

☐   Ensure all pipeline valves are closed

Warning   **High reactor pressures may cause backflow**

O   Disconnect hose and place in drain

O   Mark checkback on log book and sign log book

FIGURE 7.14. **Example of Checklist for Pumping Solvents.**

in any one decision should be available at the same time in one VDU page. Apart from deciding the content of each VDU page, these methods can be used to design a hierarchical structure for the total display system.

Knowledge-Based Tasks

Process transients and equipment failures may require workers to develop a new strategy to control the process. Detection, diagnosis, and fault-compensation are tasks in which workers may have little experience and the information needs may be different from those of familiar tasks. Again, methods of task and error analyses, particularly those concerned with human cognitive functions, may be useful in deciding what information should be displayed to help workers detect process transients, diagnose their causes and develop new strategies.

To illustrate how this framework can be realized in the design of computer-based displays, a case study is presented which was conducted for a refinery.

7.5.2. Process Details

The case study described here concerns a human factors audit of a computer controlled process system which was being introduced in a distillation unit of a chemical plant. The unit was in transition from replacing its pneumatic panel instrumentation with the new system. However, control had not yet been transferred and the staff were still using the panel instrumentation. The role of the project was to evaluate a preliminary design of the computer-based display system and provide recommendations for future development.

The description will focus on the presentation of information concerning the operation of furnaces which heat crude oil before being delivered to the distillation columns. Figure 7.15 shows a graphic display of one the furnaces used for monitoring the process only. To adjust the process parameters and control the process, another display was used which will be referred to as the "control display" from now on. The control display contained all the automatic controllers and other manual control elements used for the furnace operations. Because of the large number of process parameters to be monitored, the graphic display was supplemented with a "textual display" on which additional information was available in the form of tables. However, no clear distinction existed as to what information should be assigned to the graphic or textual displays, and both were used for monitoring the process.

The discussion below will focus briefly on the design of the graphic displays in order to illustrate the methodology used. The aim of the furnace operation (see Figure 7.15) is to achieve a specified output temperature of the crude oil. This is done by means of a master temperature controller which regulates the pressures of the fuels used. An air/fuel ratio controller regulates the flow of the combustion air, receiving as input the flow rates of the fuels

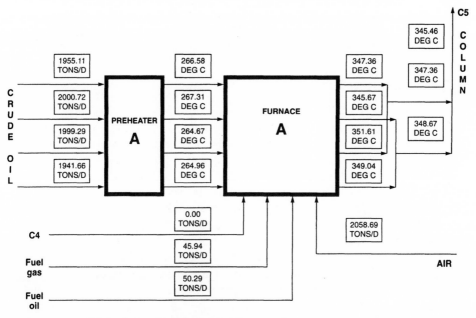

FIGURE 7.15. **Original Graphic Display for Furnace A.**

and the combustion air. The process is on automatic control most of the time. However, there are tasks such as responding to burner failures, changing types of firing or type of crude oil which require manual control of the furnace. Complex skills are required to carry out these operations safely. The SRK framework was used to evaluate the computer-based displays of this process.

7.5.3. Skill-Based Tasks

A highly desirable feature of display design is that skill-based tasks such as identification of parameters, interpretation of their meaning, and navigation through various VDU pages should be executed at high speed and without errors. A limitation of the graphic display (Figure 7.15) is that the various flow and temperature parameters are not identified. The worker has to go to a text display to identify the process parameters, an activity which can give rise to errors if other tasks require attention as well. This also increases memory load.

Another limitation concerns the duplication of process information on different VDU pages. Specifically, most of the data on the graphic displays were presented again in a different format on the text displays. This increased the number of VDU pages used in total to perform a task. This secondary navigation task undoubtedly increased the workload of the worker who had to control the process. Within the text displays themselves, information was not structured in a functional manner. That is, information required by workers was embedded in data on productivity and quality analysis, the latter being mainly displayed for the benefit of technologists and production managers whose information needs were essentially different from those of the workers.

7.5.4. Rule-Based Tasks

To determine the content of each VDU page, a number of furnace operations were analyzed in terms of operator decisions and information needs. Specifically, a hierarchical task analysis (HTA) was performed for the following operations: starting-up the furnace from cold, changing the type of fuel firing, changing the type of crude oil, and responding to burner failures. Figure 7.16 shows part of an HTA of the task of manually increasing the furnace load to the specified target. The task analysis shows that the oxygen content in the flue gases is functionally related to the output temperature of the crude, and the flow rates of the fuels and combustion air. It was concluded, therefore, that these parameters should be presented on the same graphic display. Error analysis was used to examine the errors made during a number of incidents recorded from other refinery plants. Errors associated with responding to burner failures, for instance, were made because workers did not set the automatic controllers to the appropriate mode. It was decided, therefore, to incorporate data about the status of these controllers on the graphic displays.

Figure 7.17 shows the recommended graphic display for the same furnace. Apart from labeling the process parameters for easy identification, the graphic display includes information about the efficiency of the furnace, the oxygen content and smoke of the fuel gases, and the automatic controllers used. Although personnel cannot adjust the controllers from the graphic display, they can monitor important information about the set-points (SP) and the current values (CV). The units of measurement for each process parameter were omitted in order to economize on the available screen space. This should not create difficulties, provided that the same units were used consistently across different areas, for example, tons per day for flow rates, bars for pressures, or degrees Celsius for temperatures. The recommended display also presents some alarm information about the status of the six burners. This information was included because it would facilitate on-the-job learning of the process dynamics. The operator would be able to see, for instance, how the pattern of temperatures of the four exit coils may change in response to different burner failures.

Because the staff were consulted extensively during the application of task analysis and error analysis methods, the information presented on the graphic display in Figure 7.17 corresponds with their own information needs.

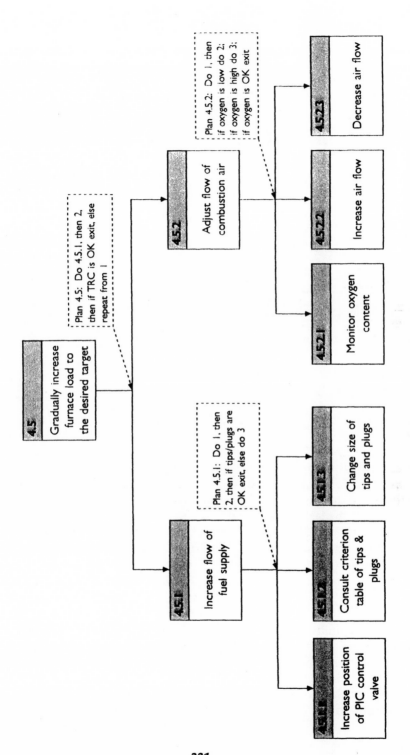

FIGURE 7.16. Hierarchical Task Analysis of the Task of Increasing Furnace Load.

331

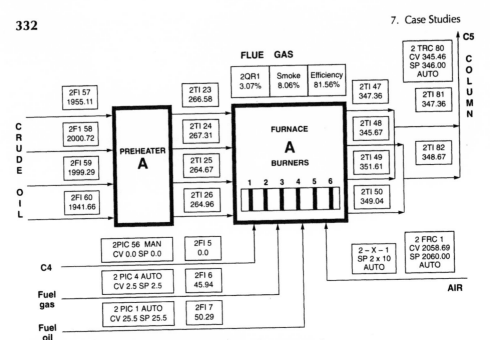

FIGURE 7.17 **Recommended Graphic Display for Furnace A.**

7.5.5. Knowledge-Based Tasks

Process transients and plant failures present workers with critical tasks which they may not have practiced thoroughly in the past or during refresher training exercises. Timely detection and diagnosis of such transients is very important in the management of emergencies of many chemical processes. To facilitate the performance of these tasks, a concise overview display of the four furnaces was developed containing all critical parameters for each furnace. To some extent, the task and error analyses performed on various process operations were useful in identifying the critical parameters. Interviews with the staff were used to verify the recommended display with the actual users. This display is presented in Figure 7.18.

As with the pneumatic panels, the display capitalizes on human pattern recognition capabilities. Should a furnace not be operating efficiently or a failure occur, this can be quickly detected by observing deviations from the standard symmetrical shape. In practice, the extent of any such deviations will be proportionally equivalent to the actual process parameter deviation.

Additionally, the shape of the deviation will prompt the operator to search for more detailed information upon which to act. An example of a process deviation as represented by an asymmetrical display might be the low flow of crude through the coils due to a blockage. This may be represented by a decrease in crude supply and fuel supply and an increase in inlet temperature. This type of overview display has the following advantages:

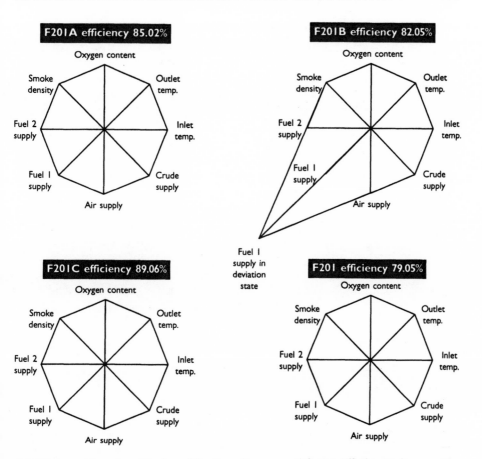

FIGURE 7.18. **Overview Display of the Four Furnaces of the Distillation Unit.**

Supports the Early Detection of Abnormal Process States

Although some diagnostic information is available in the activated alarms, the workers still need to know the size of any deviations from the target states and require a concise picture of those critical parameters which are in alarm and those which are nearing it. This information is available on the overview display and facilitates the early detection of process deviations.

Facilitates Fault Compensation and Elimination

Although the information on the overview display does not explicitly specify what actions should be taken to rectify the abnormal process state, it does suggest the goal priorities the workers should set in advance in order to protect the integrity of the system.

Enhances on-the-Job Learning of the Process Dynamics

The overview display can be used as a learning tool for the process dynamics because it provides three valuable types of information: the correlations among process variables; the size of changes and effects; and the response times and lags. When the staff are able to learn the precise dynamic response of the system to their actions, they are in a better position to develop effective fault compensation strategies in the medium to long term. Careful consideration of the changes of the key process variables during fuel or crude changes can be a great source of learning for the less experienced operator. In the case of furnace start-up operations, the overview display may help even the more experienced worker to promptly detect any process deviations and compensate effectively before critical consequences ensue.

Once the worker has a display which enables the whole furnace unit to be monitored at a glance, it is then necessary to supply more detailed information on each individual furnace. This display should represent the process in a way which facilitates comprehension of the heat transfer stages and provides all relevant information required to fulfill the objectives of safe and efficient firing. Figure 7.17 presents a recommended graphic display at a more detailed level of system description.

7.6. CASE STUDY 5: AUDIT OF OFFSHORE EMERGENCY BLOWDOWN OPERATIONS

7.6.1. Introduction

This study was concerned with improving reliability in fire and gas emergencies by assessing and improving workers' performance with respect to emergency blowdown operations offshore. An audit was carried out in order to identify and evaluate the factors affecting the decision making process and actions relevant to blowdown during fire and gas alerts.

Offshore oil platforms are highly automated, requiring little direct operator input to maintain production. In the event of a serious abnormality such as a fire or a gas escape, the control room worker is required to make decisions as to whether to depressurize one or more systems and which systems to blowdown. Other workers have the facility to depressurize some systems at a local control room.

The costs of a wrong decision are loss of production, on the one hand, and failure to respond to a real emergency on the other. In order to improve his or her decision basis, the control room worker will usually dispatch others to the source of the emergency signal to declare whether it is spurious (false alarm) or whether it is real, but it is containable without need for depressurizing. This takes up valuable time, during which the situation could escalate dangerously.

Activation of the blowdown, however, will not depressurize a system fully for a considerable length of time. One of the reasons for considering retaining the possibility of human intervention was that the automated blowdown system was not considered completely reliable at the time because of the limitations of the fire and gas detection hardware. This would have the effect of resulting in increasing the likelihood of spurious blowdown production losses.

There had been three incidents of gas/oil escapes over prolonged periods on various offshore platforms, and this had prompted the human factors investigation.

The human factors audit was part of a hazard analysis which was used to recommend the degree of automation required in blowdown situations. The results of the human factors audit were mainly in terms of major errors which could affect blowdown success likelihood, and causal factors such as procedures, training, control room design, team communications, and aspects of hardware equipment. The major emphasis of the study was on improving the human interaction with the blowdown system, whether manual or automatic. Two specific platform scenarios were investigated. One was a significant gas release in the molecular sieve module (MSM) on a relatively new platform, and the other a release in the separator module (SM) on an older generation platform.

7.6.2. Method of Audit

In order to identify the major factors affecting the decision to resort to blowdown and the execution process of this operation, a number of data collection techniques were utilized, as shown in Table 7.1.

Based on the collected information, a decision/action (DA) chart was developed to provide an overview of the main decisions involved in the blowdown operation and the main influential factors such as time stress, conflicting responsibilities, risk of gas ignition etc. Task Analysis and Error Analysis of the blowdown operation were subsequently carried out to obtain a description of the sequence of tasks steps and the likely human error modes which could occur.

7.6.3. Findings of Audit

The main factors found to be affecting the decision to blowdown manually were the blowdown philosophy, procedures, and training. Factors affecting the efficiency of the decision making process and the execution of blowdown were the ergonomics of the information presentation in the control room communications, and various aspects of the hardware.

TABLE 7.1	
Data Collection Techniques in the Human Factors Audit	
Interview:	Individual personnel at the following levels from at least two shifts on each platform were interviewed: offshore installations manager, operations supervisor, shift supervisor, control room technician, area technician.
Walk-throughs:	During the interview, personnel were asked to "walk through" a blowdown scenario from first-up yellow alarm to blowdown or the decision not to blowdown.
Observation:	Normal operations and the response to alarms.
Review of alarm logs:	The alarm log, gas alarm start times and time of return to green status were reviewed.
Procedural Review:	Any relevant procedures/instructional documents were reviewed (there were virtually none concerning blowdown).
Incident Analysis:	A small number of blowdown incident reports were reviewed, but these were found to contain very little information relevant to manual blowdown initiation.
Critical Incident Technique:	Personnel were asked to recall any incidents or near misses. This produced far more useful information for this study than the incident reports.
Training Review:	Offshore personnel were questioned about their training and also asked questions to determine their depth of knowledge of the blowdown system. Training personnel onshore were briefly interviewed.
Ergonomics Audit:	An ergonomics audit of main and auxiliary control rooms was carried out on both platforms.

7.6.3.1. Blowdown Philosophy

The major finding of the study was that the manual blowdown philosophy, particularly with respect to gas situations, was not clearly defined. This was most apparent in the offshore attitudes and perceptions regarding the importance of blowdown as a safety system. No decision criteria specifying when blowdown should or should not be activated were provided for the support of control room staff. Blowdown was essentially left to the discretion of the workers. Consequently, the offshore interpretation of this vagueness and ambivalence amounted to a perceived low priority of blowdown. It was concluded that this perception would probably lead to a significant delay in blowdown or possibly the omission of blowdown when it was actually required.

7.6.3.2. Training and Procedures

Training and procedures were not available for blowdown situations. Blowdown was dealt with only briefly in training, and was either ambiguously referred to in procedures or did not appear at all. Workers viewed blowdown as a last resort option. This view was reinforced by the absence of procedures or formal training. This view could only be changed by a combination of training in the functioning and safety value of blowdown, and by a philosophy and procedures which justified its use and specified how and when to use it.

7.6.3.3. Ergonomics of the Control Room
On both platforms the ergonomics of layout and instrumentation would hinder rapid and effective response to a significant fire or gas release. The overall ergonomics in both control rooms betrayed the lack of a coherent human–machine interface design philosophy being implemented within the design process.

One of the most noticeable ergonomic deficiencies in both control rooms was the number of panels that had to be scanned and monitored during the scenarios, and the number of rapid actions required at opposite ends of the control room. The need for virtually simultaneous actions and movements would have been discovered and resolved in the design stage had a task analysis and human error analysis been carried out on an emergency operation.

The VDUs (visual display units) in both control rooms were under-utilized for response to and control of abnormal events, in comparison to other industries. The control room computer had the capability to give the technicians considerable flexibility and power in controlling both productivity and safety factors offshore. However, this power was not utilized.

7.6.3.4. Communications
The communications systems on both platforms seemed to be prone to error and overload, although much reliance was placed on them. In particular, the control room technician had a great deal of communicating to do since he was responsible for coordinating the activity of a large number of workers on the platform, while at the same time monitoring panels, accepting alarms, and carrying out other duties.

7.6.3.5. Hardware Aspects
The most important hardware items appeared to be the detectors themselves. The gas detection system gave frequent spurious alarms, and on both platforms the ultraviolet (UV) fire detectors were also prone to spurious activation from distant hot work for example, and had a limited ability to detect real fires. The unreliability of these systems had a general effect on response time and would, overall, lengthen the time to respond. The second aspect which was related to hardware was function and performance testing of the emergency blowdown systems. It is critical that the workers believe the systems will work when required, and this can only be achieved by occasional use or at least function testing.

7.6.3.6. Decision-Making Aspects of Blowdown Activation

The overall decision-making process is shown schematically in the decision/action (DA) chart (see Chapter 4) of Figure 7.19. In the case of a verified fire the decision was quickly and unequivocally made to manually blowdown the system. In the case of an unignited gas release however, other actions were seen as being of more urgency. The main reason for a delay in manual blowdown on both platforms was not due to any inhibition in activating the system but the time required for the identification of the source of the leak and its subsequent isolation.

In the molecular sieve module (MSM) scenario, early blowdown could cause significant production delays, and in many cases might be unnecessary if the leak could be quickly isolated or was in fact caused by something trivial (e.g., an oil spillage or a relatively small momentary release). In the MSM there were many potential leak sources due to the complexity of pipework arrangements, and once the system was blown down the identification of the leak became more difficult. There were also possible start-up problems once the system was blown down. Therefore isolation was seen as the primary objective, as long as the probability of ignition was perceived to be low and the gas cloud was drifting away from the platform. In the separator scenario on the other platform there were fewer production penalties following blowdown. However, isolation was still perceived to be the primary objective if possible. On this platform the blowdown function was seen as a last resort, and the fact that blowdown had never been functionally tested, combined with anxieties expressed over secondary risks to the flare system, reinforced this perception.

Thus, on both platforms, in the case of an unignited gas release, the cause of the leak would be investigated and isolated if possible. If the source was discovered and it could not be isolated, then the last resort action of blowdown became applicable, unless it was felt that (in the case of the MSM scenario) it could be allowed to depressurize through the leak itself (as this would be quicker than blowdown). However, at some stage the perceived dangers of ignition and/or gas migration would eventually outweigh other considerations, and a decision to blowdown would be made.

The DA flow chart in Figure 7.19 shows the dynamic nature of the decision process, in as much as the inventory of gas will probably be slowly depleting itself, while the size of the gas cloud and dangers of ignition may be increasing. In fact the workers will continually ask themselves whether ignition is likely, as this determines to a certain extent the urgency of blowdown activation. As time passes, stress levels will increase since they are responsible for safe shutdown, and if the cloud should ignite prior to blowdown, they will be asked why blowdown was not implemented. Maintenance of production reasons would be unlikely to satisfy any subsequent accident investigation, especially if fatalities occurred. Nevertheless it is likely, particularly in the case of the MSM scenario where production problems will be more pronounced,

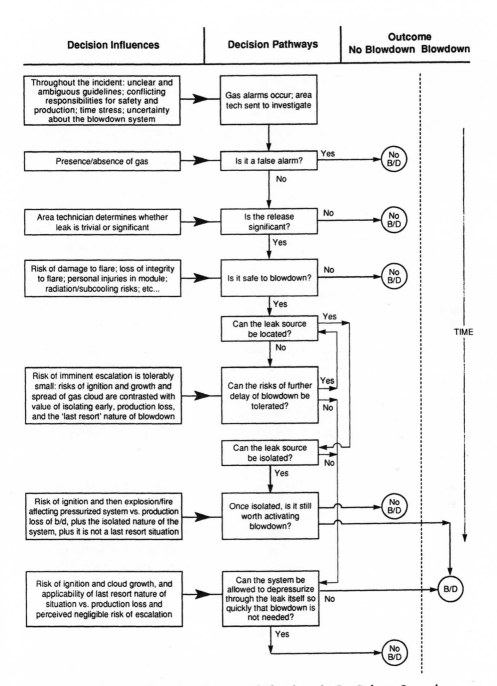

Decision Influences	Decision Pathways	Outcome
		No Blowdown Blowdown

Throughout the incident: unclear and ambiguous guidelines; conflicting responsibilities for safety and production; time stress; uncertainty about the blowdown system → Gas alarms occur; area tech sent to investigate

Presence/absence of gas → Is it a false alarm?   Yes → No B/D   No

Area technician determines whether leak is trivial or significant → Is the release significant?   No → No B/D   Yes

Risk of damage to flare; loss of integrity to flare; personal injuries in module; radiation/subcooling risks; etc... → Is it safe to blowdown?   No → No B/D   Yes

Can the leak source be located?   Yes   No

Risk of imminent escalation is tolerably small: risks of ignition and growth and spread of gas cloud are contrasted with value of isolating early, production loss, and the 'last resort' nature of blowdown → Can the risks of further delay of blowdown be tolerated?   Yes   No

Can the leak source be isolated?   No   Yes

Risk of ignition and then explosion/fire affecting pressurized system vs. production loss of b/d, plus the isolated nature of the system, plus it is not a last resort situation → Once isolated, is it still worth activating blowdown?   → No B/D

Risk of ignition and cloud growth, and applicability of last resort nature of situation vs. production loss and perceived negligible risk of escalation → Can the system be allowed to depressurize through the leak itself so quickly that blowdown is not needed?   No → B/D   Yes → No B/D

TIME

FIGURE 7.19 **Decision Flow Chart for Manual Blowdown in Gas Release Scenario.**

that reasonable attempts will be made to locate and isolate the leak source before blowdown is activated.

In summary, there was little support from the blowdown philosophy, or from procedures and training, to assist in the making of the decision to manually blowdown in a significant gas release scenario. The decision making was made more difficult by the very high workload levels placed on the control room personnel, and the delays and possible errors in critical communications which served as inputs to the decision making process. Furthermore, the decision was complicated by other conflicting considerations, some partially related to safety and others to production, and the general uncertainty about the untested blowdown system and its "last resort" character. Lastly, the workers in the control room and in the affected modules were in a highly stressful situation.

Overall therefore, the workers who were required to blowdown manually were in a difficult situation, and tended to delay manual blowdown until it was absolutely necessary.

7.6.3.7. Task Analysis and Error Analysis of the Blowdown Operation

Task analysis was carried out in order to organize all the performance data about the way that workers process information, the nature of the emergency and the way that decisions are made. Figure 7.20 shows a tabular task analysis of the workers' response to a significant unignited gas leak in MSM. The analysis was a combination of a tabular HTA and a CADET analysis (see Chapter 4). Human error analysis identified the major human failure modes which could affect time to blowdown (see Table 7.2).

7.6.3.8. Levels of Automation

It was concluded from the analysis that blowdown response time was affected more by the attitude of the platform personnel toward the system than the reaction times of the system components. Therefore the implementation of semi or fully automatic blowdown on the platforms would not necessarily enhance performance unless the workers had support in terms of philosophy, training, procedures, and hardware reliability.

Implementation of the recommendations of this study (see next section) would improve the response of both manual and automatic blowdown. The automatic blowdown would be quicker and more reliable than manual blowdown. However, this advantage would be a matter of a few minutes only, and any automatic system will inevitably cause spurious blowdowns.

From the ergonomics perspective, some level of automation was desirable. This was because the worker was in a high stress situation and had to make a relatively difficult decision. A semiautomatic system in which blowdown would occur automatically after a fixed time period following coincident red alarms (given no worker veto) would appear to have the advantages

TIME	OP	SYSTEM STATUS	INFO AVAILABLE	OPERATOR EXPECTATION	PROCEDURE (WRITTEN, MEMORIZED)	DECISION/ COMMUNICATION ACT	EQUIPMENT/ LOCATION	FEEDBACK	SECONDARY DUTIES AND DISTRACTIONS	COMMENTS
0:0	CR	Yellow Alert; Gas leak in MSM	Audible alarm, yellow flashing light on fire & gas panel	Uncertain: could be real	1. Accept alarm 2. Call area technician 3. Make PA announcement 4. Determine which gas detector is in alarm (its location)	Suspend current operation. Scan panels for flashing yellow or red light. Turn around to MSM Fire & Gas panel. Press "accept" button	CCR Layout	Visual and audible	Whatever op is doing when alarm occurs	Initially disorientating because MSM does not have its own sound source. Alarm could be missed if second simultaneous alarm occurs on main bank of Fire & Gas panels
							MSM Fire & Gas panel			
						Go to radio. Radio Area Technician (AT) to go to MSM and verify gas leak.	Radio channel availability		Noise, Interference. AT may be busy.	Channels may be unavailable
						Make PA announcement for personnel to stop hotwork and stay away from MSM	PA systems			This message takes up time and distracts operator from watching Fire & Gas panel. Consider taped messages.
					Gas detector location manual	Check location of gas detector in book.	Gas detector book			Diagrammatic book used on Platform A is very good. Consider equivalent for Platform B.
						Radio AT. Tell AT location of gas detector in alarm.	Radio		Noise, interference	

FIGURE 7.20 **Task Analysis of Operator Response to a Significant Unignited Gas Leak in MSM.**

TABLE 7.2
Major Human Errors Affecting Time to Blowdown
1. Delays are caused by gas detector inhibition and panel alarm detection problems.
2. Delays are caused by workload and communication load problems.
3. Decision is made to try and locate source of leak.
4. Decision is made to continue to attempt source location identification for some time and hence delay blowdown.
5. Workers carry out alternative (higher priority) operations (e.g., load shedding).
6. Decision is made that blowdown is unnecessary if isolated.
7. Decision is made that blowdown is unnecessary even if not isolated.
8. Forget to blowdown (due to workload, distractions, or low prominence of information).
9. Procrastinate—maintain blowdown as last resort.
11. Allocation of function problem (personnel responsible believe other personnel will activate blowdown).
12. Execution fails (fail to turn key on Y platform, and leave turned; or fail to isolate before attempting blowdown on X platform (separator blowdown only).
13. Believe someone has already operated it—fail to verify blowdown is occurring.
14. Awaiting high level order.
15. Stress reaction—fail to activate due to high stress experienced in scenario.

of being relatively quick and robust, while allowing the operator flexibility to override if necessary. It also made the difficult decision of blowdown into a decision of whether or not to intervene and stop blowdown. In cases of uncertainty, workers may be less likely to intervene and override a semiautomatic system, whereas in a manual situation uncertainty may cause them to hesitate and fail to activate the blowdown system. Therefore, because the semiautomatic system would require a positive response and justification to prevent blowdown, it would tend to be more reliable than a manual system.

7.6.4. Conclusions and Recommendations

The major conclusion was that if blowdown is in fact of primary importance, then the current manual system must be improved to enhance blowdown response time. The estimated average activation time of 20 minutes would have to be brought closer to a matter of a few minutes. The basic problem underlying the long response time was a lack of a clear statement of blowdown philosophy and any effective procedural/training support. These aspects were therefore the root causes of the problem, and it was these which had to be corrected, irrespective of the decision to maintain a manual system or move to a higher level of automation.

A selection of the recommendations which were produced are presented below and cover most aspects of the blowdown operation. However, their common underlying theme is to provide optimum decision support in a high stress, high workload emergency situation such as a significant fire or gas release.

1. Issue a simple, clear management statement on blowdown, specifying the criteria for its activation and nonactivation.
2. Train all personnel in the functioning and value of blowdown during an emergency. This may be achieved by safety specialists explaining the various hazards and relative risks of blowing down and failing to depressurize.
3. Train operating personnel in recognizing criteria for blowdown activation and verification, and what to do if it fails. This should consist of a mixture of drills, scenarios, walkthroughs and, if possible, simulator trials.
4. A brief checklist should be drawn up based on an explicit platform-specific procedure for the control room in the event of an emergency. The checklist should emphasize any checks that must be made before blowdown is initiated, and what should be observed during blowdown. This should be in the form of a flow chart.
5. The ergonomics inadequacies identified in both control rooms show that while operational experience is being fed into the design process, there is still significant room for improvement. A design philosophy, based on current ergonomics practice in a range of industries should be developed onshore (but with offshore inputs), for the purpose of being used in future designs and for any retrofits that take place on existing platforms. The development of this philosophy will require specialized expertise.
6. The most likely means of rectifying the very high workload and error rates which are likely to occur, would be to make use of the VDUs in the control room to provide important information in a centralized location, perhaps also in certain cases to effect control actions. This would be especially useful in one of the platform control rooms, since it was never intended to be a centralized control room, and there were therefore many information limitations which could probably only be compensated for by providing VDU information.

7. The molecular sieve module fire and gas panel should be given its own alarm sound source and alarm accept control, since there is currently a risk of an alarm being missed.

8. It is also recommended that a more comprehensive and accurate incident reporting system be set up. The incident reporting system is the only means of assembling and using data from actual incidents that have occurred. However, the level of detail in incident reports gives very little information which could be utilized to improve worker performance. Also, no near miss data were available. It is suggested that the incident reporting and analysis system be further developed for specifically enhancing human factors and thus reliability.

It has been demonstrated that the company-wide blowdown philosophy (or lack of) was the main root cause of the problem. As such it can be seen that operational areas addressed by the audit were essentially management controllable.

8

Implementing an Integrated Error and Process Safety Management System at the Plant

8.1. INTRODUCTION

The purpose of this chapter is to set out in concrete terms the ways in which the techniques described in the previous sections can be implemented in the chemical process plant environment. The integration of error and process safety management systems is also described.

Three main opportunities for applying the techniques to minimize error and maximize performance can be identified:

Optimization
This is the situation where the company has decided to implement a specific program to reduce error. There are two main opportunities for optimization activities:

- During the design of a new process plant, in order to ensure that when the plant is operational its systems (hardware, procedures, training, etc.) will provide the maximum support for error-free operation
- Where it has been decided (possibly because of an unacceptable level of human error leading to safety and production problems) to implement a comprehensive error management program at an existing plant

Evaluation
This is a situation where a plant appears to be operating successfully, without a major human error problem. However, management are interested in assessing the systems in the plant from the point of view of minimizing the error potential. This type of exercise is particularly relevant for plants dealing with substances or processes with high hazard potential, for example, in terms of

the toxicity and energy of the materials. In such plants, errors, although rare, may have severe consequences for individuals or the environment if they do occur. Management may therefore wish to evaluate or audit the factors that can directly affect error potential, and take appropriate action if these factors are found to be less than adequate.

The areas that may be considered as part of such an evaluation are the performance-influencing factors (PIFs) described in Chapter 3. The human factors assessment methodology (HFAM) method described in Chapter 2, provides a systematic framework for the evaluation process.

Problem Solving

Where a specific incident leading to safety, quality or production problems has occurred, the plant management may wish to perform a very focused intervention. This will be directed at identifying the direct and underlying causes of the problem, and developing an appropriate remedial strategy. The process for performing an analysis of this type is described in the incident analysis section of Chapter 6.

Risk Assessment

This application is similar to evaluation except that it may be performed as part of an overall qualitative or quantitative risk assessment. In the latter case, quantitative assessment techniques such as those described in Chapter 5 may be applied.

In subsequent sections the specific procedures for implementing a program to optimize human performance and minimize human error will be described.

8.2. MANAGING HUMAN ERROR BY DESIGN

The following sections describe a design process based on the CCPS approach to human factors in chemical process safety management, which addresses a wide variety of issues relevant to reducing error potential. Many of these issues can be considered both during the process of designing a new plant and also for an existing operation. The design process addresses both management level factors (e.g., objectives and goals) and also operational level factors (e.g., training and procedures).

The CCPS approach to chemical process safety management has been described in a number of documents:

- *A Challenge to Commitment* (CCPS, 1989a)
- *Guidelines for Technical Management of Chemical Process Safety* (CCPS, 1989c)
- *Plant Guidelines for Technical Management of Chemical Process Safety* (CCPS, 1992a)

- *Guidelines for Auditing Process Safety Management* (CCPS, 1993)
- *Guidelines for Investigating Chemical Process Incidents* (CCPS, 1992d)

These publications contain information on twelve key elements of chemical process safety management. In this section, seven of those elements that can be significantly impacted by paying careful attention to human factors principles are addressed. These are:

- Accountability, Objectives, and Goals
- Capital Project Review and Design Procedures
- Process Risk Management
- Management of Change
- Process and Equipment Integrity
- Training and Performance
- Incident Investigation

8.2.1. Accountability, Objectives, and Goals

Accountability is the obligation to answer for one's performance with respect to expectations, goals, and objectives. It is an important element of an effective process safety management system. To improve safety, the risk associated with human errors must be reduced. The work situation is the predominant cause of human errors and management has control over the work situation.

For reduction of human errors to be a top priority, management must convey this priority throughout the organization by administering policies that

- Demonstrate commitment
- Establish a blame-free atmosphere
- Provide resources
- Promote understanding
- Eliminate error-likely situations

Demonstrate Commitment
The likelihood of success of any endeavor is largely dependent on the commitment to that success. This is especially true with improving process safety and reducing human errors. That commitment must start at the very top and flow strongly through all levels of the organization.

Establish a Blame-free Atmosphere
The people most knowledgeable about a particular task are the people who perform it every day. Their help is essential for reducing the associated risks. Continuous feedback from the worker will provide the framework for improvements to the job. This feedback can only be fostered in an atmosphere of trust. If an incident occurs in which human error is a suspected cause, man-

agement response is critical. If management realizes that it is really the work situation that is at fault and that eventually a similar incident would have occurred no matter who the worker was, then a blame-free atmosphere exists. If instead, management lashes out and seeks to blame individuals for the incident, then trust will vanish. In such a negative environment, meaningful communication cannot take place.

Provide Resources
It is readily acknowledged that resources such as manpower, equipment, and training are generally provided by management. What may not be recognized is that a human factors science exists which is capable of improving the power of these resources. Managers need to make use of this science by incorporating human factors expertise into their organizations to improve their process design, operation, and maintenance.

Promote Understanding
Promoting understanding throughout the organization is an essential part of management's human error communication. By having persons in the organization well versed in human factors principles, management can greatly enhance the understanding of these principles by everyone in the organization. This allows incorporation of these ideas into all aspects of process design, operation, and maintenance.

Eliminate Error-Likely Situations
The final element in management's communication of a desire to reduce human error is the identification and elimination of error-likely situations. Every task is an opportunity for a human error, but some situations represent greater risks than others. Identifying these high-risk situations is not easy and an expertise in applying human factors principles to the workplace is an essential prerequisite for this identification. Eliminating these hazardous situations is often relatively simple once they have been identified. In some cases it may be appropriate to provide error-tolerant systems, which are those that facilitate identification of and recovery from the errors.

8.2.2. Process Safety Review Procedures for Capital Projects

The "Capital Projects Design Review Procedures" element of process safety management assures that the design, the equipment and the construction are properly reviewed for all new projects. Process safety review procedures should be involved with the project from its inception. One method of illustrating the various phases of a project is shown in Figure 8.1 (Figure 5-1 from CCPS, 1989a).

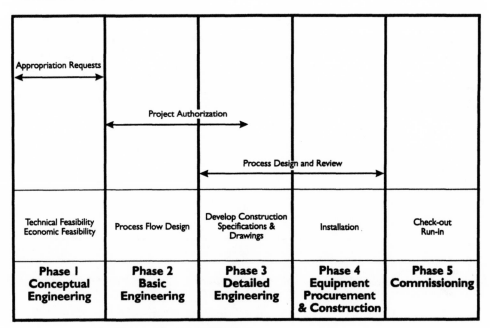

FIGURE 8.1. **The Phases of a Capital Project.**

While this process implies an ordered, structured process, it should be noted that the various stages overlap and it is frequently necessary to return to an earlier step in the process to modify or clarify information or decisions made in an earlier phase. The influence of the human factors aspects on design needs to be similarly integrated into the process design procedure. The particular human factors elements to be addressed at each phase are discussed below. These phases can be directly related to the human factors engineering and ergonomics (HFE/E) design approach described in Section 2.2.

Phase I—Conceptual Engineering
During this phase the objectives of the system as well as the system performance specifications are identified. In addition, the technical and economic feasibility of the project is evaluated.

Human factors considerations include determining the characteristics of the population in the area where the facility will draw its labor force. Factors such as educational levels, languages, norms, and environmental conditions should be considered because they can have a significant impact on later stages of the design process.

Phase II—Basic Engineering
This phase is where the basic design work, such as process flow design, is performed. Piping and instrumentation diagrams (P&IDs) and equipment

data sheets used for design are some of the products of this phase. The human factors issues include the allocation of functions between human and machine, what is best to automate and let a machine (computer) perform and what should be assigned to the human. This process needs to be given considerable thought and effort, and should include normal as well as unusual plant conditions. The strategy of simply automating what can easily and cheaply be automated and relegating the remainder to the human should be avoided. The intent should be to maximize system performance by the logical and proper distribution of functions.

Knowing what functions or tasks are to be performed by humans permits the development of initial staffing strategies. Consideration should also be given to work load. It is also necessary to begin to examine the information and response requirements needed to operate the system.

The system requirements will influence such human factors issues as

- Personnel skill requirements
- Physical demands placed on the operator (e.g., temperature, noise, vibration)
- Mental demands (e.g., the number and complexity of demands and the response time for decision making)

Since most systems are based on existing systems it is wise to begin to evaluate the existing systems to gain insight into potential problems and improvement opportunities. While these insights can be gained through observations, interviews or questionnaires, the most effective method is the participation of operators, mechanics, engineers, and supervisors on design review teams. This is an excellent technique to get the user involved in the design process and to tap into the resources that will not only identify potential problem areas but often will provide good methods for determining cost effective ways to address these problems. Those participating also develop a sense of ownership with the design and become stake holders in the process. They also develop a good basic understanding of the system and how it works before they receive formal training on its operation and maintenance.

Phase III—Detailed Engineering
In this phase the specifications and drawings for construction are being developed and issued. Attention should be given to human–machine and human–computer interface which includes such issues as software design, general work space design, controls, displays (including computer displays), communication and information requirements, maintainability, labeling, and handling tasks. The intent should be to incorporate good human factors principles into the design in an effort to eliminate error-likely or accident prone situations. Once again reviewing past incidents may prove to be beneficial at identifying areas to address. It may be appropriate to analyze specific human interface designs

utilizing mock-ups or other simulations. The application of information from reference guidelines, mock-ups and simulations should be used to assist in the development of the detailed design. Depending on the performance results, it may be necessary to alter the allocation of the function or task.

Phase IV—Equipment Procurement and Installation

This phase involves the purchasing of fabricated and bulk materials and installations on site. As equipment is fabricated it is necessary to review its design to ensure that the design has not been modified and to detect any unforeseen problems that might negatively impact human and therefore system performance. Further evaluations and testing may be necessary on simulators. Planning should begin for preparing operating procedures and training manuals and devices (see Section 8.2.5). The need for additional performance aids such as computer-assisted decision making tools should be considered. The implications of the layout of plant equipment such as valves and pumps should be reviewed from a human factors standpoint to ensure that issues such as accessibility are addressed.

Quality of installation and the adherence to design specifications of the equipment should be evaluated to ensure that errors during shipping and installation were not made. Often overlooked at this phase are human factors considerations for the construction crew, such as selection of the contractor, training of the crew, lighting, shift work, procedures, and supervision.

Phase V—Commissioning

This phase involves the performance of check-out and run-in activities to ensure that equipment and piping are mechanically integrated, functionally located and free of obstructions It is also necessary to ensure that instruments and controls are functioning properly and that all previously identified problems have been addressed. All maintenance and operating procedures need to be verified as correct.

The impact of the system design on human performance should be examined during this phase. The designer should consider whether or not the operator will be able to keep the system operating correctly under normal conditions and is he or she able to effectively handle unusual conditions, returning them to normal operating conditions. Observation of personnel and discussions with them are effective ways that should be employed in this phase. Where necessary, appropriate modifications should be made to the system to ensure proper performance.

8.2.3. Process Risk Management

Hazard identification is the first step in process risk management. In order for this procedure to be adequately utilized, the hazard identification team must know what to look for. This requires that participants be trained in the capabilities of people. This training should include biomechanical information for tasks such as lifting; display and control design factors such as colors, directions, and order; and man–machine interface issues such as feedback skill level, complexity, and environment. Operating Input from operating personnel through a blame-free culture must be sought for effective hazard identification. The operators must be encouraged to point out problems that have occurred. Upon raising the human factors issues, the recommendations for change must be addressed by the design engineer.

Management must modify the culture and develop human factors awareness in the hazard identification teams so that they will be capable of identifying the potential for human error. A good practice is to involve operators in the hazard identification team.

In order to do a risk analysis properly, the analyst must understand human capabilities, including the capability to recover from errors. Human error data may be available from generic sources or from in-plant studies done previously.

Effective process management during emergencies requires that emergencies must be visualized before they occur and that the capabilities of the human be taken into account when designing response mechanisms. Procedures should be developed to assist the operators and the emergency response team. Simulation exercises designed to test and evaluate emergency response capability should be conducted, critiques held, and corrections made. These will provide insight into the effectiveness of the humans' role.

8.2.4. Management of Change

Management of change procedures must be simple yet effective in capturing changes. Human nature being what it is, many changes will be deemed too trivial to qualify as a change to avoid the extra effort of following the management of change procedures (documentation, approval reviews, etc.)

Many changes that are not normally recognized as such, can significantly affect people's performance. Examples are changes in:

- Management style or philosophy
- Organizational structure or reporting relationships
- Company and union relationship
- Shift regimes
- Staffing (at either the worker or supervisor level)
- Work environment (lighting, noise level, temperature, etc.)
- Educational level of employees (either higher or lower)
- Work organization
- Work flow
- Authority and/or responsibilities of first-line supervisors or operators

- Production rate
- Skill level required to perform tasks
- Information feedback
- Process control philosophy (e.g., computerization that removes supervisory control from the operator and may lead to inability of the operator to respond effectively to an abnormal situation)
- Stress level (e.g., more complicated or less complicated processes, layoffs)

8.2.5. Process and Equipment Integrity

Equipment integrity is primarily achieved by good design and installation and the proper consideration of human factors should be an integral part of the design phases (see Section 8.2.2 on Capital Project Review and Design Procedures and the earlier sections of this chapter.

Maintenance, Inspection, and Testing
Maintaining equipment integrity throughout its lifetime is achieved by continual inspection, testing, and maintenance; equipment therefore should be designed and installed to provide a good working location for operation and maintenance. This requires considering lighting, shelter, humidity, noise, temperature, access, personnel working position, availability of correct tools, and the like. Clear responsibilities for isolating, verifying, and sign off are necessary. Labeling exactly what must be worked on, and explicit, accurate, unambiguous communications on what must be done are essential. Frequent reviews of work permitting systems are required to ensure that specified procedures continue to be followed.

Stress from working against a deadline, working in danger, long working hours and hot conditions, must be considered. Workshops should be designed with human factors principles in mind. The environment, both physical and managerial, working positions, and tools should be designed to facilitate good error-free maintenance.

Procedures
Correct procedures for maintenance, inspection, and testing will be needed and should be written with human factors ideas in mind. Some considerations for good procedures are:

- The operators should be intimately involved in writing procedures. This will both make use of their extensive knowledge of the process and help to ensure their commitment to following those procedures.
- Procedures should specify the process "windows" for safe operation.
- The procedure should show how to do the task and not be confused with a training manual. The latter, but not the former, should include

the reasons for the task and explanations of the results or objectives of each step.

- They must be kept up-to-date and be easy to revise. A system needs to exist so that all outdated copies are immediately replaced with current editions.
- Diagrams and examples should supplement text for clarity and understanding.
- The best presentation will have considered text font, color, columns, relationship of text to diagrams, single-sided printing, etc.
- The text should be written at the appropriate reading level. The active tense should be used and complex sentences avoided.
- Procedures should be easy to use on the job, portable, and resistant to abuse.
- Steps that include some feedback to verify their completion are desirable.
- The whole procedure should be validated to ensure that it is practicable and is indeed followed in practice.
- It should be easier to obtain and follow the correct procedure than to improvise.
- Procedures should follow a standard format, facility-wide. The objective is to avoid errors by ensuring consistency and ease of use.

8.2.6. Training and Performance

Training

Training requires much more than simply following and practicing the procedures. It requires understanding the reasons for the procedures, the consequences of deviations from these procedures and recognition and accommodation for the fact that actual performance will differ from that observed in training sessions. The training should, as far as is possible, reduce there differences, or at least the significance of such differences. It is essential that operating procedures and training be closely integrated.

Training should in no way be considered as a substitute or remedy for poor design. Although poor training is frequently given as the reason for people making mistakes, we must emphasize that working in an error-likely situation is probably a more valid reason. Good training will have considered the relevant human factors elements.

Training can be considered to take place at two levels:

- The general (appreciation, overview, awareness), education level
- The detailed level of being able to perform tasks, functions or acquire skills

While it is impractical to provide all employees with detailed human factors training, process safety, and management training should provide an overview of the subject. The detailed training in operations, maintenance, inspections, and tests, must include reference to specific and readily applicable human factors issues. An integral part of training is the existence of quality procedures to follow. Writing quality procedures is not a casual skill, it requires proper consideration of all the ways the user interacts with the procedure. Discussion of the human factors aspects of preparing procedures is given in Section 8.2.5, Process and Equipment Integrity. A more detailed case study of procedure development is provided in Chapter 7.

In all training programs, the importance of identifying, reporting, and eliminating error-likely situations, reporting mistakes and near misses and looking for ways to prevent errors should be emphasized. Once again this requires a blame-free environment.

The following are some human factors aspects relevant to training but they are meant only as a stimulus to further study and consideration and not as an authoritative and exhaustive checklist.

- The trainer must understand human factors principles.
- The training environment should be noise-free, with proper lighting and a comfortable temperature provided. Good audio-visual aids should be used.
- Lines of communication should be kept short. Transfer of information is probably least accurate by word of mouth. Avoid superfluous information. Show is better than tell.
- Understanding the objectives of a procedure and the consequences of deviation are a vital part of learning how to perform the procedure. Knowing process limitations and tolerances is also important.
- Choosing the tasks best suited to equipment and allowing people to do the task they are best suited to, is better than training people to do jobs for which they are ill-suited. Practice needs to be carried using training situations similar to the active tasks and the differences between the real and artificial situations need to be appreciated.
- All activities are based on various levels of ability. Skill based actions are virtually automatic and may fail due to distractions and change. Rule based actions require good procedures and adequate time. Knowledge based actions require technical knowledge and organized thinking. Different types and levels of training and practice are required for these different abilities. Full discussions of these concepts and their implications for training are provided in Chapter 3.
- Recognize the effects of stress in case of emergencies. The actual emergency probably will differ considerably from those practiced. The reluctance to acknowledge that the emergency exists is a well-known cause of delayed response. Complex decisions are doomed to failure

under stress; anticipate as many decisions as possible and make them simple by providing decision aids such as those described in Chapter 4.
- People develop bad habits and take things for granted; processes, equipment, and technology change, so retraining is essential at appropriate intervals.
- Develop realistic evaluation exercises to verify that effective learning of skills has occurred.

Performance

In addition to proper training and quality procedures, good performance requires a supportive culture and working environment. The procedures provide the "how to," the training reinforces this with the background, the understanding, and the practice to develop the necessary skills; the environment must support their quality execution.

An encouraging, motivational environment is likely to evoke better performance than a threatening one. It is generally accepted that results are better where employees are empowered to contribute and participate rather than just follow instructions.

8.2.7. Incident Investigation

The most important thing that must be in place to have a successful incident investigation program is a blame-free culture. Unless this is in place, the only incidents that will be reported are those with outcomes that are difficult to cover-up such as serious injuries and large spills. Even first-aid cases will go unreported, if there is a penalty for having an injury (or a prize for not having one). Once the culture is at least partially in place, *all incidents must be reported, including near misses. It is important to have all these incidents in your data base for a complete analysis of your program needs.*

Investigating these reports then becomes paramount. However investigation takes know-how. A training program that focuses on finding the root, secondary, underlying, or hidden causes is required. Superficial investigations that stop when unsafe acts and conditions are identified and only blame the person add nothing to incident investigation. The objective is to discover the real causes of the incident, so that they can be corrected. After investigations are started, a data bank for the information extracted from the exercise should be initiated. The data can be used for risk analysis (see Section 8.2.3 on Process Risk Management and Chapter 5), giving direction to your training programs (see Section 8.2.6 on Training and Performance), and providing design feedback for future plant changes and new plant designs (see Section 8.2.2 on Process Safety Review Procedures).

Sometimes special expertise may be needed when incidents are investigated. Human factors professionals may be needed to do a Task Analysis to determine if a task exceeds the capabilities of the human. Assessments of the factors in a situation that would have affected the likelihood of the error (performance-influencing factors—see Chapter 3) may also need to be made by the investigator.

In order to accurately describe the actual sequence of events that make up the accident techniques such as STEP (sequential timed event plotting—see Chapter 6 and the case study in Chapter 7) can be used by the investigator.

8.3. SETTING UP AN ERROR MANAGEMENT SYSTEM IN AN EXISTING PLANT

Some of the elements of an error management system were described in the Section 8.2. In this section, the process of setting up an error management program in an existing plant will be described. The components of the error management program have been discussed in previous chapters and are summarized in Figure 8.2.

This indicates that error management comprises two strategies: proactive methods are applied to prevent errors occurring, and reactive strategies are used to learn lessons from incidents that have occurred and to apply these lessons to the development of preventive measures. Both proactive and reactive methods rely on an understanding of the courses of human error based on the theories and perspectives presented in this book. The tools and tech-

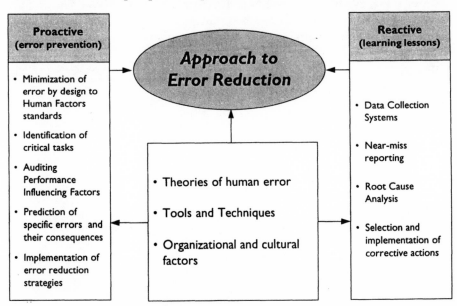

FIGURE 8.2. **General Error Management Structure.**

niques that need to be applied, such as task and error analysis, have been described and illustrated in Chapters 4 and 7. The PIFs that need to be evaluated and controlled are described in Chapter 3. In setting up a successful error management program, it is vital to address organizational and cultural issues as well as technical methods. The overall structure of the error management program is illustrated in Figure 8.3.

8.3.1. Stage 1: Obtain Senior Management Commitment to Support the Program

Since senior management will ultimately control the resources required to implement a program, they need to be convinced that the benefits which are likely to arise will provide a reasonable return on their investment. Evidence presented in Chapter 1 of this book and from other sources can be used to provide a convincing argument that investment in improving human performance and reducing error will produce a rate of return which is at least equal to and will probably exceed that obtained from the same investment in hardware. It should be emphasized, however, that any program which requires a change of attitudes and culture, particularly in a sensitive area such as human error, will required time for its effects to be felt. Management commitment must therefore be reasonably long term.

The decision to initiate a human error management program will normally be taken by senior plant management or at a corporate level. The reasons for setting up a program may be the occurrence of significant losses that are clearly attributable to human error, or from regulatory pressures to produce improvements in this area.

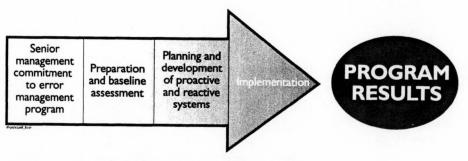

FIGURE 8.3. **Stages in Setting Up an Error Management Program.**

8.3.2. Stage 2: Evaluate the Current Situation

An evaluation of the nature and magnitude of the current human error problem in the organization is an essential step in the program. This assessment identifies the existing concerns and perceived problem areas and sets the initial agenda for the error reduction program. Typical information sources that will be utilized are plant incident records, problem reports and near miss reports if such a system exists. Structured interviews will also provide an important source of data concerning problem areas. The initial baseline evaluation will normally be carried out by a human reliability specialist appointed by senior management.

The main output from this stage is a detailed assessment of the human error problem areas. If possible this should include quantitative data on the incidence of errors and the significance of their consequences. This will provide a valuable baseline against which to evaluate the success of the error management program.

In addition to these assessments of the causes and consequences of errors, the baseline assessment should also include an evaluation of the systems that are currently in place to control errors. This will include an evaluation of the current state of a representative set of the PIFs discussed in Chapter 3. Typical factors would include

- Training policies
- Procedures
- Work organization (e.g., shift work)
- Design of human–machine interfaces (e.g., process information displays, alarm systems, plant labeling)
- Communications systems
- Work control systems (e.g., work permits)

8.3.3. Stage 3: Set Up a Program Development and Coordination Committee

The purpose of this stage is to set up a group of plant personnel who will be responsible for initiating and sustaining the effectiveness of the error management program. The composition of this group should include individuals from all areas of the plant (e.g., operations, line management, engineering) who may be affected by the error management process when it is implemented. It is also desirable that a human factors professional is on the committee in order to provide specialist knowledge as required. However, the most important requirement of the committee is that it is enthusiastic and strongly committed to the systems approach to error reduction.

In order to obtain this commitment, training in the systems view will be necessary, preferably using examples from the plant that the group can readily relate to. It will normally be the responsibility of the human factors specialists who have conducted the baseline analysis to provide this training.

8.3.4. Stage 4: Program Strategy Development

The committee will formulate an error management strategy to address the specific needs of the plant based on the information collected during the baseline exercise. The strategy will address the following areas:

- Specific interventions to address the problem areas identified during the baseline study
- Setting up proactive error management systems
- Development of reactive programs including data collection and root cause analysis systems
- Development of programs to create an appropriate culture to support the error management system
- Assessment of policy needs to support these programs

The objective of initially focusing on the problem areas identified during the baseline assessment is to enable the error management program to demonstrate its capability to solve problems as quickly as possible. By providing solutions to perceived problem areas, the program will gain credibility which will considerably enhance its acceptance and support. The results of the initial interventions can also be used as case studies as part of the consciousness raising process that will be necessary in order to successfully implement the program in the plant. It is recommended that the problem solving exercise is performed for one or two of the problem areas identified during the baseline study **before** the program is launched. This will enable tangible benefits to be described during the launch of the program.

8.3.5. Stage 5: Program Launch

The launch should ensure that it is perceived to be an important and significant innovation. Major factors that will enhance the success of the launch are listed below.

Publicity
The program launch should be prominently featured and explained in company communication channels such as newsletters and in-house magazines.

Senior Management Support
Senior management should attend the launch and endorse its importance.

Demonstrated Credibility
The case studies obtained from the pre-launch error reduction interventions should be used to demonstrate the practical value of the program.

8.3.6. Stage 6: Implementation

The program plans for the five areas addressed during the program strategy development phase are implemented at this stage.

Specific Interventions
The error reduction activities that were begun in the pre-launch phase should be completed at this stage.

Setting Up the Proactive Error Management System
This system will include the following topics:

- Identification of critical tasks. This is an important initial step as it identifies the areas where application of proactive error reduction approaches will produce the greatest benefits (see Chapter 5)
- Prediction of specific errors and their consequences by using task analysis and error prediction techniques (Chapters 4 and 5)
- Auditing PIFs (Chapter 3)
- Implementation of error reduction measures

Setting Up Reactive Error Management Systems
This involves the development of data collection and root cause analysis systems as described in Chapter 6.

Development of a Plant Culture to Support
the Error Management Program
This aspect of the program may be difficult to achieve quickly, especially in an environment where a blame culture has predominated. A supportive culture can only be achieved in a situation where there is a clear policy direction from senior management, and this policy is implemented in a day to day basis by line managers and supervisors. In addition, a continuing program of education is necessary to overcome skepticism and to reinforce the practical benefits of the program. In order to gain acceptance from the workforce the program has to demonstrate that active participation produces tangible benefits in terms of enhanced safety and a reduced burden of blame from errors that are due to causes outside the workers' control.

*Development of Policies to Provide Long-Term Support
for the Program*

The various elements of the error management program such as the development of high quality procedures and training and effective feedback and communications systems need to be supported by policies and standards to implement these policies. The development of these policies is an important strategic aspect of the implementation process.

8.3.7. Stage 7: Evaluation and Improvement

Following the implementation of the error management program, evaluations of its effectiveness must be made on a continuing basis. Measures of effectiveness can be gained partly by direct discussions with individuals at all levels of the plant, using a predesigned evaluation checklist to ensure that all the evaluation dimensions are assessed in a systematic way. Examples of evaluation questions follow.

- Since the program began are people generally more willing to admit that they have made errors?
- Has the program made it easier to understand the causes of errors?
- Are people still blamed or punished if they admit to making errors?
- Have you noticed any changes in the conditions that may have caused you to make errors in the past (e.g., improved procedures, better plant labeling, more effective sharing of experience)?
- Have changes been made that are likely to reduce the incidence of errors?

In addition to these types of evaluations, the incident reporting and near miss systems should be monitored for improvements in performance that can be attributed to the program. On the basis of feedback from these and other sources such as changes in quality levels or efficiency, modifications should be made to the program to enhance and sustain its effectiveness. Concrete evidence regarding the benefits arising from the program should be communicated to those involved.

8.3.8. Stage 8: Maintenance and Ownership

The key to maintaining the long-term effectiveness of the program lies in ensuring that it provides tangible benefits for its stake-holders. These include the plant management as well as the various levels of the workforce. Because the plant worker will have the greatest day-to-day contact with the factors that impact directly on the likelihood of errors, it is vital that they eventually come to own and operate the error management system.

However, this will only occur if their efforts in identifying the factors that give rise to errors is matched by evidence that their contributions lead to tangible changes being made that improve their jobs. If this is achieved then the error management program will in the long run become self-sustaining, and self-financing through reductions in accident rates, and improved quality and efficiency.

8.4. SUMMARY

The general approach that has been advocated in this chapter is that it is the responsibility of an organization, through its safety management policies, to create the systems, environment, and culture that will minimize human error and thereby maximize safety.

The potential benefits to be gained are considerable. In addition to reducing the burden of death, injury, and ill health that arises from human caused accidents, enhancing safety through the reduction of error has the potential for direct financial payback. A recent U.K. study (HSE, 1994) indicated the annual cost of accidents to employers was equal to between 5% and 10% of all U.K. companies gross trading profits (between $6 and $12.5 billion). Even if human error is conservatively estimated at being the direct cause of only 50% of industrial accidents, the savings in reducing human error by 50% could be estimated as being of the order of $15 billion if these results were translated to the U.S. economy. The benefits of investing in human error prevention programs are therefore difficult to challenge. It is hoped that the contents of this book will provide the chemical process industry with the knowledge and tools it needs to begin to achieve these benefits.

References

Abe, N. (1978). Levels of Trust and Reactions to Various Sources of Information in Catastrophic Situations. In E. A. Quarantelli (Ed.), *Disasters: Theory and Research.* Beverly Hills: Sage Publications.

Ainsworth, L. K., & Whitfield, D. (1984). "The Use of Verbal Protocols for Analyzing Power Station Control Skills." Report No. 114, Applied Psychology Dept., Aston University, Birmingham, U.K.

Allusi, E. A., & Morgan, B. B. (1982). Temporal Factors in Human Performance and Productivity. In E. A. Allusi & E. A. Fleishman (Eds.), *Human Performance and Productivity.* Hillsdale, NJ: Erlbaum.

Annett, J., Duncan, K. D., Stammers, R. B., & Gray, M. J. (1971). *Task Analysis.* London: Her Majesty's Stationery Office.

Armstrong, M. E., Cecil, W. L., & Taylor, K. (1988). "Root Cause Analysis Handbook." Report No. DPSTOM-81, E. I. DuPont De Nemours & Co., Savannah River Laboratory, Aiken, SC 29808.

Armstrong, M. E. (1989). Human Factors in Incident Investigation. In *Proceedings of the 33rd Human Factors Society Annual Meeting.* Santa Monica, CA: Human Factors Society, Inc., pp. 1024–1028.

Astley, J., Shepherd, A., & Whitfield, D. (1990). A Review of UK and International Knowledge and Practice in the Selection of Process Control Operators. In E. J. Lovesey (Ed.), *Ergonomics Setting Standards for the 90's. Contemporary Ergonomics, 1990.* London: Taylor and Francis.

Bainbridge, L. (1987). Ironies of Automation. In J. Rasmussen, K. Duncan, & J. Leplat (Eds.), *New Technology and Human Error.* New York: Wiley.

Bainbridge, L. (1974). Analysis of Verbal Protocols from a Process Control Task. In E. Edwards & F. P. Lees (Eds.), *The Human Operator in Process Control.* Washington, DC: Taylor & Francis.

Banks, W., & Wells, J. E. (1992). A Probabilistic Risk Assessment Using Human Reliability Analysis Methods. In *Proceedings of the International Conference on Hazard Identification and Risk Analysis, Human Factors, and Human Reliability in Process Safety.* New York: American Institute of Chemical Engineers, CCPS.

Barnes, R. M. (1980). *Motion and Time Study, Design and Measurement of Work* (7th ed.). New York: Wiley.

Bartlett, F. C. (1943). Fatigue following Highly Skilled Work. *Proceedings of the Royal Society (Series B)* 131, 247–257.

Beishon, R. J. (1967). Problems of Task Description in Process Control. *Ergonomics* 10, 177–186.

Beishon, R. J. (1969). An Analysis and Simulation of an Operator's Behavior in Controlling Continuous Baking Ovens. In A. de Brisson (Ed.), *The Simulation of Human Behavior.* Dunuod: Paris.

Bellamy, L. J., Kirwan, B., & Cox, R. A. (1986). Incorporating Human Reliability into Probabilistic Safety Assessment. *5th International Symposium on Loss Prevention and Safety Promotion in the Process Industries*, Cannes, France. Paris: Société de Chimie Industrielle.

Bellamy, L. J., & Geyer, T. A. W. (1988). Addressing Human Factors Issues in the Safe Design and Operation of Computer Controlled Process Systems. In B. A. Sayers (Ed.), *Proceedings of SARSS '88. Human Factors and Decision Making: Their Influence on Safety and Reliability.* 19–20 October, Altrincham, Manchester, U.K. London: Elsevier Applied Science.

Bellamy, L. J., Geyer, T. A., Wright, M. S., & Hurst, N. W. (1990). The Development in the U.K. of Techniques to Take Account of Management, Organizational and Human Factors in the Modification of Risk Estimates. *Proceedings of the 1990 Spring National Meeting, American Institute of Chemical Engineers*, Orlando.

Berkun, M. M. (1964). Performance Decrement Under Psychological Stress. *Human Factors* 6, 21–30.

Bishop, J., & Larhette, R. (1988). *Managing Human Performance—INPO's Human Performance Enhancement System.* Conference Record for the 1988 IEEE Fourth Conference on Human Factors and Power Plants 88CH2576-7. Institute of Electrical and Electronic Engineers: New York. Pp. 471–474.

Blackman, H. S., Gertman, D. I., & Gilmore, W. E. (1983). "CRT Display Evaluation: The Checklist Evaluation of CRT-Generation Displays." Report no. NUREG/CR-3357. Washington, DC: Nuclear Regulatory Commission.

Blum, M. L., & Naylor, J. C. (1968). *Industrial Psychology.* New York: Harper and Row.

Bridges, W. G., Kirkman, J. Q., & Lorenzo, D. K. (1994). Include Human Errors in Process Hazard Analyses. *Chemical Engineering Progress*, May.

Butikofer, R. E. (1986). *Safety Digest of Lessons Learned.* Washington, DC: American Petroleum Institute, API Publication 758.

Cakir, A., Hart, D. J., & Stewart, T. F. M. (1980). *Visual Displays Units.* New York: Wiley.

Center for Chemical Process Safety (CCPS) (1985). *Guidelines for Hazard Evaluation Procedures* (1st ed.). New York: American Institute of Chemical Engineers, CCPS.

Center for Chemical Process Safety (CCPS) (1989a). "A Challenge to Commitment." New York: American Institute of Chemical Engineers, CCPS.

Center for Chemical Process Safety (CCPS) (1989b). *Guidelines for Chemical Process Quantitative Risk Analysis.* New York: American Institute of Chemical Engineers, CCPS.

Center for Chemical Process Safety (CCPS) (1989c). *Guidelines for Technical Management of Chemical Process Safety.* New York: American Institute of Chemical Engineers, CCPS.

Center for Chemical Process Safety (CCPS) (1992a). *Plant Guidelines for Technical Management of Chemical Process Safety*. New York: American Institute of Chemical Engineers, CCPS.

Center for Chemical Process Safety (CCPS) (1992b). *Guidelines for Hazard Evaluation Procedures* (2nd ed.). New York: American Institute of Chemical Engineers, CCPS.

Center for Chemical Process Safety (CCPS) (1992c). *International Conference on Hazard Identification and Risk Analysis, Human Factors and Human Reliability in Process Safety*. New York: American Institute of Chemical Engineers, CCPS.

Center for Chemical Process Safety (CCPS) (1992d). *Guidelines for Investigating Chemical Process Incidents*. New York: American Institute of Chemical Engineers, CCPS.

Center for Chemical Process Safety (CCPS) (1993). *Guidelines for Auditing Process Safety Management Systems*. New York: American Institute of Chemical Engineers, CCPS.

Cohen, A. (1977). Factors in Successful Occupational Safety Programs. *Journal of Safety Research* 9(4), 168–178.

Colquhoun, W. P., Blake, M. J., & Edwards, R. S. (1969). Experimental Studies of Shift Work III: Stabilized 12-hour Shift Systems. *Ergonomics* 12, 865–882.

Crosby, P. B. (1984). *Quality without Tears: The Art of Hassle-free Management*. New York: McGraw-Hill.

Crossman, E. R .F. W., Cooke, J. E., & Beishon, R. J. (1974). Visual Attention and the Sampling of Displayed Information in Process Control. In E. Edwards & F. P. Lees (Eds.). *The Human Operator in Process Control*. Washington, DC: Taylor & Francis.

Cullen, the Hon. Lord (1990). *The Public Inquiry into the Piper Alpha Disaster*. UK. Dept. of Energy. London: Her Majesty's Stationery Office.

Deming, W. E. (1986). *Out of the Crisis*. Cambridge, MA: MIT Center for Advanced Engineering Study.

Dixon, N. F. (1976). *On the Psychology of Military Incompetence*. London: Jonathan Cape.

Dixon, N. F. (1987). *Our Own Worst Enemy*. London: Jonathan Cape.

Dougherty, E. M., & Fragola, J. R. *Human Reliability Analysis: A Systems Engineering Approach with Nuclear Power Plant Applications*. New York: Wiley.

Dumas, R. (1987). Safety and Quality: The Human Dimension. *Professional Safety* December 1987, pp. 11–14.

Duncan, K. D. (1974). Analytic Techniques in Training Design. In E. Edwards & F. P. Lees (Eds.), *The Human Operator in Process Control*. Washington, DC.: Taylor & Francis.

Duncan, K. D., & Gray, M. J. (1975). Scoring Methods for Verification and Diagnostic Performance in Industrial Fault-Finding Problems. *Journal of Occupational Psychology* 48, 93–106.

Edwards, E., & Lees, F. (1974). *The Human Operator in Process Control*. Washington, DC: Taylor and Francis.

Embrey, D. E. (1985). Modelling and Assisting the Operator's Diagnostic Strategies in Accident Sequences.

Embrey, D. E. (1986). Approaches to aiding and training operators' diagnoses in abnormal situations. *Chemistry and Industry* 7 July, 454–459.

Embrey, D. E (1986). SHERPA: A Systematic Human Error Reliability Prediction Approach. *Proceedings of the American Nuclear Society International Topical Meeting on Advances in Human Factors in Nuclear Power Systems*.

Embrey, D. E. (1992). Incorporating Management and Organizational Factors into Probabilistic Safety Assessment. *Reliability Engineering and System Safety* 38, 199–208.

Embrey, D. E. (1993). *THETA (Top-Down Human Error and Task Analysis). Software Manual.* Human Reliability Associates, 1 School House, Higher Lane, Dalton, Wigan, Lancs. WN8 7RP, England.

Embrey, D. E. (1994). *Software Guides for Human Reliability Quantification.* Human Reliability Associates Ltd., 1, School House, Higher Land, Dalton, Wigan, Lancs, England.

Embrey, D. E., & Humphreys, P. (1985). Support for Decision-Making and Problem-Solving in Abnormal Conditions in Nuclear Power Plants. In L. B. Methlie & R. H. Sprague (Eds.), *Knowledge Representation for Decision Support Systems.* Amsterdam: North Holland.

Embrey, D. E., Kirwan, B., Rea, K., Humphreys, P., & Rosa, E. A. (1984). *SLIM-MAUD. An approach to Assessing Human Error Probabilities Using Structured Expert Judgment* Vols. I and II. Washington, DC: NUREG/CR—3518 US Nuclear Regulatory Commission.

Ferry, T. S. (1988). *Modern Accident Investigation and Analysis* (2nd edition). New York: Wiley.

Fitts, P. M., & Jones, R. E. (1947). Analysis of Factors Contributing to 460 "Pilot Error" Experiences in Operating Aircraft Controls. Reprinted in H. W. Sinaiko (Ed.) (1961), *Selected Papers on Human Factors in the Design and Use of Control Systems.* New York: Dover.

Flanagan, J. C. (1954). The Critical Incident Technique. *Psychology Bulletin* 51, 327–358.

Folkard, S., Monk, T. H., & Lobban, M C. (1979). Towards a Predictive Test of Adjustment to Shift Work. *Ergonomics* 22, 79–91.

Folkard, S., & Monk, T. H. (1985). *Hours of Work: Temporal Factors in Work Scheduling.* New York: Wiley.

Friedman, M., & Rosenman, R. (1974). *"Type A" Behavior and Your Heart.* New York: Knopf.

Froberg, J. E. (1985). In S. Folkard & T. H. Monk (Eds.), *Hours of Work: Temporal Factors in Work Scheduling.* New York: Wiley.

Garrison, W. G. (1989). *Large Property Damage Losses in the Hydrocarbon-Chemical Industries: A Thirty Year Review* (12th ed.). Chicago: Marsh and McLennan Protection Consultants.

Gertman, D. I., Haney, L. N., Jenkins, J. P., & Blackman, H. S. (1985). Operator Decision Making Under Stress. In G. Johannsen, G. Mancini, & L. Martensson (Eds.). *Analysis Design and Evaluation of Man–Machine Systems* (Proceedings of the 2nd IFAC/IFIP/IFORS Conference, Varese, Italy).

Geyer, T. A., Bellamy, L. J., Astley, J. A., & Hurst, N. W. (1990). Prevent Pipe Failures Due to Human Errors. *Chemical Engineering Progress,* November.

Goodstein, L. P. (1982). "Computer-Based Operating Aids." Report NKA/LIT—3. 2(82)111, RISØ National Laboratory: Roskilde, Denmark.

Goodstein, L. P., Anderson, H. B., & Olsen, S. E. (1988). *Tasks, Errors and Mental Models.* Washington, DC: Taylor and Francis.

Greenwood, M. Woods, H. M., & Yule, G. U. (1919). "The Incidence of Industrial Accidents upon Individuals with Special Reference to Multiple Accidents." Industrial Fatigue Research Board Report 4. London: Her Majesty's Stationery Office.

Griew, S., & Tucker, W. A. (1958). The Identification of Job Activities Associated with Age Differences in the Engineering Industry. *Journal of Applied Psychology* 42, 278.

Hale, A. R., & Glendon, A. I. (1987). *Individual Behavior in the Control of Danger.* Amsterdam: Elsevier.

Hall, R. E., Fragola, J., & Wreathall, J. (1982). *Post Event Decision Errors: Operator Action Tree/Time Reliability Correlation.* NUREG/CR-3010 U. S. Regulatory Commission: Washington, DC.

Hartley, L. R., Morrison, D., & Arnold, P. (1989). Stress and Skill. In A. M. Colley & J. R. Beech (Eds.). *Acquisition and Performance of Cognitive Skills.* New York: Wiley.

Hendrick, K., & Benner, L. Jr. (1987). *Investigating Accidents with STEP.* New York: Marcel Dekker.

Herzberg, F., Mausner, B., & Snyderman, B (1959). *The Motivation to Work.* New York: Wiley.

Hockey, G. R. J., Briner, R. B., Tattersall, A. J., & Wiethoff, M. (1989). Assessing the Impact of Computer Workload on Operator Stress: The Role of System Controllability. *Ergonomics* 32(2), 1401–1418.

Hollnagel, E., & Woods, D. D. (1983). Cognitive Systems Engineering: New Wine in New Bottles. *International Journal of Man–Machine Studies* 18, 583–600.

Hollnagel, E. (1993). *Human Reliability: Analysis, Context and Control.* San Diego, CA: Academic Press.

Hurst, N. W., Bellamy, L. J., & Wright, M. S. (1992). Research Models of Safety Management of Onshore Major Hazards and Their Application to Offshore Safety. Proceedings of a Conference on Major Hazards: Onshore and Offshore. *IChemE Symposium Series No. 130.* Rugby, UK: Institution of Chemical Engineers.

Hurst, N. W., Bellamy, L. J., Geyer, T. A., & Astley, J. A. (1991). A Classification Scheme for Pipework Failures to Include Human and Socio-Technical Errors and their Contribution to Pipework Failure Frequencies. *Journal of Hazardous Materials* 26, 159–186.

Janis, I. L. (1972). *Victims of Groupthink.* Boston: Houghton Mifflin.

Johnson, W. G. (1980). *MORT Safety Assurance Systems.* New York: National Safety Council and Marcel Dekker.

Joschek, H. I. (1981). Risk Assessment in the Chemical Industry. *Proceedings of the International ANS/ENS Topical Meeting on Probabilistic Risk Assessment, September 20–24, 1981, New York.* American Nuclear Society: 555 North Kensington Avenue, LaGrange Park, IL 60525.

Joyce, R. P., Chenzoff, A. P., Mulligan, J. F., & Mallory, W. J (1973). *Fully Proceduralized Job Performance Aids: Vol. II, Handbook for JPA Developers* (AFHRL-TR-73- 43(II)) (AD775-705). Wright Patterson Air Force Base, OH: U.S. Air Force Human Resources Laboratory.

Juran, J. M. (1979). *Juran's Quality Control Handbook* (3rd ed.). New York: McGraw-Hill.

Kahn, R. L., & French, J. P. R. (1970). *Status and Conflict: Two Themes in the Study of Stress.* In J. E. McGrath (Ed.), *Social and Psychological Factors in Stress.* New York: Holt, Rinehart and Winston.

Kahn, R. L. (1974). *Conflict, Ambiguity and Overload: Three Elements in Job Stress.* In A. Maclean (Ed.), Occupational Stress. Springfield, IL: Charles C Thomas.

Kahn, R. L., Wolfe, D. M., Quinn, R. P., Snoek, J. D., & Rosenthal, R. A. (1964). *Organizational Stress: Studies in Role Conflict and Ambiguity.* New York: Wiley.

Kantowitz, B. H., & Fujita, Y. (1990). Cognitive Theory, Identifiability and Human Reliability Analysis. *Journal of Reliability Engineering and System Safety* 29, 317–328.

Kantowitz, B. H., & Sorkin, R. D. (1987). Allocation of Function. In G. Salvendy (Ed.), *Handbook of Human Factors.* New York: Wiley.

Kasl, S. V. (1974). Work and Mental Health. In J. O'Toole (Ed.). *Work and the Quality of Life.* Cambridge, MA: MIT Press.

Kepner, C. H., & Tregoe, B. B. (1981). *The New Rational Manager.* Princeton, NJ: Kepner-Tregoe, Inc.

Kirwan, B. (1990). Human Reliability Assessment. In J. R. Wilson & E. N. Corlett (Eds.). *Evaluation of Human Work, a Practical Ergonomics Methodology.* Washington, DC: Taylor and Francis.

Kirwan, B. (1990). Human Error Analysis. In J. R. Wilson & E. N. Corlett (Eds.). *Evaluation of Human Work, a Practical Ergonomics Methodology.* Washington, DC: Taylor and Francis.

Kirwan, B. (1992). Human Error Identification in Reliability Assessment. Part 2: Detailed comparison of techniques. Applied *Ergonomics* 23(6), 371–381.

Kirwan, B., & Ainsworth, L. K. (1993). *A Guide to Task Analysis.* Washington, DC: Taylor & Francis.

Kirwan, B., Embrey, D. E., & Rea, K. (1988). *Human Reliability Assessors Guide* P. Humphreys (Ed.). Report no. RTS 88/95Q, Safety and Reliability Directorate, United Kingdom Atomic Energy Authority, Wigshaw Lane, Culcheth, Warrington, WA3 4NE, England.

Kletz, T. A. (1991). *An Engineer's View of Human Error* (2nd ed.). Rugby, UK: Institution of Chemical Engineers.

Kletz, T. A. (1992). HAZOP and HAZAN. *Identifying and Assessing Process Industry Hazards* (3rd ed.). Rugby, UK: Institution of Chemical Engineers.

Kletz, T. A. (1993). *Learning from Disaster: How Organizations Have No Memory.* Rugby, UK: Institute of Chemical Engineers.

Kletz, T. A. (1994a). *Learning from Accidents* (2nd ed.), Stoneham, MA: Butterworth-Heinemann.

Kletz, T. A. (1994b), *What Went Wrong? Case Histories of Process Plant Disasters* (3rd ed.). Gulf Publishing Co., Houston.

Kontogiannis, T. K., & Embrey, D. E. (1990). Assessing and Improving Decision Making in the Management of High Stress Situations. Paper presented at a conference on Emergency Planning, IBC Technical Services Ltd., Gilmoora House, Mortimer Street, London, W1N 7TD, UK.

Kontogiannis, T., & Lucas, D. (1990). "Operator Performance Under High Stress: An Evaluation of Cognitive Modes, Case Studies and Countermeasures." Report No. R90/03 prepared for the Nuclear Power Engineering Test Center, Tokyo. Human Reliability Associates, Dalton, Wigan, Lancashire, UK.

Kossoris, M. O., & Kohler, R. F. (1947). *Hours of Work and Output.* Washington, DC: U.S. Government Printing Office.

Leplat, J. (1982). Accidents and Incidents in Production: Methods of Analysis. *Journal of Occupational Accidents* 4(2–4), 299–310.

Leplat, J., & Rasmussen, J. (1984). Analysis of Human Errors in Industrial Incidents and Accidents for Improvements of Work Safety. *Accident Analysis and Prevention* 16(2), 77–88.

Lorenzo, D. K. (1990). *A Manager's Guide to Reducing Human Errors: Improving Human Performance in the Chemical Industry*. Washington, DC: Chemical Manufacturers Association, Inc. .

Lucas, D. (1987). Mental Models and New Technology. In J. Rasmussen, K. Duncan, & J. Leplat (Eds.). *New Technology and Human Error*. New York: Wiley.

Lucas, D., & Embrey, D. E. (1987). "Critical Action and Decision Evaluation Technique." Internal report, Human Reliability Associates, 1 School House, Higher Lane, Dalton, Wigan, Lancashire., UK.

Marcombe, J. T., Krause, T. R., & Finley, R. M. (1993). Behavior Based Safety at Monsanto's Pensacola plant. *The Chemical Engineer*, April 29.

Marshall, E. C., Duncan, K. D., & Baker, S. M. (1981). The Role of Withheld Information in the Training of Process Plant Fault Diagnosis. *Ergonomics* 24, 711–724.

Maslow, A. H. (1954). *Motivation and Personality*. New York: Harper.

McCormick, E. J., & Sanders, M. S. (1982). *Human Factors in Engineering and Design* (5th ed.). New York: McGraw-Hill.

McCormick, E. J., & Sanders, M. S. (1983). *Human Factors in Engineering and Design*. New York: McGraw-Hill.

McKenna, F. P. (1985). Do Safety Measures Really Work? An Examination of Risk Homeostasis Theory. *Ergonomics* 28, 489–498.

McKenna, F. P. (1989). Altered Egos. *Occupational Safety and Health*, September, 40–42.

McSween, T. E. (1993). Improve Your Safety Program with a Behavioural Approach. *Hydrocarbon Processing*, August.

Meister, D. (1966). Applications of Human Reliability to the Production Process. In W. B. Askren (Ed.) Report No. AMLR-TR-67-88, Symposium on Human Performance in Work.

Meister, D. (1977). Methods of Predicting Human Reliability in Man–Machine Systems. In S. Brown & J. Martin (Eds.). *Human Aspects of Man–Machine Systems*. Milton Keynes, UK: Open University Press.

Meister, D. (1979). *Human Factors in System Development*. New York: Wiley.

Meister, D. (1984). Human Reliability. In F. A. Muckler (Ed.), *Human Factors Review: 1984*. Santa Monica, CA: Human Factors Society Inc.

Meister, D. (1985). *Behavioral Analysis and Measurement Methods*. New York: Wiley.

Mill, R. C. (Ed.) (1992). *Human Factors in Process Operations*. Rugby, UK: Institute of Chemical Engineers.

Miller, D. P., & Swain, A. D. (1987). *Human Error and Human Reliability*. In G. Salvendy (Ed.), *Handbook of Human Factors*. New York: Wiley.

Monk, T. H., & Embrey, D. E. (1981). A Field Study of Circadian Rhythms in Actual and Interpolated Task Performance. In A. Reinberg, N. Vieux, & P. Andlauer (Eds.), *Night and Shift Work: Biological and Social Aspects*. Oxford, UK: Pergamon Press.

Moran, J. B., & Ronk, R. M. (1987). Personal Protective Equipment. In G. Salvendy (Ed.), *Handbook of Human Factors*. New York: Wiley.

Moray, N. (Ed.) (1979). *Mental Workload, Its Theory and Measurement*. New York: Plenum Press.

Moray, N. (1988). Mental Workload Since 1979. In D. J. Oborne (Ed.), *International Reviews of Ergonomics*. London: Taylor and Francis. Pp. 125–150.

Munger et al. (1962). *An Index of Electronic Equipment Operability: Data Store.* Report AIR-C43-1/62-RP(1). Pittsburgh PA: American Institute for Research.

Murgatroyd, R. A., & Tait, J. F. (1987)."A Case Study of the Application of the SHERPA Technique to the Dinorwig Spiral Casing Ultrasonic Testing System." Human Reliability Associates, 1 School House Higher Lane, Dalton, Wigan, Lancs., UK.

Murrell, K. F. H., Powesland, P. F., & Forsaith, B. (1962). A Study of Pillar-Drilling in Relation to Age. *Occupational Psychology* 36, 45.

Murrell, K. F. H. (1965). *Ergonomics: Man in His Working Environment.* London: Chapman and Hall.

Norman, D. A. (1981). Categorization of Action Slips. *Psychological Review* 88, 1–15.

Norros, L., & Samatti, P. (1986). *Nuclear Power Plant Operator Errors during Simulator Training.* Research reports 446, Technical Research Centre of Finland.

Oil Insurance Association (1971). *Report on Boiler Safety.*

Orpen, C. (1982). Type A Personality as Moderator for the Effects of Role Conflict, Role Ambiguity and Role Overload on Individual Strain. *Journal of Human Stress* 8(2), 8–14.

Ozog, H. (1985). Hazard Identification, Analysis and Control. *Chemical Engineering,* February 18, 161–170.

Paradies, M., Unger, L., & Ramey-Smith, A. (1992). Development and Testing of the NRC's Human Performance Investigation Process (HPIP). In *Proceedings of the International Conference on Hazard Identification and Risk Analysis, Human Factors and Human Reliability in Process Safety,* New York: American Institute of Chemical Engineers, Center for Chemical Process Safety. Pp. 253–260.

Patternotte, P. H., & Verhagen, L. H. J. M. (1979). Human Operator Research with a Simulated Distillation Process. *Ergonomics* 22, 19–28.

Pedersen, O. M. (1985). *Human Risk Contributions in the Process Industry: Guides for their Pre-Identificationx in Well-Structured Activities and for Post-Incident Analysis.* Report No. RISØ-M-2513, Risø National Laboratories, Roskilde, Denmark.

Pennycook, W. A., & Embrey, D. E. (1993). An Operating Approach to Error Analysis. In *Proceedings of the First Biennial Canadian Conference on Process Safety and Loss Management.* Edmonton, 24th April. Waterloo, Ontario, Canada: Institute for Risk Research, University of Waterloo.

Peters, T. J., & Waterman, R. H. (1982). *In Search of Excellence.* New York: Harper and Row.

Petersen, P. (1984). *Human Error Reduction and Safety Management.* Goshen, New York: Aloray Inc.

Pew, R. W., Miller, D. C., & Feeher, C. E. (1981). *Evaluation of Proposed Control Room Improvements Through Analysis of Critical Operator Decisions.* EPRI/NP-1982. Cambridge MA: Bolt Beranek & Newman, Inc.

Phillips, L. D., Embrey, D. E., Humphreys, P., & Selby, D. L. (1990). A Sociotechnical Approach to Assessing Human Reliability. In R. M. Oliver & J. A. Smith (Eds.), *Influence Diagrams, Belief Nets and Decision Making: Their Influence on Safety and Reliability.* New York: Wiley.

Pirani, M., & Reynolds, J. (1976). Gearing Up for Safety. *Personnel Management,* February, 25–29.

Pontecorvo, A. B. (1965). A Method of Predicting Human Reliability. *Annals of Reliability and Maintenance* 4, 337–342.

Poulton, E. (1976). Continuous Noise Interferes with Work by Masking Auditory Feedback and Inner Speech. *Applied Ergonomics* 7, 79–84.

Poulton, E. (1977). Continuous Intense Noise Masks Auditory Feedback and Inner Speech. *Psychological Bulletin* 84, 977–1001.

Poulton, E. (1978). Increased Vigilance with Vertical Vibration at 5 Hz: An Alerting Mechnism. *Applied Ergonomics* 9, 73–76.

Price, H. E. (1985). The Allocation of System Functions. *Human Factors* 27, 33–45.

Rasmussen, J. (1979). Reflection on the Concept of Operator Workload. In N. Moray (Ed.), *Mental Workload—Its Theory and Measurement*. New York: Plenum Press.

Rasmussen, J. (1981). Models of Mental Strategies in Process Control. In J. Rasmussen & W. Rouse (Eds.), *Human Detection and Diagnosis of System Failures*. New York: Plenum Press.

Rasmussen, J. (1982). Human Errors: A Taxonomy for Describing Human Malfunction in Industrial Installations. *Journal of Occupational Accidents* 4, 311–333.

Rasmussen, J. (1986). *Information Processing and Human–Machine Interaction*. Amsterdam: North Holland.

Rasmussen, J. (1989). Chemical Process Hazard Identification. *Reliability Engineering and System Safety*, Vol. 24, pp. 11–20. New York: Elsevier Science Publishers Ltd.

Rasmussen, J. (1990). Human Error and the Problem of Causality in Analysis of Accidents. In D. E. Broadbent, J. Reason, & A. Baddeley (Eds.). *Human Factors in Hazardous Situations*. Oxford, U.K.: Clarendon Press.

Reason, J. T. (1988). The Chernobyl Errors. *Bulletin of the British Psychological Society*, 40 (June).

Reason, J. T. (1990). *Human Error*. Cambridge: Cambridge University Press.

Reason, J. T., & Mycielska, K. (1982). *Absent Minded? The Psychology of Mental Lapses and Everyday Errors*. Englewood Cliffs, NJ: Prentice Hall.

Reinartz, S. (1989). Activities, Goals and Strategies: Team Behavior during Simulated Nuclear Power Plant Incidents. In *Conference Proceedings, Human Reliability in Nuclear Power*. London: International Business Communications.

Report of the Presidential Commission on the Space Shuttle *Challenger* Accident. (1986). Washington, DC: U.S. Government Printing Office.

Roger, D., & Nesshoever, W. (1987). The Construction and Preliminary Validation of a Scale for Measuring Emotional Control. *Personality and Individual Difference* 8, 527–534.

Rosa, R. R., Colligan, M. J., & Lewis, P. (1986). Extended Workdays: Effects of 8-hour and 12-hour Rotating Shift Schedules on Test Performance, Subjective Alertness, Sleep Patterns and Psychosocial Variables. *Proceedings of the Human Factors Society 30th Annual Meeting*, Vol. 2, 882–889.

Salvendy, G. (1987). *Handbook of Human Factors*. New York: Wiley.

Sanderson, P. M., & Harwood, K. (1988). The Skills Rules and Knowledge Classification: A Discussion of Its Emergence and Nature. In L. P. Goodstein, H. B. Anderson & , S. E. Olsen (Eds.). *Tasks, Errors and Mental Models*. Washington, DC: Taylor and Francis.

Shaw, L. S., & Sichel, H. S. (1971). *Accident Proneness*, Oxford: Pergamon.

Shepherd, A. (1985). Hierarchical Task Analysis and Training Decisions. *Programmed Learning and Educational Technology* 22, 162–176.

Shepherd, A., Marshall, E. C., Turner, A., & Duncan, K. D. (1977). Diagnosis of Plant Failures from a Control Panel. *Ergonomics* 20, 347–361.

Simpson, G. C. (1988). Hazard Awareness and Risk Perception. In A. S. Nicholson & J. E. Ridd (Eds.). *Health Safety and Ergonomics.* Stoneham, MA: Butterworth-Heinemann.

Slovic, P., Fischhoff, B., & Lichenstein, S. (1981). Perceived Risk: Psychological Factors and Social Implications. *Proceedings of the Royal Society of London* 376, 17–34.

Small, A. M., Sr. (1987). Design for Older People. In G. Salvendy (Ed.), *Handbook of Human Factors.* New York: Wiley.

Smillie, R. J. (1985). Design Strategies for Job Performance Aids. In T. M. Duffy & R. Waller (Eds.), *Designing Usable Text.* New York: Academic Press.

Smith, M. J., Colligan, M. G., & Tasto, D. L. (1982). Health and Safety Consequences of Shift Work in the Food Processing Industry. *Ergonomics* 25, 133–144.

Stammers, R. B. (1981). Theory and Practice in the Design of Training Simulators. *Programmed Learning and Educational Technology* 18, 67–71.

Suokas, J. (1989). The Identification of Human Contribution in Chemical Accidents. *6th International Symposium on Loss Prevention and Safety Promotion in the Process Industries.* Oslo, 19–22 June, 1989.

Swain, A. D. (1987). *Accident Sequence Evaluation Program Human Reliability Analysis Procedure.* NUREG/CR-4772. Washington, DC: US Nuclear Regulatory Commission.

Swain, A. D (1989). "Comparative Evaluation of Methods for Human Reliability Analysis." Report No. GRS-71, Gesellschaft für Reaktorsicherheit MbH, Shwertnergasse 1, 500 Köln 1, Germany.

Swain, A. D., & Guttmann, H. E. (1983). *Handbook of Human Reliability Analysis with Emphasis on Nuclear Power Plant Applications.* NUREG/CR-1278. Washington, DC: US Nuclear Regulatory Commission.

Taylor, F. W (1911). *The Principles of Scientific Management.* New York: Harper & Row.

Uehara, Y., & Hasegawa, H. (1986). Analysis of Causes of Accident at Factories Dealing with Hazardous Materials. *5th International Symposium on Loss Prevention and Safety Promotion in the Process Industries,* Vol. 1, Ch. 23. Paris: Société de Chimie Industrielle.

United Kingdom Atomic Energy Authority (1987). *Short Guide to Reducing Human Error in Process Operations.* Warrington, UK: AEA Technology Ltd.

United Kingdom Atomic Energy Authority (1991). *Long Guide to Reducing Human Error in Process Operations.* Warrington, UK: AEA Technology Ltd.

Verhaegen, P., Strubbe, J., Vonck, R., & van der Abeele, J. (1985). Absenteeism, Accidents and Risk-Taking. *Journal of Occupational Accidents* 7, 177–186.

Vernon, H. M. (1918). "An Investigation of the Factors Concerned in the Causation of Industrial Accidents." Memo R1, Health of Munitions Workers Committee: London.

Volle, M., Brisson, M., Perusse, M., Tanaka, M., & Doyon, Y. (1979). Compressed Work Week: Psychophysiological and Physiological Repercussions. *Ergonomics* 22, 1011–1010.

Wagenaar, W. A. (1992). Influencing Human Behavior: Toward a Practical Approach for Exploration and Production. *Journal of Petroleum Technology,* November.

Wagenaar, W. A., & Groeneweg, J. (1987). Accidents at Sea: Multiple Causes and Impossible Consequences. *International Journal of Man–Machine Studies* 27, 587–598.

Wagenaar, W. A., Hudson, P. T., & Reason, J. T. (1990). Cognitive Failures and Accidents. *Applied Cognitive Psychology* 4, 273–294.

Warm, J. S. (1984). (Ed.). *Sustained Attention in Human Performance.* New York: Wiley.

Warr, P. B. (1978). *Psychology at Work.* Middlesex, UK: Penguin Books.

West, B., & Clark, J. A. (1974). Operator Interaction with a Computer-Controlled Distillation Column. In E. Edwards & F. P. Lees (Eds.), *The Human Operator in Process Control.* London: Taylor & Francis.

Whiston, J., & Eddershaw, B. (1989). Quality and Safety—Distant Cousins or Close Relatives? *The Chemical Engineer: U.K.,* June 1989.

Wickens, C. D. (1984). *Engineering Psychology and Human Performance.* Columbus, Ohio: Charles Merrill.

Wiener, E. L. (1985). Beyond the Sterile Cockpit. *Human Factors* 27(1), 75–90.

Wierwille, W. W., & Eggemeier (1993). Recommendations for Mental Workload Measurement in a Test and Evaluation Environment. *Human Factors* 35(2), 263–281.

Wilde, G. J. S. (1984). Evidence Refuting the Theory of Risk Homeostasis? A Rejoinder to Frank P. McKenna. *Ergonomics* 25, 879–890.

Wilkinson, R. T. (1964). Artificial "Signals" as an Aid to an Inspection Task. *Ergonomics* 7, 63.

Wilson, J. R., & Corlett, E. N. (1990). *Evaluation of Human Work, A Practical Ergonomics Methodology.* Washington, DC: Taylor and Francis.

Woods, D. D. (1982). "Operator Decision Behavior during the Steam Generator Tube Rupture at the Ginna Nuclear Power Station." Research Report 82-1057-CONRM-R2, Westinghouse Research and Development Centre: Pittsburgh, PA.

Woods, D. D., & Roth, E. M. (1990). Cognitive Systems Engineering. In Helander, M. (Ed.), *Handbook of Human–Computer Interaction,* Amsterdam: North-Holland.

Woods, D. D., Wise, J. A., & Hanes, L. F. (1981). An Evaluation of Nuclear Power Plant Safety Parameter Display Systems. In *Proceedings of Human Factors Society 25th Annual Meeting.* Santa Monica, CA: Human Factors Society, Inc.

Wright, L. (1977). Presenting Technical Information: A Survey of Research Findings. *Instructional Science* 6, 93–134.

Wright, P. (1987). *Writing Technical Information. Review of Research in Education,* 14. Itasca, IL: E. Peacock.

Wright, P., & Barnard, D. (1975). Just Fill in This Form—A Review for Designers. *Applied Ergonomics* 6(4), 2.3–2.20.

Zorger, F. H. (1966). Section 5. In W. G. Ireson (Ed.). *Reliability Handbook,* New York: McGraw-Hill.

Bibliography

BOOKS

Cushman, W. H., & Rosenberg, D. J. (1991). *Advances in Human Factors/Ergonomics, Human Factors in Product Design.* New York: Elsevier.

Eastman Kodak Company (1983). *Ergonomics Design for People at Work,* Volume 1. New York: Van Nostrand Reinhold.

Eastman Kodak Company (1986). *Ergonomics Design for People at Work,* Volume 2. New York: Van Nostrand Reinhold.

Fleger, S. A., Permenter, K. E., Malone, T. B., & Carlow Associates Inc. (1988), *Advanced Human Factors Engineering Tool Technologies,* Technical Memorandum 2-88, U. S. Army Engineering Laboratory

Gilmore, W. E., Gertman, D. I., & Blackman, H. S. (1989), *User–Computer Interface in Process Control—A Human Factors Engineering Handbook.* New York: Academic Press.

Grandjean, E (1988). *Fitting the Task to the Man.* Philadelphia, PA: Taylor and Francis.

Kinkade, R. G., & Anderson, J. (1984). *Human Factors Guide for Nuclear Power Plant Control Room Development,* Electric Power Research Institute NP-3659.

Kletz, T. (1994a). *Learning from Accidents* (2nd ed.). Stoneham, MA: Butterworth-Heinemann.

Konz, S. (1979), *Work Design.* Columbus, OH: Grid Inc.

Lorenzo, D. K. (1990). *A Manager's Guide to Reducing Human Errors.* Washington, DC: Chemical Manufacturers Association, Inc.

Norman, D. A. (1988). *The Design of Everyday Things.* New York: Basic Books.

Norman, D. A. (1993), *Things That Make Us Smart,* Addison-Wesley Publishing Co.

Pelsma, K. H. (1987), *Ergonomics Source Book—A Guide to Human Factors Information.* Report. Store-Ergosystems Assoc., Inc.

Pheasant, S (1988) *Body Space: Anthropometry, Ergonomics and Design.* Philadelphia, PA: Taylor and Francis.

Rasmussen, J., Duncan, K., Leplat, J. (1987), *New Technology and Human Error.* New York: Wiley.

Reason, James (1990). *Human Error.* Cambridge: Cambridge University Press.

Salvendy, Gavriel (1987), *Handbook of Human Factors.* New York: Wiley.

Sanders, M. S., & McCormick, E. J. (1987). *Human Factors in Engineering and Design.* New York: McGraw-Hill.
Van Cott, H. P., & Kinkade, R. G. (1972). *Human Engineering to Equipment Design.* New York: McGraw-Hill.
Woodson, W. E. (1981). *Human Factors Design Handbook.* New York: McGraw-Hill.

PERIODICALS

Applied Ergonomics, published quarterly by IPC House, Surrey, England.
Ergonomics, published monthly by Taylor and Francis Ltd.
Ergonomics in Design the Magazine of Human Factors Applications, published quarterly by the Human Factors and Ergonomics Society.
Human Factors, the Journal of the Human Factors and Ergonomics Society, published quarterly by the Human Factors and Ergonomics Society.

Index

A

Accident causation analysis. *See also*
  Root cause analysis system
  (RCAS)
  data collection systems and, 255–258,
    261, 262–265
  human error concept of, 41–43
  onion model of, 262–263
Accountability, design process implementation guidelines, 347–348
Action oriented techniques, 162–179
  decision/action flow diagram, 169–
    171
  hierarchical task analysis (HTA), 162–
    167
  operational sequence diagram (OSD),
    172–176
  operator action event tree (OAET),
    167–169
  signal-flow graph analysis, 176–179
Active human error
  accident causation analysis, 41
  defined, xvii, 43
Activity analysis, error prediction and
  reduction, data acquisition techniques, 157–158
Age, performance-influencing factors,
  141–142
Alarms, control panel design, 123
Allocation of function, human factors
  engineering/ergonomics, 63–65
Anonymity, data collection systems, cultural factors, 259–260

Atmospheric conditions, performance-
  influencing factors, 112
Audit techniques
  MANAGER, sociotechnical systems,
    90–92, 93
  offshore emergency blowdown opera-
    tions, case study, 334–344
  system-induced error approach, per-
    formance optimization, 21
  traditional safety engineering, 54–55
Authority, performance-influencing factors, 145
Automation
  human factors engineering/ergonom-
    ics, 63–65
  offshore emergency blowdown opera-
    tions, audit, case study, 340, 342

B

Behavior modification, traditional
  safety engineering, 48–49
Black box model, 67, 68
Blame and punishment
  corporate culture and, 249, 253
  design process implementation guide-
    lines, 347–348, 356

C

Case studies, 291–344
  approach of, 17–19, 22–37
  hydrocarbon leak from pipe, 292–306
  offshore emergency blowdown opera-
    tions, audit, 334–344

Case studies (*cont.*)
  overview, 291–292
  solvent mischarge in batch plant
    described, 307–314
    standard operating procedure
      design for, 314–325
  visual display unit design, 325–334
Causation. *See* Accident causation analy-
    sis; Root cause analysis system
    (RCAS)
Change analysis, root cause analysis sys-
    tem, data collection, 284–285
Checks, performance-influencing fac-
    tors, task characteristics, 127–128
Chemical process quantitative risk
    analysis (CPQRA)
  data collection systems, 253
  overreliance on, performance-influ-
    encing factors (PIFs), 146–147
  task analysis, cognitive, 181, 186
Circadian rhythm, performance-influ-
    encing factors, 116–119
Clarity of instruction, performance-in-
    fluencing factors, task
    characteristics, 126–127
Classification of error, cognitive sys-
    tems engineering, 68–70
Cognitive systems engineering, 67–85
  applications of, 78–81
    conclusions regarding, 84–85
    error prediction, 79
    error reduction, 79
    generally, 78–79
    root cause analysis using sequential
      error model, 81
    summarized, 81–84
  classification of errors, 68–69, 74–76
    error recovery, 76
    knowledge-based mistakes, 76
    rule-based mistakes, 74–76
    slips and mistakes, 74
  generic error modeling system
    (GEMS), 70–73
  human error perspective of, 46
  overview of, 67–68
  prediction, 68–69

    skill-, rule-, and knowledge-based
      classification, 69–70, 71, 72
  stepladder model
    case example, 94–96
    described, 76–78
Cognitive task analysis techniques, 179–
    187
  critical action and decision evaluation
    technique (CADET), 180–183
  influence modeling and assessment
    systems (IMAS), 183–187
  overview of, 179–180
Cognitive "tunnel vision", defined, xvii
Communication
  offshore emergency blowdown opera-
    tions, audit, case study, 337
  performance-influencing factors
    (PIFs), 142–145
Complexity, performance-influencing
    factors, 110
Consciousness raising, of human error, 3
Consequence analysis, risk assessment,
    216–217
Consultants, error prediction and reduc-
    tion, 155
Control actions, human-machine inter-
    face, 62
Control panel design, performance-in-
    fluencing factors, 121–123
Corporate culture. *See* Cultural factors
Critical action and decision evaluation
    technique (CADET), 180–183
Critical Human Action Profile (CHAP),
    282
Critical incident technique, error predic-
    tion and reduction, 156–157
Critical task identification and screen-
    ing analysis, risk assessment, 209,
    211
Cultural factors
  blame and punishment, 249, 253
  data collection, 259–260, 289
  design process implementation guide-
    lines, 356
  error-inducing conditions, 18
  rule book culture, xix, 146

D

Danger, performance-influencing factors, 110

Data acquisition techniques, 154–160. *See also* Data collection
  activity analysis, 157–158
  critical incident technique, 156–157
  documentation, 157
  expert consultation, 154–155
  observation, 156
  simulators and mock-ups, 158–160
  withholding information technique, 160

Data collection, 247–290. *See also* Data acquisition techniques
  data types, 260–265
    chemical processing industry, 260–262
    generally, 260
  defined, 247
  implementation and monitoring, 287
  interpretation, 268–269
  methods, 266–268
    major incident analysis, 267
    personnel responsible for, 266
    reporting form design, 266–267
    storage and retrieval, 267–268
  organizational factors, 255–259
  overview of, 247–249
  root cause analysis system, 270–287
    change analysis, 284–285
    evaluation, 285–287
    generally, 270
    human performance investigation process (HPIP), 278, 282–284
    Management Oversight and Risk Tree (MORT), 273, 274
    root cause coding, 275, 278, 279
    sequentially timed events plotting procedure (STEP), 273, 275–277
    tree of causes/variation diagram, 270–272
  set-up, 288–290
  systems, generally, 249–251
  system types, 251–255
    generally, 251–252, 254–255
    incident reporting systems (IRS), 252
    near miss reporting system (NMRS), 252–253

quantitative human reliability data collection system, 253–254
    root cause analysis system (RCAS), 253

Decision/action flow diagram, described, 169–171

Decision-making stage
  human-machine interface, 61
  offshore emergency blowdown operations, audit, case study, 338, 340

Demand-resource mismatch view, of human error, 15–17

Design factors
  error-inducing conditions, 18
  performance-influencing factors and, 106, 120–123
  visual display unit design, case study, 325–334

Disciplinary action, traditional safety engineering, 53

Documentation
  error prediction and reduction, data acquisition techniques, 157
  predictive human error analysis (PHEA), 192–193

E

Emergency situations, performance-influencing factors, 149–151

Emotional control, performance-influencing factors, 140

Encystment, defined, xvii

Engineering. *See* Cognitive systems engineering; Human factors engineering/ergonomics; Process safety review procedures; Reliability engineering; Traditional safety engineering

Equipment design, performance-influencing factors, 120–121

Ergonomics checklists, 196–198. *See also* Human factors engineering/ergonomics (HFE/E)
  application of, 196
  examples of, 196–198

Error. *See* Human error

Error-inducing conditions, described, 18

Error prediction and reduction, 153–
    199. *See also* Predictive human
    error analysis (PHEA)
  cognitive systems engineering, 68–69,
    79, 81, 83–84
  data acquisition techniques, 154–160
    activity analysis, 157–158
    critical incident technique, 156–157
    documentation, 157
    expert consultation, 154–155
    observation, 156
    simulators and mock-ups, 158–160
    withholding information technique,
      160
  ergonomics checklists, 196–198
  human error analysis techniques, 189–
    195
    overview of, 189–190
    predictive, 190–194
    work analysis, 194–195
  overview of, 153–154
  program benefits, 10–12
  risk assessment and, 201–245. *See also*
    Risk assessment
  system-induced error approach, per-
    formance optimization, 20–21, 21
  task analysis, 161–189
    action oriented techniques, 162–179
    applications of, 161–162
    cognitive techniques, 179–187
    evaluation of, 187–189
    purpose of, 161
Error recovery, cognitive systems engi-
    neering, 76
Events and Causal Factors Charting
    (ECFC), 282
Event tree, THERP case study, 229–233
Experience, performance-influencing
    factors, 133–135
Expert consultation, error prediction
    and reduction, 154–155
External error mode, defined, xvii
Externals, defined, xvii

F
Failure model, risk assessment, quantifi-
    cation process, 225

Failure modes effects and criticality
    analysis (FMECA)
  overreliance on, management policies,
    146–147
  performance-influencing factors and,
    109
False assumptions, case study illustrat-
    ing, 27–30
Fatigue, performance-influencing fac-
    tors, 115
Fault diagnostic support, performance-
    influencing factors, 128
Fault tree construction
  human reliability role, 202–204
  representation, 219–222
Feedback
  cognitive systems engineering, step-
    ladder model, 76–78
  data collection, 259–260, 289
  human error, 8

G
Generic error modeling system (GEMS),
    70–73
Goals, design process implementation
    guidelines, 347–348
Groups, performance-influencing fac-
    tors, 145

H
Hazard and operability (HAZOP) study
  hazard identification by, 19
  overreliance on, performance-influ-
    encing factors (PIFs), 146–147
  performance-influencing factors and,
    109
  risk assessment
    critical task identification and
      screening analysis, 211
    human reliability role, 202, 205, 206
Hazard identification, HAZOP, 19
Health & Safety Executive Research Pro-
    gram (U.K.), 90–92
Hierarchical task analysis (HTA)
  case study, solvent mischarge, 315–319
  described, 162–167

Hours of labor, performance-influencing factors, 113–116
Human error, 1–37. *See also* entries under Error
  benefits of elimination program for, 10–12
  case studies of, 22–37
  concepts of, 39–44
    accident causation analysis, 41–43
    generally, 39–40
    reliability engineering, 40–41
  costs of, 5
  demand-resource mismatch view of, 15–17
  neglect of, rationale for, 10
  overview of, 1–4
  perspectives on, 44–47
    cognitive systems engineering, 46, 67–85. *See also* Cognitive systems engineering
    human factors engineering/ ergonomics, 44–46, 55–67. *See also* Human factors engineering/ ergonomics
    sociotechnical systems, 46–47, 85–93. *See also* Sociotechnical systems
    traditional safety engineering, 44, 47–55. *See also* Traditional safety engineering
  probability, defined, xvii
  role of, in accidents, 4–10
  system-induced error approach
    case study approach illustrating, 17–19
    overview of, 12–15, 20
    practical applications, 19–22
Human error analysis techniques, 189–195
  overview of, 189–190
  predictive, 190–194
  work analysis, 194–195
Human factors assessment methodology (HFAM)
  data collection systems and, 265
  design process implementation guidelines, 356–357

performance-influencing factors and, 105, 109
  sociotechnical systems, 87–90, 93
Human factors engineering/ergonomics (HFE/E), 55–67. *See also* Ergonomics checklists
  automation and allocation of function, 63–65
  evaluation of, 67
  human error perspective of, 44–46
  human-machine interface
    control actions, 62
    decision-making stage, 61
    generally, 56–60
    perception stage, 60–61
  information processing and mental workload, 62–63
  overview of, 55–56
  system-induced error approach, performance optimization, 19
  system reliability assessment and error, 65–67
Human information-processing
  defined, xvii
  human factors engineering/ergonomics, 62–63
Human-machine interface
  defined, xvii
  ergonomics checklists, 197
  human factors engineering/ergonomics, 56–60
    control actions, 62
    decision-making stage, 61
    generally, 56–60
    perception stage, 60–61
  inadequate, case study illustrating, 24–27
Human performance evaluation system (HPES)
  cultural factors, 259–260
  personnel responsible, 266
Human performance investigation process (HPIP), 278, 282–284
Human reliability
  assessment methodology for, SPEAR, 207–209, 210
  defined, xvii
Hydrocarbon leak, case study, 292–306

I

Identifications, inadequate, case study illustrating, 27
Implementation, data collection system, 287
Implementation guidelines, 345–363
  design process, 346–357
    accountability, objectives, and goals, 347–348
    change management, 352–353
    generally, 346–347
    incident investigations, 356–357
    process and equipment integrity, 353–354
    process safety review procedures, 348–351
    risk management, 351–352
    training and performance, 354–356
  overview, 345–346
  set-up, 357–363
Incident reporting systems (IRS), described, 252
Influence diagram approach (risk assessment)
  calculations for, 241–245
  quantification techniques, 238–240
Influence modeling and assessment systems (IMAS), described, 183–187
Information processing, human factors engineering/ergonomics, 62–63
Inspection, design process implementation guidelines, 353
Instruction, clarity of, performance-influencing factors, 126–127. See also Training
Instrumentation scales, inappropriate, case study illustrating, 27
Interface. See Human-machine interface
Internal error mechanism, defined, xviii
Internal error mode, defined, xviii
Internals, defined, xviii
International Safety Rating System (ISRS), 54
Investigation management
  data collection types, 262
  design process implementation guidelines, 356–357

J

Job aids and procedures, performance-influencing factors, 123–128

K

Knowledge-based classification. See Skill-, rule-, and knowledge-based classification (SRK)
Knowledge-based level of control, defined, xviii
Knowledge-based tasks
  cognitive systems engineering, classification of errors, 76
  visual display unit design, case studies, 328, 332–334

L

Labeling
  equipment design, performance-influencing factors, 120
  inadequate, case study illustrating, 26
Latent human error
  accident causation analysis, 41, 42
  data collection and, 257
  defined, xviii, 43
Leadership, performance-influencing factors, 145. See also Management policies
Lighting, performance-influencing factors, operating environment, 111–112
Locus of control
  defined, xviii
  performance-influencing factors (PIFs), personality factors, 140

M

Maintenance, design process implementation guidelines, 353
Major incident analysis, data collection methods, 267
Management Oversight and Risk Tree (MORT), root cause analysis system, data collection, 273, 274, 278, 284
Management policies

design process implementation guidelines, 352–353
error-inducing conditions, 18
implementation guidelines, set-up, 358
performance-influencing factors (PIFs), organizational and social factors, 145–148
root cause analysis system, data collection, 286–287
MANAGER
performance-influencing factors, 105
sociotechnical systems, 90–92, 93
Manual variability, defined, xviii
Mental model, human factors engineering/ergonomics, 61, 64–65
Mental workload, human factors engineering/ergonomics, 62–63
Mindset syndrome
cognitive systems engineering, 76
defined, xviii
Mistakes
cognitive systems engineering, 74
defined, xviii
Mock-ups, error prediction and reduction, 158–160
Monitoring
data collection system, 287
human factors engineering/ergonomics, automation, 64
Motivation, performance-influencing factors, 136–137

N

Near miss reporting system (NMRS)
data collection systems, 257–258
described, 252–253
Night work, performance-influencing factors, 116–119
Noise, performance-influencing factors, 111–112

O

Objectives, design process implementation guidelines, 347–348
Observation, error prediction and reduction, 156

Offshore emergency blowdown operations, case studies, 334–344
Operating environment, performance-influencing factors, 109–119
Operating procedures, flawed, case study illustrating, 30
Operational sequence diagram (OSD), described, 172–176
Operator action event tree (OAET), described, 167–169
Operator characteristics, 133–142
experience, 133–135
personality factors, 135–141
physical condition and age, 141–142
Organizational factors
data collection
chemical processing industry, 265
generally, 255–259
error-inducing conditions, 18
failures, case study illustrating, 35–37
human error and, 5–10
ineffective, case study illustrating, 32–34
performance-influencing factors (PIFs), 142–148
system-induced error approach and, 22
Organizational learning, performance-influencing factors, 147–148

P

Partitioned operational sequence diagram (OSD), described, 175–176
Perception stage, human-machine interface, 60–61
Performance-influencing factors (PIFs), 103–152
applications of, 105–107
classification structures for, 107–109
data collection systems and, 264–265
defined, xviii
interactions of factors in, 148–149
operating environment, 109–119
chemical process environment, 109–110
physical work environment, 111–112
work pattern, 112–119

Performance-influencing factors *(cont.)*
  operator characteristics, 133–142
    experience, 133–135
    personality factors, 135–141
    physical condition and age, 141–142
  organizational and social factors, 142–
    148
    teamwork and communication,
      142–145
  overview of, 103–105
  risk assessment
    qualitative human error analysis,
      212–213
    success likelihood index method
      (SLIM), 234–235
  task characteristics, 120–133
    control panel design, 121–123
    equipment design, 120–121
    job aids and procedures, 123–128
    training, 128–133
  variability of performance, normal
    and emergency situations, 149–151
Performance optimization, system-induced error approach, 19
Performance-shaping factors (PSFs),
  103. *See also* Performance-influencing factors (PIFs)
Personality, performance-influencing
  factors, 135–141
Personal protective equipment (PPE)
  performance-influencing factors, 120–
    121
  safety campaigns, traditional safety
    engineering, 51–52
Physical condition and age, performance-influencing factors, 141–142
Physical work environment, performance-influencing factors, 111–112
Plant investigation, human performance
  investigation process (HPIP), 282
Policy factors, error-inducing conditions, 18. *See also* Management
  policies
Population stereotype, defined, xviii
Prediction. *See* Error prediction and reduction
Predictive human error analysis (PHEA)

case study, solvent mischarge, 319–
  320, 321
described, 190–194
risk assessment
  qualitative analysis methods, 218, 219
  qualitative human error analysis,
    213–216
Prevention. *See* Error prediction and reduction
Probabilistic safety analysis (PSA), 104
Problem solving, implementation guidelines, 346
Procedures, performance-influencing
  factors, 123–128
Process safety review procedures, design process implementation
  guidelines, 348–351
Production requirements, safety requirements versus, 129–130
Protocol analysis, expert consultation, 155
Psychology
  personality, performance-influencing
    factors, 135–141
  root cause analysis system, 286
Punishment. *See* Blame and punishment

Q
Qualitative human error analysis, 211–
  219
  case study, in SPEAR, 217–219
  consequence analysis, 216–217
  error reduction analysis, 217
  overview of, 211–212
  performance influencing factor analy-
    sis, 212–213
  predictive human error analysis
    (PHEA), 213–216
  task analysis, 212
Quality assurance, Total Quality Management (TQM), 253
Quantification, risk assessment, human
  reliability role in, 206–207. *See also*
  Risk assessment: quantification
Quantitative human reliability data collection system, 253–254
Quantitative risk assessment (QRA),
  103–104

R

Reactance, defined, xviii
Recovery analysis, predictive human error analysis, 216
Recovery error, defined, xviii, 43
Reliability engineering, human error concept of, 40–41
Reporting form design, data collection methods, 266–267
Representation, risk assessment, 219–222
Responsibility
    failures in, case study illustrating, 34–35
    performance-influencing factors, 144
Rest pauses, performance-influencing factors, 113–116
Retrieval methods, data collection, 267–268
Risk assessment, 201–245
    critical task identification and screening analysis, 209, 211
    defined, xviii
    design process implementation guidelines, 351–352
    human reliability role in, 202–207
        case study, 202–204
        error implications for, 205–206
        quantification aspects, 206–207
    implementation guidelines, 346
    overview of, 201–202
    qualitative human error analysis, 211–219
        case study, in SPEAR, 217–219
        consequence analysis, 216–217
        error reduction analysis, 217
        overview of, 211–212
        performance influencing factor analysis, 212–213
        predictive human error analysis (PHEA), 213–216
        task analysis, 212
    quantification, 222–240
        human reliability role in, 206–207
        overview of, 222–224
    quantification process, 224–225
    quantification techniques, 225–240

influence diagram approach, 238–240, 238–245
success likelihood index method (SLIM), 233–238
technique for human error rate prediction (THERP), 226–233
representation, 219–222
system for predictive error analysis and reduction (SPEAR), 207–209, 210
Risk homeostasis, defined, xviii
Risk homeostasis theory (RHT), performance-influencing factors, 138–140
Risk-taking, performance-influencing factors, 137–138
Role ambiguity, defined, xix
Role conflict, defined, xix
Root cause
    analysis of, using sequential error model, 81
    case study, hydrocarbon leak, STEP, 297–299
    defined, xix
Root cause analysis system (RCAS). *See also* Accident causation analysis
    data collection, 264, 270–287
        change analysis, 284–285
        evaluation, 285–287
        generally, 270
        human performance investigation process (HPIP), 278, 282–284
        Management Oversight and Risk Tree (MORT), 273, 274
        root cause coding, 275, 278, 279
        sequentially timed events plotting procedure (STEP), 273, 275–277
        tree of causes/variation diagram, 270–272
    described, 253
Root cause coding, 275, 278, 279
Routine violations, case study illustrating, 30–32
Rule-based classification. *See* Skill-, rule-, and knowledge-based classification (SRK)

Rule-based level of control, defined, xix
Rule-based tasks
  error classification, cognitive systems
    engineering, 74–76
  visual display unit design, case stud-
    ies, 326, 328, 330–332
Rule book culture
  defined, xix
  performance-influencing factors, 146
Rules and procedures, performance-in-
    fluencing factors, 123–128

S
Safety campaigns, traditional safety en-
    gineering, 50–53
Safety management system audits, tradi-
    tional safety engineering, 54–55
Safety requirements, production require-
    ments versus, 129–130
Scheduling, performance-influencing
    factors, 116–119
Sequential error model. See Skill-, rule-,
    knowledge-based classification
    (SRK)
Sequential Timed Event Plotting (STEP)
  case study, hydrocarbon leak, 292–306
  data collection system and, 264
  design process implementation strate-
    gies, 357
  root cause analysis system, 273, 275–
    277
Shift rotation, performance-influencing
    factors, 116–119
Signal-flow graph analysis, described,
    176–179
Simulators and mock-ups, error predic-
    tion and reduction, 158–160
Skill(s)
  deterioration of, human factors engi-
    neering/ergonomics, 63–64
  level of, performance-influencing fac-
    tors, 133–134
Skill-, rule-, knowledge-based classifica-
    tion (SRK)
  case examples, 96–101
  cognitive systems engineering, 69–70,
    71, 72

root cause analysis using, 81
Skill-based level of control, defined, xix
Skill-based task, visual display unit de-
    sign, case studies, 326, 329–330
Sleep deprivation, performance-influ-
    encing factors, 113
Slips
  cognitive systems engineering, 74
  defined, xix
Social factors, performance-influencing
    factors, 142–148
Sociotechnical approach to human reli-
    ability (STAHR). See Influence
    diagram approach (IDA)
Sociotechnical systems, 85–93
  comparisons among approaches of, 93
  human error perspective of, 46–47
  human factors analysis methodology,
    87–90
  overview of, 85–86
  performance-influencing factors and,
    105
  TRIPOD approach, 86–87
  U.K. Health & Safety Executive Re-
    search Program, 90–92
Solvent mischarge, in batch plant, case
    studies, 307–325
Spatial operational sequence diagram
    (OSD), described, 176
Standard operating procedures (SOPs),
    solvent mischarge in batch plant,
    case studies, 314–325
Stepladder model
  case example, 94–96
  described, 76–78
Stereotype fixation, defined, xix
Stereotype takeover, defined, xix
Stimulus operation response team
    (SORTM), 282
Storage methods, data collection, 267–
    268
Stress
  case study illustrating, 23–24
  performance-influencing factors
    emergency situations, 149–151
    operator characteristics, 134–135
  risk assessment, 235

Success likelihood index method (SLIM), 233–238

Suddenness of onset, performance-influencing factors, 110

System factors. *See* Organizational factors

System for predictive error analysis and reduction (SPEAR)
  described, 207–209, 210
  qualitative human error analysis and, 211–219
  system-induced error approach, performance optimization, 20–21

System-induced error approach
  case study approach illustrating, 17–19
  data collection and, 256–259
  overview of, 12–15, 20
  practical applications, 19–22

System reliability assessment, human factors engineering/ergonomics, 65–67

T

Task analysis, 161–189
  action oriented techniques, 162–179
    decision/action flow diagram, 169–171
    hierarchical task analysis (HTA), 162–167
    operational sequence diagram (OSD), 172–176
    operator action event tree (OAET), 167–169
    signal-flow graph analysis, 176–179
  applications of, 161–162
  cognitive techniques, 179–187
    critical action and decision evaluation technique (CADET), 180–183
    influence modeling and assessment systems (IMAS), 183–187
    overview of, 179–180
  evaluation of, 187–189
  offshore emergency blowdown operations, audit, case studies, 340
  purpose of, 161
  risk assessment, qualitative human error analysis, 212

Task characteristics, performance-influencing factors, 120–133

Task modeling, risk assessment, quantification process, 224–225

Teamwork, performance-influencing factors, 142–145

Technique for Human Error Rate Prediction (THERP)
  case study, 229–233
  data collection systems, 254
  described, 226–228
  representation, risk assessment, 219–222

Technology, human factors engineering/ergonomics, overreliance on, 65

Temporal operational sequence diagram (OSD), 172–174

Testing, design process implementation guidelines, 353

Thermal conditions, performance-influencing factors, 112

Time dependency, performance-influencing factors, 110

Total Quality Management (TQM), 253

Traditional safety engineering, 47–55
  data collection and, 255–256
  defined, xix
  disadvantages of, 49–50
  disciplinary action, 53
  error prevention and, 48–49
  human error perspective of, 44
  overview of, 47–48
  safety campaigns, 50–53
  safety management system audits, 54–55
  training, 55

Training
  data collection set-up, 289
  design process implementation guidelines, 354–356
  offshore emergency blowdown operations, audit, case study, 336–337
  performance-influencing factors
    clarity of, 126–127
    task characteristics, 128–133
  traditional safety engineering, 55

Tree of causes/variation diagram, 270–272

TRIPOD approach
performance-influencing factors and, 105, 109
sociotechnical systems, 86–87, 93

Type A and B personality, performance-influencing factors, 141

U
Unforgiving environment, described, 19
United Kingdom Health & Safety Executive Research Program, sociotechnical systems, 90–92, 93

V
Vagabonding, defined, xix
Variation diagram, tree of causes/variation diagram, 271

Verbal protocol analysis, defined, xix
Violation error
accident causation analysis, 42
defined, xix, 43
Visual display unit design, case studies, 325–334

W
Warnings, performance-influencing factors, 127–128
Withholding information technique, error prediction and reduction, 160
Work analysis, human error analysis techniques, 194–195
Workload distribution, performance-influencing factors, 143
Work pattern, performance-influencing factors, 112–119